Design and Implementation of Healthcare Biometric Systems

Dakshina Ranjan Kisku
National Institute of Technology Durgapur, India

Phalguni Gupta
National Institute of Technical Teachers Training and Research, India

Jamuna Kanta Sing
Jadavpur University, India

A volume in the Advances in Medical Technologies and Clinical Practice (AMTCP) Book Series

IGI Global
DISSEMINATOR OF KNOWLEDGE

Published in the United States of America by
IGI Global
Medical Information Science Reference (an imprint of IGI Global)
701 E. Chocolate Avenue
Hershey PA, USA 17033
Tel: 717-533-8845
Fax: 717-533-8661
E-mail: cust@igi-global.com
Web site: http://www.igi-global.com

Library of Congress Cataloging-in-Publication Data

Names: Kisku, Dakshina Ranjan, editor. | Gupta, Phalguni, editor. | Sing, Jamuna Kanta, editor.
Title: Design and implementation of healthcare biometric systems / Dakshina Ranjan Kisku, Phalguni Gupta, and Jamuna Kanta Sing, editors.
Description: Hershey, PA : Medical Information Science Reference, [2019] | Includes bibliographical references.
Identifiers: LCCN 2018030564| ISBN 9781522575252 (hardcover) | ISBN 9781522575269 (Ebook)
Subjects: | MESH: Biometry | Computing Methodologies
Classification: LCC QH323.5 | NLM WA 950 | DDC 570.1/5195--dc23 LC record available at https://lccn.loc.gov/2018030564

This book is published in the IGI Global book series Advances in Medical Technologies and Clinical Practice (AMTCP) (ISSN: 2327-9354; eISSN: 2327-9370)

British Cataloguing in Publication Data
A Cataloguing in Publication record for this book is available from the British Library.

Advances in Medical Technologies and Clinical Practice (AMTCP) Book Series

Srikanta Patnaik
SOA University, India
Priti Das
S.C.B. Medical College, India

ISSN:2327-9354
EISSN:2327-9370

Mission

Medical technological innovation continues to provide avenues of research for faster and safer diagnosis and treatments for patients. Practitioners must stay up to date with these latest advancements to provide the best care for nursing and clinical practices.

The **Advances in Medical Technologies and Clinical Practice (AMTCP) Book Series** brings together the most recent research on the latest technology used in areas of nursing informatics, clinical technology, biomedicine, diagnostic technologies, and more. Researchers, students, and practitioners in this field will benefit from this fundamental coverage on the use of technology in clinical practices.

Coverage

- E-Health
- Clinical Nutrition
- Nursing Informatics
- Neural Engineering
- Biometrics
- Patient-Centered Care
- Clinical Studies
- Nutrition
- Medical Imaging
- Clinical Data Mining

IGI Global is currently accepting manuscripts for publication within this series. To submit a proposal for a volume in this series, please contact our Acquisition Editors at Acquisitions@igi-global.com or visit: http://www.igi-global.com/publish/.

The Advances in Medical Technologies and Clinical Practice (AMTCP) Book Series (ISSN 2327-9354) is published by IGI Global, 701 E. Chocolate Avenue, Hershey, PA 17033-1240, USA, www.igi-global.com. This series is composed of titles available for purchase individually; each title is edited to be contextually exclusive from any other title within the series. For pricing and ordering information please visit http://www.igi-global.com/book-series/advances-medical-technologies-clinical-practice/73682. Postmaster: Send all address changes to above address.

Titles in this Series

For a list of additional titles in this series, please visit: www.igi-global.com/book-series

Histopathological Image Analysis in Medical Decision Making
Nilanjan Dey (Techno India College of Technology, India) Amira S. Ashour (Tanta University, Egypt) Harihar Kalia (Seemantha Engineering College, India) R.T. Goswami (Techno India College of Technology, India) and Himansu Das (KIIT University, India)
Medical Information Science Reference • copyright 2019 • 340pp • H/C (ISBN: 9781522563167) • US $265.00 (our price)

Healing Through the Arts for Non-Clinical Practitioners
Jenny Bopp (Westtown School, USA) Amy M. Grebe (Zinnia Consulting and Design, USA) and Joanna Hope Denny (Pellisippi State Community College, USA)
Medical Information Science Reference • copyright 2019 • 309pp • H/C (ISBN: 9781522559818) • US $255.00 (our price)

Medical Image Processing for Improved Clinical Diagnosis
A. Swarnambiga (Indian Institute of Technology Madras, India)
Medical Information Science Reference • copyright 2019 • 313pp • H/C (ISBN: 9781522558767) • US $245.00 (our price)

Clinical Costing Techniques and Analysis in Modern Healthcare Systems
Ronald Ma (Austin Health, Australia)
Medical Information Science Reference • copyright 2019 • 265pp • H/C (ISBN: 9781522550822) • US $235.00 (our price)

Optimized Genetic Programming Applications Emerging Research and Opportunities
Bahrudin Hrnjica (University of Bihac, Bosnia and Herzegovina) and Ali Danandeh Mehr (Antalya Bilim University, Turkey)
Medical Information Science Reference • copyright 2019 • 310pp • H/C (ISBN: 9781522560050) • US $225.00 (our price)

Design and Development of Affordable Healthcare Technologies
Arindam Bit (National Institute of Technology Raipur, India)
Medical Information Science Reference • copyright 2018 • 388pp • H/C (ISBN: 9781522549697) • US $245.00 (our price)

IGI Global
DISSEMINATOR OF KNOWLEDGE

701 East Chocolate Avenue, Hershey, PA 17033, USA
Tel: 717-533-8845 x100 • Fax: 717-533-8661
E-Mail: cust@igi-global.com • www.igi-global.com

Editorial Advisory Board

Table of Contents

Detailed Table of Contents

The healthcare industry offers highly personalized services to its patients. It is necessary to correctly identify the patient and efficiently link his medical records as and when needed. It is important to note that the need of identification arises in many desperate scenarios when the patient may not be able to tell anything about himself. Biometrics can help in this scenario by using physiological or behavioral characteristics. Some of the biometrics traits could be acquired without direct participation of the person, and therefore, the patient need not provide any pin, password, or token for his identification. Biometrics can handle challenges of duplicate medical records and identity theft. However, there is an important issue that may arise when a large number of patients get registered to the system. Increase in the size of the biometric database gradually escalates the time required for identification. This calls for the need of an efficient indexing approach that can confine the search space and decrease the response time. This chapter highlights biometric indexing approaches suitable for the healthcare industry.

Mistakes in healthcare systems such as a mix-up of records or confusing medical charts lead to the wrong medications to patients. Major tasks such as administrative costs, legal expenses, and liabilities incur high cost to the healthcare industry using traditional, inaccurate patient identification processes. This can be resolved by biometric technology. Only physiological features can be used for patient identification to eliminate need of SSN, insurance card, or date of birth during registration. A biometric template can be directly mapped to an electronic health record to accurately authenticate individuals on subsequent visits. This technology ensures no medical records can be mimicked and the right care is provided to the right patient. Deep learning provides a platform to solve identification and diagnostic problems arising in medicine and can be used in healthcare biometrics to analyze clinical parameters and their combinations for disease prognosis (e.g., prediction of disease, extracting medical knowledge, therapy planning, and support).

Many healthcare providers integrate biometric recognition/verification schemes into patient identification or other information security systems. While overcoming the disadvantages of using passwords, PINs, and tokens which may be forgotten, or stolen, biometric systems are susceptible to spoofing attacks, or presentation attacks. Liveness detection is an effective mechanism used to defeat a presentation attack. This chapter focuses on voice liveness detection in automatic speaker verification (ASV) systems. The authors explain the spoofing attacks to ASV systems comprising impersonation, voice conversion, speech synthesis, and replay and then present four types of liveness detection (anti-spoofing) methods used to mitigate ASV spoofing attacks: challenge-response-based methods, acoustic feature-based methods, hardware-based methods, and multi-modal biometric-based methods. This chapter analyzes the advantages and disadvantages of each kind of liveness detection method and proposes the possible application of voiceprint-based liveness detection schemes in the insulin pump system.

With much concern over security, it has become essential to maintain the identity and track of an individual's activities in the modern healthcare sector. Although there are biometric authentication systems based on different modalities, recognition of a person using the ear has gained much attention as ears are unique. Ear localization is a first step for ear-based biometric authentication systems, and this needs to be accurate, since it plays a crucial role in the overall performance of the system. The localization of ear in the side face images captured in the wild possess great challenges due to varying angles, light, scale, background clutter, blur and occlusion, etc. In this chapter, the authors have proposed EarLocalizer model to localize the ear, which is inspired by Faster-RCNN. The model is evaluated on two wild ear databases, UBEAR-II and USTB-III, and has achieved an accuracy of 95% and 99.08%, respectively, at IOU (Intersection over Union) = 0.5. The results of the proposed model signify that the model is invariant to the environmental conditions.

This chapter outlines the current state of the art of Kinect sensor gait and activity authentication. It also focuses on emotional cues that could be observed from human body and posture. It presents a prototype of a system that combines recently developed behavioral gait and posture recognition methods for human emotion identification. A backbone of the system is Kinect sensor gait recognition, which explores the relationship between joint-relative angles and joint-relative distances through machine learning. The chapter then introduces a real-time gesture recognition system developed using Kinect sensor and trained with SVM classifier. Preliminary experimental results demonstrate accuracy and feasibility of using such systems in real-world scenarios. While gait and emotion from body movement has been researched in the context of standalone biometric security systems, they were never previously explored for physiotherapy rehabilitation and real-time patient feedback. The survey of recent progress and open problems in crucial areas of medical patient rehabilitation and rescue operations conclude this chapter.

Autism Spectrum Disorder (ASD) screening is still an ongoing process due to few objective and effective screening approaches. The consideration of facial patterns, fingerprints, and other metrics such as footprints and ear size can well explain the ASD phenotypes. These physical anomalies provide a simpler and more objective approach to screen the disorder through rather than considering the complex biological factors. Moreover, it is very easy to acquire these metrics as compared to those lengthier and restrictive procedures. The screening engine in which both the biometrics are integrated has better and reliable outcomes as compared to singular approaches. The objective of this chapter is to present an ASD screening system based on the combination of biometrics of face and fingers. The novelty of the chapter is that the classifier-based matching and fusion of the modalities has been proposed. Hence, multi-biometric-based pre-clinical system has the potential to follow ASD-affected individuals continuously, objectively, and periodically.

In order to diagnose a possible cardiac disorder, ECG (electrocardiogram) signals are usually recorded on standard grid papers in hospitals. Many efforts have been made to advance the technology in order to improve the diagnosis and management of cardiovascular disease. There is a need to convert the existing ECG records into digital forms, as it is the most efficient method to store and analyze ECG attributes for clinical uses. The main purpose of this chapter is to review the existing algorithms for digital conversion of paper ECG. It discusses the various challenges and a systematic study on different methods that have been used so far to convert paper ECG records into digitized form so that they can be retrieved efficiently. Initial challenge involved in the digitization process is gridline removal. In this process, information of ECG signal is also removed. None of the existing methods provide flawless gridline removal. The paper ECG used in hospitals differs in shape, size, formats, so the main challenge in digitization process is to achieve a worldwide ECG format.

Preface

Healthcare sectors often deal with a large amount of data related to patients' care and hospital workforce management. In recent time, healthcare biometrics has emerged as a new area of biometrics applications and as a positive effect, industry experts estimate that healthcare biometrics will drive global market with major technological transformation. However, in healthcare sectors, medical identity theft takes place when someone approaches free services and free healthcare items in exchange of medical record number or national ID card related information. Malicious persons or those who are willing to steal Medicare ID or social security number of patients, could use this information to obtain free of cost medical care, buy drugs and submit fake bills on behalf of genuine patient and could also damage the privacy of one's identity which is related to all important financial, healthcare and medical related information. In addition, it may damage the personal life of patient and it ends up with life-threatening situation while wrong information is circulated all around. Therefore, to protect the medical identity of every patient who takes benefit of healthcare and hospital services, biometrics technologies could offer to them the protection of their medical record of being theft from hackers and malicious users.

Biometrics technology brings a convenient way to safeguard the identity of individuals as well as valuable belongings associated with the subjects. In the last few decades, the world has witnessed a rapid, but consistent development of novel biometrics systems which started from nothing. With the most effective biometrics traits and computer vision techniques, these systems have been designed and thereafter, deployed to the fields where protection of identity is the biggest challenge and might be a threat to the society. Healthcare is one such area where biometrics could play a crucial role to shape up the public healthcare systems in better way. With the developments of novel biometric systems, hospitals and healthcare organizations can start adopting biometric systems for controlling logical access to centralized patients database and limit the physical movements to buildings and hospital wards. Several studies have been presented on medical biometrics in different reports where many experts have suggested to integrate entire medical facilities with biometric systems, so that better service can be provided to the patients and workforce management can be improved. Healthcare biometrics could be dealt with two issues: providing entry to access healthcare related resources and patient identification for maintaining patient privacy and safety. Healthcare biometrics could be utilized in access control, patient identification and monitoring, workforce management and patient record management. Sincere effort and effective use of healthcare biometrics might be helpful to eradicate fraudulent and susceptible happenings in healthcare services. At the same time, it can reduce the medical identity theft and manipulation of patient healthcare data and fraudulent claims made by hospital management. A careful selection of one or more biometric traits along with suitable machine vision, pattern classification, and effective soft computing techniques could serve the purpose of using biometric systems with existing medical facilities effectively.

This book has received contributions from leading educators, researchers and healthcare security experts who have proven expertise in developing high end biometrics solutions which are certainly useful for healthcare biometrics. As this book seeks to showcase the suitable biometrics technologies to be used under various healthcare environments, readers would find this book a useful asset. This book is presented in lucid manner by addressing the design issues of biometrics systems to be needed for healthcare framework. It would be a nice reference point for budding engineers, technocrats, researchers, educators, security experts, healthcare professionals and social scientists who would be interested to study healthcare issues and offer efficient solutions by implementing technologies for healthcare management.

The purpose of editing this book is to provide a collection of latest developments made on healthcare biometrics and its related areas to ensure exhibiting state-of-the-art technologies. This book discusses various issues related to healthcare biometrics and their effective solutions presented in a well-organized manner. There are a total of nine chapters in this book. Chapter 1 introduces biometrics systems for healthcare biometrics. It also presents state-of-the-art biometrics technologies, their suitability in healthcare biometrics and selection criteria of a biometrics system for healthcare biometrics. Next chapter presents a review of various performance metrics used for biometrics authentication which may be useful for healthcare biometrics because the metrics would be operating in the same environment. Chapter 3 discusses efficient indexing techniques for healthcare biometrics systems. These techniques would be found useful to reduce search space in biometrics database as well as the response time. The next chapter discusses deep learning technique for healthcare biometrics systems. It also discusses how deep learning can successfully be used for prediction of diseases, extracting medical knowledge, therapy planning and support etc. Chapter 5 presents the advantages and disadvantages of liveness detection method and proposes the possible application of voiceprint-based liveness detection schemes in the insulin pump system. Chapter 6 presents an EarLocalizer model to localize the ear, which is inspired by Faster-RCNN and it also highlights how this technology can be used in healthcare biometrics environment. Chapter 7 outlines the current state of the art of KINECT gait and activity authentication while focusing on emotional cues that could be observed from a human body. Chapter 8 presents an Autism Spectrum Disorder (ASD) screening system based on the combination of face and fingerprint biometrics. The novelty of the chapter is that the classifier based matching and fusion of the modalities has been proposed. Chapter 9 presents a review of the existing algorithms for digital conversion of Paper-ECG. It discusses the various challenges and a systematic study on different methods that have been used to convert Paper-ECG records into digitized form so that they can be retrieved efficiently.

Lastly, it is believed that the book will certainly be helpful to the communities composed of educators, researchers, faculty members, industry practitioners, graduate students, etc. Besides, the same would be proved to be beneficial to the professionals working with healthcare services, government policy making tasks and planning, government enforcement agencies, legal and regulatory services and with business enterprises in order to understand the practical essence of next-generation healthcare biometrics systems and future research directions. Further, it would help them in installing and commissioning of the healthcare biometrics systems in the healthcare sectors where susceptible activities of attackers are detected. The overall nature of the proposed book is to attract a large number of enthusiastic people who are actively involved in biometrics research and would like to move forward into the direction of healthcare biometrics. Hope, this little effort would be appreciated to bring together some significant research works on healthcare biometrics in this edition.

Dakshina Ranjan Kisku
National Institute of Technology Durgapur, India

Phalguni Gupta
National Institute of Technical Teachers Training and Research, India

Jamuna Kanta Sing
Jadavpur University, India

Acknowledgment

Firstly, we would like to thank all the authors for their outstanding contributions to this book.

Secondly, we would like to express our heartfelt gratitude to all the reviewers who provided insightful and constructive feedback on the contributed chapters. Thank you for making the time in your busy schedule.

Finally, but not the least, a special note of thanks to the colleagues at National Institute of Technology Durgapur, National Institute of Technical Teachers' Training and Research Kolkata, and Jadavpur University, Kolkata.

Chapter 1
Biometric Technologies in Healthcare Biometrics

Rinku Datta Rakshit
Asansol Engineering College, India

Dakshina Ranjan Kisku
National Institute of Technology Durgapur, India

ABSTRACT

The aim of this chapter is to introduce biometrics systems and discuss the essential components of biometrics technologies in the healthcare system. The discussion also includes the state-of-the-art biometrics technologies, selection criteria of a suitable biometrics system, biometrics identity management, and multi-biometrics fusion for healthcare biometrics system.

INTRODUCTION

Recently the concernment for biometrics is increasing swiftly to safeguard a patient's secrecy and the confidentialness of patient's actuality in the healthcare system. Identifying a patient and protecting a patient file is a ticklish task in any healthcare system. Due to improper identification, patients may lose their life in some cases. An unconscious patient cannot present an ID. The traditional patient tagging or mismatch between patient tag and hospital records are not free from flaws due to clerical mistakes or errors. This type of error can raise the inaccurate identification, wrong blood transfusion, or deceitful medicine infliction, which may be very dangerous for a patient's life. Therefore, there is a need for robust identification using biometrics in the healthcare system.

The term biometrics is a combination of two terms – 'bios' means life and 'metrics' means assessment. It refers to the metrics related to the physical and behavioral traits of human beings. The biometrics technology is implemented to analyze people for their identification, access control or surveillance by means of their biological information. Every person is unique and comprises a separate identity in the form of physical traits like fingerprint, face, iris, hand geometry and behavioral traits like voice, typing rhythm, and gait. Biometrics technologies are defined as automated way of recognizing or ascertaining the identity of an individual arisen on an idiosyncratic biological (anatomical or physiological) or be-

DOI: 10.4018/978-1-5225-7525-2.ch001

havioral characteristics. It is very ticklish to feign a biometrics authentication system as biometrics can give invulnerable and competent recognition of a person and they cannot be plundered or unremembered. A biometrics system comprises four major modules –

- A device is used to take sample of a particular biometrics trait of an individual. These devices capture biometrics sample in some digital form.
- A feature extraction software module is used to transform the input biometrics sample into a template that is the fundamental component of the matching phase.
- A matching software module perpetrates comparison of an input biometrics sample with previously stored biometrics templates.
- An interface facilitates the communication of a person with the application system.

The choice of the applicable biometrics technology is a crucial task. It depends on the circumstance where the identification or verification system is enacted, the user scheme, needs for matching precision, and the implementation cost of the system and strengths, and cultural convention that could influence user adoption. To choose the proper biometrics technology, the prime factor which is used is its accuracy.

Biometrics identity is the physical or behavioral traits of a human being. Using biometrics identity of a patient, a healthcare system can simplify patient admission and immediately get entree to previous medical files. Biometrics identity also facilitates to recognize senseless or wretched patients easily, find out abuse of medical services, and establish an unprecedented way of identification across all medical services.

Biometrics identity management in healthcare is a combination of computer security and business management to ascertain that resources are to be accessed by patients, physicians, and the right healthcare personnel must be available for the right reasons. As the healthcare systems are growing exponentially and related records are accessed by several healthcare personnel concurrently, therefore the records must be protected securely. Resources in a healthcare system are very important artifacts for patients, doctors, and other healthcare personnel, so assuring that resources are protected against unauthorized access become crucial. Biometrically authenticated identity ensures that only authorized user is accessing the resources, not by imposters using fabricated identity token. Biometrics identity management eliminates the likelihood of spoofing.

Sometimes unimodal biometrics identification systems fail to produce the expected level of accuracy due to image quality, intra class variability, sensitivity to noise, non-universality, and various factors. To enhance the perfection of the biometrics system, multibiometrics system is required, which required many biometrics samples as a proof of the same individual. The fusion in biometrics system refers to the process of combining two or more biometrics modalities together. Multimodal biometrics fusion is possible using different fusion strategies and various levels of fusions. The fusion of separate modalities can ameliorate the overall perfection of the biometrics system in the different challenging environments. This provides additional tools to healthcare to accurately identify patients, physicians, and other healthcare related personnel.

Correct identification of physicians, supporting staffs and patients along with patient record maintenance, assignment of access permission for healthcare providers are major challenges in any healthcare system. These challenges can be overcome by integration of biometrics technology with healthcare system, due to its ability to provide a mechanism for verification or identification of individuals. However, biometrics technologies are vulnerable to spoofing from the faked copies of different biometrics samples.

The development of biometrics authentication systems will recline on its efficiency to refuse fraud biometrics samples. Therefore liveness detection could be an important addition to biometrics technologies.

Integration of biometrics technology in healthcare system has closed doors for offenders to thieve the most worthy health information and stopped unauthorized access to laboratory apparatus in healthcare. Biometrics is the application of automated ways to identify an individual by using physiological or behavioral characteristic of that person. Biometrics systems are fitting the ground of an expansive array of extremely invulnerable identification and verification resolutions. Different instances of biometrics systems include fingerprint, face, retinal, palm print, foot print, iris recognition and hand geometry. As the security requirements are very high in healthcare sectors, therefore biometrics are to be used to control physical as well as logical access to restricted area and resources. There is no other worthy option than biometrics for person authentication in healthcare which is more safe, secure, affordable and efficient.

The aim of this chapter is to introduce biometrics and its propriety in healthcare system. This chapter comprehends the need of biometrics technologies in healthcare, state-of-the-art biometrics technologies, selection criteria of a biometrics technology, biometrics identity, and biometrics identity management in healthcare system, and multi-biometrics fusion for healthcare biometrics. Finally, this chapter discusses the various challenges of biometrics technology.

BACKGROUND

Biometrics recognition (Jain, Ross, & Prabhakar, 2004) is the way of identifying a person by his/her biometrics traits. Necessity of improving performance of person authentication by biometrics is increases day by day. A lot of exploration has been done on the improvement and deployment of biometrics authentication systems based on fingerprint, face, iris, retina, voice, palm print, hand geometry, voice, signature, ear etc. These biometrics technologies are already used for border control, passport verification, and attendance system and in some other sensitive places. However, recently it felt that biometrics could be a value added service to the healthcare system as it needs protection of patient identity, patient record, patient insurance documents and other valuable belongingness along with personal identification and access control.

In these days, society needs a faultless security measurement that will be very natural, serviceable and favorable at any costs. Biometrics based authentication systems make this requirement possible. Biometrics is a good means of automatic recognition of an individual based on their physiological and behavioral traits. These systems give safety founded on "what you are" rather than "what you possess" and "what you remember". Therefore, such systems are very useful in healthcare system where patients need not to get anything like passwords and is permanently with a person unlike smart-card etc. This type of security measure cannot be shared or used by any other person. Different systems based on various physiological and behavioral traits have been raised which cover fingerprint (Maltoni, Maio, Jain, & Prabhakar, 2009), face (Chellappa, Wilson, & Sirohey, 1995), iris (Bowyer, Hollingsworth, & Flynn, 2008; Ng, Tay, & Mok, 2008), retina (Tabatabaee, Milani-Fard, & Jafariani, 2006), voice (Kinnunen, & Li, 2010; Campbell, 1997), hand geometry (Sanchez-Reillo, Sanchez-Avila, & Gonzalez-Marcos, 2000) (Sidlauskas & Tamer, 2008), ear (Choras, 2007), Palm print (Harb, Abbas, Cherry, Jaber, & Ayache, 2015), palm vein (Han & Lee, 2012) (Mirmohamadsadeghi & Drygajlo, 2011), and signature (Faundez-Zanuy, 2005). To exceed the difficulty of traditional biometrics measures some new biometrics measures have been introduced like ECG (Shen, 2005; Riera, Soria-Frisch, Caparrini, Grau, & Ruffini,

2008), foot-print (Jia et al., 2012; Kumar & Ramakrishnan, 2010), and tongue print (Zhang, Liu, & Yan, 2010; Zhang, Liu, Yan, & Shi, 2007; Diwakar & Maharshi, 2013)etc. Some of these have been already used in forensic applications. Deployment of biometrics solution in healthcare is the best alternative for security measure.

Safety and privacy are the utmost stringent issues for data transmission in healthcare system. A biometrics based security resolution for encryption and authentication is proposed by (Zhang, Poon, & Zhang, 2009) for Tele-Healthcare systems. Without an authentication mechanism, the distant physicians will not know who sends the physiological signals. An invader can pose to be a genuine user and send false data which leads to a wrong diagnosis. A survey (Chandra, Durand, & Weaver, 2008) is carried out on healthcare consumers and providers. According to this paper, biometrics attained excellent adoption from providers than consumers. A web-based, biometrics-enabled, patients and Electronic Medical Record (EMR) management system BEPOC is presented in (Nwosu & Igbonagwam, 2014), which practically and efficiently addresses the issue of patient identification during medical mission. (Leonard, Pons, & Asfour, 2009) proposed the fingerprint, iris, retina scan and DNA (FIRD) framework that take advantage of a patient's biometrics prominences to uniquely connect them to their medical data. The framework makes an infrastructure that will distinctively recognize a patient by his or her complete Electronic Health Record (EHR) with accurate precision. A face recognition system is discussed by (Sumathi & Malini, 2010) to enhance E-health. (Kumar et al., 2017) proposed an anonymous and biometrics based authentication scheme for the Naked hospital environment. This paper considers a hospital scenario and show how patients can get authenticated without using gadgets to entree services from their circumstance. Medical sensors embedded in the environment will deliver them the requisite digital services after invulnerable authentication. Also, this work can resists various well known attacks such as insider attack, replay attack and identity privacy attack. (Jahan, Chowdhury, & Islam, 2017) presents extensive overview of biometrics technology in addressing E-Health security challenges. This paper proposed a robust and efficient biometrics scheme for user verification and authentication using fingerprint biometrics to enhance privacy and security in E-Health system. (Mohammedi et al., 2018) proposed a new biometrics-based authentication scheme for mobile healthcare system using mobile devices, biometrics inputs and Elliptic Curve Cryptography (ECC). The security solution and simulation outcomes exhibit that the proposed approach is applicable for use in practical healthcare applications.

NEED OF BIOMETRICS IN HEALTHCARE

Most of healthcare systems are maintaining Electronic Health Records (EHR) of their patients to provide fast, secure, and cost-effective care. Mainly passwords are used to safeguard computer systems in healthcare from unauthorized users. A patient health records may be mixed up or misplaced which might aftereffect in wrong medication. Moreover, medical information of one patient could be shifted to another person's record or to a deceitful person which causes medical identity theft. To keeping EHRs data integrity is a stringent feature. To secure critical health records of patients, passwords are not the best solution because often people forget or share their passwords. Password needs memorization. Biometrics offers a powerful solution for replacing password based authentication. Biometrics based authentication approach help healthcare workers to provide timely and quality care to their patients.

Also in healthcare sector healthcare professional have to give a lot of time on patient identification. But this worthy time could be spent on attention and treatment of other patients, if conventional patient identification system can be replaced by biometrics based patient identification. Identifying a patient correctly at healthcare facility is a crucial job. Producing an ID for patient identification is very difficult for senseless patients. Conventional patient tagging is not free from deficiency. Other important needs of biometrics in healthcare sector are to maintain security and access control because healthcare frauds like duplicate claims, fake billing and medical identity theft or theft of patient records are increasing day by day.

The acceptance of a particular biometrics depends on the adroitness of that biometrics technology to evolve the challenges of healthcare personnel identification as well as patient identification effectively with the help of their unique biometrics trait. The real-time patient identification as well as healthcare personnel identification decreases amount of stored data those are used for patient identification and healthcare personnel identification and related cost of data storing. Healthcare biometrics enables only authorized healthcare personnel to access patient's health record which eliminates the chances of medical identity theft and other healthcare insurance related thefts.

Integration of a biometrics trait with healthcare system boosts in different ways –

- Biometrics solution helps a healthcare system to verify a patient on each visit reliably.
- It dispels duplicate medical record keeping.
- It facilitates instant and secured access of patient records.
- It grants permission to only authorized healthcare personnel to access secure health records of patients.
- Accurately connect patients with their medical records.
- Maintain data integrity of patient record.
- It grants permission to only authorized persons to access a room or building physically.

BIOMETRICS TECHNOLOGIES

Biometrics technologies are widely used in different fields, mainly for secure access control and true person identification. Already biometrics technologies are used in a number of airports, for well protected building access, in cars (for unlocking the vehicle), in schools, at border crossing. Biometrics based systems use biometrics traits as powerful authentication tool. Biometrics traits are metrics related to human prominences. Biometrics systems are mainly based on two types of human characteristics – physiological and behavioral. Physiological characteristics comprise fingerprint, face, palm print, hand geometry, iris, retina etc. Whereas behavioral characteristics use pattern of human behavior like typing rhythm, gait, voice and signature. Biometrics identifiers of every person are unique. These are more feasible in identity verification than knowledge-based and token-based systems because biometrics traits are unique for a person. To ensure high security, biometrics based authentication systems are acceptable technology because every human being possesses the biometrics traits. The block diagram of a biometrics authentication system is shown in Figure 1.

Every biometrics system contains four main modules – sensor module, preprocessing module, feature extraction module, and matching module.

Figure 1. The block diagram of a biometrics authentication system

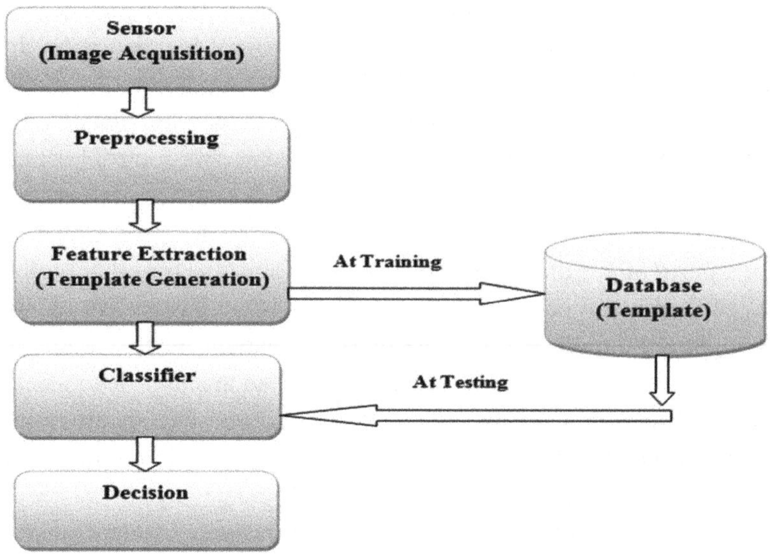

1. **Sensor Module**: This module provides an interface through which an individual interact with the system. This module captures all necessary data for identification or verification purpose. In most biometrics systems, this module is also known as image acquisition system.
2. **Preprocessing Module**: This module dispels artifacts from the data captured by a sensor to enhance the quality of the input biometrics sample, e.g. removing background noise or adding up missing information to the biometrics template. Image normalization, de-noising, filtering, histogram equalization are some of preprocessing techniques.
3. **Feature Extraction Module**: This module is very significant module of a biometrics system. Any extracted characteristic from the image is referred to a feature. Invariant features from biometrics samples are extracted using some algorithms and then templates are generated from these extracted features in this module.
4. **Matching Module**: In this module, probe sample is send to the classifier that compares it with stored templates and produces the decision.

There are two phases of every biometrics system – enrollment and testing. When an individual uses a biometrics system for first time, he/she enrolled himself/herself by his/her biometrics traits. In posterior uses, biometrics sample is captured from an individual and compared with the template stored at the time of enrollment.

Every biometrics based authentication system works either in identification mode or in verification mode. In identification mode the biometrics system compares a probe sample with all templates (one-to-many) present in the biometrics database. In verification mode the biometrics system compares a probe sample with a particular template (one-to-one) stored in the database in order to ascertain the identity.

Finger Print Recognition

Fingerprint recognition (Maltoni, Maio, Jain, & Prabhakar, 2009) is the automated process of assimilating the pattern of furrows, ridges and minutiae points of the fingers to identify a person based on the comparison of two fingerprints. Fingerprint is very effective biometrics to restrain unauthorized personnel from accessing patient's records and used for true healthcare personnel as well as true patient identification. A fingerprint pattern of any human consists of dark lines of ridges along with white lines or valleys between them on the skin surface of a finger. The flow of ridges and their features are used for fingerprint recognition. Research reveals that fingerprints of two persons were not found to be the same, not even for the identical twins and fingerprint patterns remains same for a person throughout his/her life. The fingerprint images those are used for automatic identification can be acquired using different sensors. Actually digital imaging technology is used in capturing, storing and analyzing the fingerprint data.

There is a concern to search an extremely accurate and feasible way to secure critical healthcare records and restrain any faults which could lead the dangerous problem for life. Fingerprints which are accurate form of identification can be used to identify a person uniquely. The fingerprint biometrics mainly needed in healthcare systems to prevent unauthorized access of patient record, true patient identification as well as healthcare personnel identification. The use of fingerprint biometrics in healthcare system will raise secrecy and indemnity of patients by assuring unerring identification.

Fingerprint biometrics facilitate hospitals to ascertain consent to government rules such as the Health Insurance Portability and Accountability Act (HIPAA) and other such laws worldwide that decree safeguard of patient's privacy. Fingerprint authentication allows only authorized persons to access patient's data. Therefore, with fingerprint authentication patients and healthcare professionals can grope very sure and certain regarding the privacy of their healthcare data.

Hospitals and healthcare organizations can implement fingerprint technology effectively for registration of new patients, controlling visitors, proper distribution of medicines, executing payments for medical claims, maintaining the privacy of patient's information and impenetrable entree to the premises. Also fingerprint readers can be used in hospitals to access medical cabinets for maintaining patient's security. This facility will help to manage both medical supplies and medications efficiently. This process has been implemented in a number of major US hospitals and used by millions of healthcare related personnel.

Fingerprint systems have several advantages. Fingerprints are always undeniably tied to the actual user and always on them. A fingerprint system relieves the user from the liability of recalling to bearing their smart cards or tokens or to remember passwords. Fingerprint system gives very high accuracy and it is very cheap in terms of cost. This system is very simple to use. Fingerprint system requires small storage space to store biometrics templates. However, this system is not also free from flaws and safe to use. Temporary or permanent injury on finger, dryness or dirty of the finger's skin can cause some serious problem for fingerprint system. In addition, fingerprint spoofing is also considered as a major concern for successful identity verification.

Already the fingerprint identification system has been effectively implemented in different departments such as law enforcement and government agencies. It is an existing system which shows a high level of validity that can be used for strengthening physical as well as logical entree within a healthcare facility. Fingerprint systems are accepted extensively by hospitals because of its credibility, favor and the efficiency to be spread out in a wide range of circumstances.

Advantages

- Accuracy is very high.
- It is most economical biometrics authentication technique.
- It is a developed biometrics technology
- Easy to use.
- The memory required to store template in database is less.

Disadvantages

- Most devices are incapable to enroll a little percentage of users.
- Perfection of the system can come off over time
- It can make mistakes with the dryness or dirty of the finger's skin.

Face Recognition

Face is another physiological modality which can be used to identify and authenticate a person in healthcare. Face is an individual characteristic of a person. In our daily life, a person is identified by his/her face. Remembering and distinguishing different faces is possible because human brains have natural ability to do this. It is important application to have a human visual system in machines where machines can behave like a human brain and have the capability to recognize different faces with different identity.

Face recognition (Chellappa, Wilson, & Sirohey, 1995) system is an automated technology which is able to identify or verify a person by processing their digital image or a video frame from a video source if their face recognition individuality has been pre-introduced. The face recognition algorithm extracts discriminating facial features like shape and position of nose, eyes, jaw and cheekbones stored these extracted features in the form of templates in a database. This type of system can be feasible to identify a patient at the time of registration, at the time of admission and at the time of medication. Face recognition systems are also very useful to identify healthcare personnel and restrict the access of patient's health record as well as cabinet of the patient.

The social acceptability of face recognition technology is very high and implementation cost of this system is low. Automated face recognition technology can identify a person in the crowd without letting him know, if it installed properly in airports, multiplexes and other public places. The system does not crave the assistance of the probe subject to work which makes it most suitable for healthcare system because in healthcare system, in some critical cases, patient can not cooperate with the system. Face recognition is a clean process because the patient never needs to touch it or have direct contact with anything, which could create a major problem in healthcare due to deadly infection.

Further, face recognition is considered a complicated problem in computer vision due to changes in pose, illumination, occlusion, facial expression, head rotation, and scale. Face image quality is very significant in face recognition systems as great degrees of changes are feasible in face images. Among all biometrics systems, facial recognition has the highest false acceptance and rejection rates, which raise questions on the usefulness of face recognition software in cases of high security application area. Face recognition system performs well in case of full frontal faces along with neutral facial expression and capturing of full frontal faces with neutral facial expression of patients and healthcare personnel are possible because overt (user aware) application of face recognition system might be useful in healthcare.

Health record privacy is the main disquiet when it is secured by storing biometrics data tag with health records. Data stored by face biometrics can be attained by third party if not stored properly. Face recognition technology can be spoofed by intruders by using facial mask, digital images or printed images of a person.

Advantages

- No direct contact is required.
- Implementation cost is less.
- Easy for people to verify.

Disadvantages

- Perfection of the system is deteriorated by changing the image acquirement environment.
- Recognition rate is also decreased by changing in physiological characteristics.
- Propensity of privacy misuse is high due to non-cooperative enrollment and identification ability.

Retinal Scan

A retinal scan (Tabatabaee, Milani-Fard, & Jafariani, 2006) is a biometrics authentication technique that uses the unique patterns of retina blood vessels to identify or verify a person. The human retina is located in the rear portion of the eye. This is thin tissue made of neural cells. Each person's retina is unique due to the depth form of the capillaries that supply the blood to retina. Even identical twins do not have an identical pattern. The retinal pattern typically remains unaltered from birth until death. Due to the permanence of unaltered pattern of retina, it is the most accurate and feasible biometrics, that can be used for authentication. Retinal scan can be used in healthcare system for true patient identification, healthcare personnel identification and physical as well as logical access control.

A retinal scan is made possible by using an unperceived beam of infrared ray having low energy into an individual's eye when they look via the scanner's eyepiece. This ray of low energy infrared light quests a standardized path on the retina as retinal blood vessels imbibe light more spontaneously than the neighboring tissue and the extent of reflection varies at the time of scan. These variations of pattern is digitized and stored as templates in database.

Retinal scanners are mainly used for authentication and identification purposes. Retinal scanning has been exploited by different government agencies including the FBI, CIA, and NASA. Recently, retinal scanning gains major popularity in commercial fields.

Retinal scanning has several advantages. Biometrics authentication using retinal scanning has low occurrence of false positives. False negative rate is extremely low (almost 0%). Retinal scanning is a speedy approach which identifies a person very quickly and it is highly feasible because no two people have the identical retinal pattern even also in identical twins. This is very hard to replicate a retina. The eye from a dead person would collapse very fast, therefore in retinal scanning liveness detection is not required. Retinal scanning also has some medical application. Retina scan based biometrics is more hygienic because the patient never has to touch anything. Though retinal scanning gives very high accuracy, it can be affected by several factors.

Advantages

- False positive rate is very low.
- False negative rate is extremely low (almost 0%).
- It is highly reliable.
- Result produced very fast by this technology.
- Highly accurate.
- Replicate a retina is impossible.

Disadvantages

- Measurement accuracy can be affected by several diseases like cataracts and severe astigmatism.
- Retinal scanning is not very user friendly because it requires very close contact of the person with the camera while scanning is done.
- Equipment cost of retinal scanning is very high.
- Regular uses of this technology leads to serious eye problems.

Iris Recognition

Iris recognition (Bowyer, Hollingsworth, & Flynn, 2008; Ng, Tay, & Mok, 2008) is an automated method to identify an individual based on one or both of the irises of a person's eyes by using mathematical pattern-recognition techniques. The complex patterns of irises are permanent, unique, and can be observed from some distance.

Iris recognition uses fine near infrared illumination to capture complex patterns of iris images by video camera technology. Digital templates are generated from iris images by computational techniques and stored in database. These stored templates are further used for identification of an individual. The speed of iris recognition is very high with low false match rates.

Iris recognition technology is already implemented in several countries around the world for the use of some national ID programs and passport-free automated border-crossings. Besides the speed of matching and utmost counteraction to false matches of iris recognition technology, another advantage of it is the permanence of the iris as a protected, internal and externally visible part of the eye.

As people's eyes are unique, iris scanners are more suitable in high-security applications. The uniqueness and permanence nature of iris and the technology's contactless and non-intrusive nature make it more suitable for healthcare system to safely recognize a patient at the reception desk, blood drawing room or doctor's room, or to identify a doctor and other healthcare personnel. Also iris recognition technology can be used by people after eye surgery because eyes also are same after surgery, and iris scanners can be used by blind people as long as their eyes have irises. Generally, eyeglasses and contact lenses do not cause improper readings.

Advantages

- **Accuracy:** Iris recognition is the ideal biometrics recognition technology with respect to accuracy while compared to other biometrics. The false acceptance rate as well as false rejection rate is extremely less in this modality which ascertains high recognition accuracy in its results.

- **Scalability:** This technology can be used for both large and small scale applications because it is highly scalable.
- **Distance:** Iris scanning can be possible from a usual distance.
- **Stable:** Iris patterns are permanent in a person's whole life.
- **Easy to Use:** Iris recognition system is very easy to use. A person only need to stand still in front of the camera and the task of capturing image is done immediately. It is a comfortable process for everyone.

Disadvantages

- **Expensive:** Iris scanners are very expensive in terms of cost.
- **Distance:** Iris is very tiny in size and can't be captured from a several meters space.
- **Movement:** A person has to be standing steady in front of the scanning instrument to be enrolled by iris scanners and movement cause serious problem.
- **Reflection:** The presence of reflections sometimes causes a problem for iris scanning. Mainly it could occur due to lenses, eyelashes that would cause a reflection.
- **Infrared Light:** The constant use of this system may damage the iris due to the scanning with infrared light.
- **Storage Requirement:** A large memory space is required for storing the data.
- **Eyelids:** It is very difficult to control frequent blinking of eyelids.
- **Transformation:** Iris may disfigure non-elastically as its size may be changed due to some medical conditions or other conditions.

Hand Geometry Recognition

Hand geometry (Sanchez-Reillo, Sanchez-Avila, & Gonzalez-Marcos, 2000; Sidlauskas & Tamer, 2008) is a biometrics cue that might be used to recognize users by the appearance of their hands. Hand geometry scanner measure different dimensions of a hand and compare those measurements to the stored templates in database.

Hand geometry devices have been made since the early 1980s, making hand geometry the first biometrics to trace extensive computerized use. It mainly includes applications like time-and-attendance management and access control.

Hand geometry is not thought to be as unique as other biometrics technology for high-security applications. Hand geometry can give increased accuracy when combined with other forms of biometrics traits or other forms of identification, such as identification cards or PIN. In large demography, hand geometry is not a good choice for one-to-many applications, where a user is recognized from his biometrics except any other identification.

Hand geometry is mainly feasible for practical usage in real-world applications where security in an important concern. A hand geometry recognition process mainly captures geometry of the hand like length, width and other qualities of the finger. In any hand geometry based recognition system, hand image segmentation is an important step. The accuracy of hand geometry based system is mainly depends on the quality and exactness of the segmented hand image.

Advantages

- Its cost is medium as it only requires a medium resolution CCD camera and a platform.
- Mainly algorithms having low computational cost are used for this system which increases the speed.
- The template size is very small (9-25 bytes) which reduces storage requirement.
- It is very simple and striking to users which leads to a null user rejection
- It can be easily combined into other systems.

Disadvantages

- It has limited accuracy.
- It is not competent for arthritic person, as placing the hand on the scanner is very difficult.

Palm Print Recognition

Palm print (Harb, Abbas, Cherry, Jaber, & Ayache, 2015) recognition is an authentication process founded on the distinctive patterns of palm prints of a person's hands. Palm print recognition systems usage a scanner or a camera to capture the palm print images, and software that processes image of biometrics sample of a person's palm and assimilates it to a template of that person stored in database. Palm print scanners use thermal, optical or tactile processes to express the details in the pattern of ridges and bifurcations in an image of a human palm, also details like, scars, creases and texture are also considered. These three systems lean on heat-emission analysis, visible light analysis and pressure analysis, respectively. To capture images of palm print persons must touch their hands to a screen of palm print scanner or may be contactless.

Fingerprints and Palm prints are often used simultaneously to improve the exactness of identification. A handprint covers more skin area which includes more details, since making false positives is improbable and simultaneously generating intended falsification is very difficult. In other circumstances, such as criminal inquisition, a partial or full palm print may be acquired when fingerprints are not present. For example, a criminal might endue gloves to avoid giving up fingerprints but unknowingly leave a partial palm print when a glove slips at the time of a crime.

The handprint recognition was first used in 1858, when Sir William Herschel records the prints of Indian civil service employees working under him and assimilated them to new handprint samples taken on paydays to ascertain identification.

Friction ridge impression is used in both palm and finger print biometrics to represent the useful information. This information assembles ridge structure, ridge flow, and ridge characteristics of the raised part of the epidermis. The data illustrated by these friction ridge impressions authorizes a decision that similar areas of friction ridge impressions either risen from the same person or could not have been made by the same person.

Advantages

- The palm area is much larger than fingerprint, hence it generate more distinguishable features which makes it more worthy in identification process.

- Palm prints are easily taken by low resolution devices

Disadvantages

- Generally the palm print scanners are heavy and costly as they required to take a larger area of the hand.
- It requires more processing time.

Footprint Recognition

Footprint recognition (Jia et al., 2012; Kumar & Ramakrishnan, 2010) is an automated system which measures the features of footprint for knowing the identity of an individual. A footprint is a natural measure for taking a particular identifier which does not vary much over time. Footprint identification is relatively a new technology which is a substitute to access control in health domains such as thermal baths and spas. This technology can be facilitating healthcare personnel to identify new born babies at hospitals. Different research groups from all over world proposed different novel approaches for footprint identification. Footprint technology utilizes texture, foot shape, foot silhouette and friction ridge to identify a person.

Infant fingerprints are not suitable for identifying a baby, but infant footprints are more suitable for scanning and captured footprints are further processed by algorithms to extract invariant features from the footprint which are used for identification.

In many countries recording footprint of infant are mandatory. Currently maximum hospitals use ink for the foot to capture footprint on a paper. This method has various restrictions and well trained healthcare personnel are required for this.

Though several biometrics applications are available in the market, there is a concernment of low cost authentication technology that can authenticate a low quality test image which is captured in low resolution. Still footprint is a forlorn field of research. But footprint biometrics has both the behavioral and physiological characteristics to identify a person. The footprint recognition needs removal of shoes and socks to capture the image of bare foot which increases the extra concern of the user. Also dirty feet or cracked heels make the image noisy which reduces the accuracy of this technology. Physiological and behavioral biometrics characteristics makes footprint more suitable for authentication. Foot print has geometric features which helps the system at the time of recognition. Minutia features of foot biometrics are considered completely unique to an individual. The minutiae characteristic is already experimented in fingerprint analysis for its uniqueness. A healthcare system can use footprint system for authentication in those places where people strictly enter in bare foot to maintain cleanliness.

Advantages

- Footprint recognition is very attractive for newborn babies due to its easy applicability.
- Implementation cost is low.
- This technology is widely accepted.
- Image acquisition is very fast.
- Very fast and accurate identification.
- Using high resolution scanner better accuracy can be obtained.

- It is a non-invasive method.

Disadvantages

- Performance of footprint system is affected due to dirty feet and cracked heels that make the image noisy which reduces the accuracy of this technology

Palm Vein Recognition

Palm vein (Han & Lee, 2012; Mirmohamadsadeghi & Drygajlo, 2011) recognition is a biometrics authentication tool which is used to recognize a person uniquely based on vascular patterns of his/her palm. A palm has an extensive and more intricate vascular pattern which comprises enormous discriminating features for personal identification. An infrared ray is used to capture the palm vein image of the user's. The lighting of the infrared ray is limited by the illumination on all direction of the sensor, and the sensor is capable to take the palm image irrespective of the location and stroll of the palm. The palm vein recognition system matches the transformed vein pattern with the enrolled vein pattern, while measuring the location and orientation of the palm by a pattern matching algorithm. The palm is an ideal portion of the body for this system; it usually does not have any obstruction (like hair) for taking the image of blood vessel pattern. Also the palm vein pattern does not affected by the skin color. In vein vessels the deoxidized hemoglobin takes up light having a wavelength of about 7.6×10^{-4} mm within the near-infrared area.

Palm vein biometrics-based authentication is much more advantageous because it can identify a patient accurately as well as increase patient safety in healthcare. Palm vein biometrics for patient identification eliminates false medical records and protects medical identity theft and cheat at the point of health service by attaching a patient's biometrics profile to true Electronic Health Record (EHR) to ascertain accurate and safe treatment at healthcare organization. At first palm vein based biometrics authentication generates a palm vein map from the image of a patient's palm which is captured by using near-infrared light. Then using this palm vein pattern unique electronic medical record of a patient is extracted. Actually, a palm vein recognition reader is an automated system which scans palm of a patient and automatically fetch their electronic medical record. This technology is 100% safe. The palm vein recognition is safe for a user as near-infrared light used in the scanner is not so harmful. Already palm vein biometrics has been experimented and certified to be out of danger for use in healthcare as an access control and identification tool.

Advantages

- **Security:** Palm vein patterns are very secure and tenacious to imitate as palm vein patterns subsist inside the body.
- **Usability:** The palm vein biometrics reader is very simple to use for patients. People needs to only place a flattened hand on the palm vein hand guide and deploy their fingers. The palm vein sensor then takes the image of the biometrics sample.
- **User Acceptance:** User acceptance of palm vein biometrics is very high.
- **Increasing Data Integrity:** At appropriate patient care touch points palm vein biometrics accurately links a patient with his/her electronic health records and maintains patient data integrity.

- **Accuracy:** The FAR and FRR are generated by this system is very low, hence it is deliberated as a more accurate biometrics modality.
- **Long Term Stability:** Palm vein biometrics is not changed over a long period of time.

Disadvantages

- **Cost:** Implementation cost of palm vein technology is very high.
- Palm vein print can be affected by the humidity, body temperature, heat radiation, nearness of vein to surface, focus and camera calibration.

DNA Recognition

DNA profiling (Gehring et al., 1995) is also known as DNA testing, DNA fingerprinting or DNA typing is the method of deciding an individual's DNA features, which are unique to a person. DNA testing desired to identify a species, more than an individual, is known as DNA bar coding.

DNA recognition is generally used as a forensic tool in criminal inquisitions, for example assimilating one or more individuals' profiles to DNA found at a crime scene so as to measure the possibility of their presence in the crime. Also, it is used in medical research and parentage testing.

DNA profiling can also be used to prevent medical identity theft and patient identification. DNA profiling is highly accurate. Recently, DNA records are maintaining in file by most countries. DNA profiling has various practical uses. It can used to match tissues of organ donors with those of people, who need transplants. It helps to identify diseases that are passed down through a family, and helps to find medication for some hereditary diseases.

Advantages

- It is an easy and painless method for the person being tested.
- It is an affordable and unfailing technique.
- It is less time consuming.
- Anyone at any age can be tested with this method without any major concerns.
- This can be used for several cases like legal claims, missing person's cases, identification for the military, and paternity and prenatal testing.
- The technique is highly developed and improved.
- Its accuracy is very high.
- Error in this system is impossible.
- It is a standardized method.

Disadvantages

- The sample of DNA can easily be ruined before testing.
- This process is very complex and tiring, and results produced by it may be hard to interpret.
- Multiple samples are required to test. Therefore much more time is required for ideal accuracy.
- This approach is very expensive.

ECG Recognition

The electrocardiogram (ECG) is a pathology tool that is regularly used to measure the muscular and electrical operations of the heart. The heart is an electrical pump and the heart's electrical alacrity can be assessed by electrodes given on the skin. The electrocardiogram can assessed the rhythm and rate of the heartbeat, and also give imperceptible proof of blood flow to the heart muscle.

In healthcare system the electrocardiogram (ECG) biometrics process can authenticate the patient's identity as well as protect the privacy of their sent electrocardiogram signal. In ECG recognition discriminating features are extracted from heartbeat waveforms.

The ECG can be used as a biometrics tool (Shen, 2005; Riera, Soria-Frisch, Caparrini, Grau, & Ruffini, 2008) because it satisfies the requirements of biometrics characteristics. The ECG pattern of each individual is unique, it can easily measure, and this characteristic of a person does not change over time. The use of a biometrics reclines on the purpose and application of the system.

Advantages

- ECG recognition is a non-expensive approach.
- It is safe, and easy to perform.
- The necessary equipments of this technology are easily available.

Disadvantages

- It doesn't constantly give the proper diagnosis.
- ECG signals are influenced by different noise sources such as power line or electromyographic noise and motion artifacts.

Tongue Print Recognition

Tongue print (Zhang, Liu, & Yan, 2010; Zhang, Liu, Yan, & Shi, 2007; Diwakar & Maharshi, 2013) is a new biometrics tool that can be used for user authentication. Tongue print is unique to a person and cannot be easily counterfeit because no two tongue prints are same. Tongue contains several identifiers such as ridges, wrinkles, seams, and marks. Tongue print recognition captured digital image of the tongue and matched with a database for verification. Research on tongue prints is at an initiatory stage.

The presence of fissures was the most common morphological characteristic noticed in the surface of the tongue. The fissures were mainly located in the central region of the tongue. The predominant shape of the tongue in both males and females was "U shape." V-shaped tongue with a sharp tip was also observed in a substantial sample of females. Due to uniqueness of tongue prints, it can be used as a powerful biometrics tool in healthcare organization for person identification and access control.

Advantages

- The shape and surface textures of tongue is unique to each person.
- This organ that can be freely and completely exposed for test.

- **Inner part:** Tongue is a vital internal organ that is well protected from environment and enclosed in mouth.
- **Liveness:** Tongue is an important biometrics by which liveness of a person is detected easily. Sticking out one's tongue is an undeniable proof of life.

Disadvantages

- Tongue is affected by diseases, which reduce the accuracy of person identification.

Ear Recognition

Automatic identity recognition by ear biometrics (Choras, 2007) is a new biometrics research field, but has attained a rising amount of effort over the past few years. The ear images can be easily captured from a distance without the knowing of a person. This advantage makes ear recognition competent for security and surveillance applications. The ear biometrics can be implemented via ear photographs for automated identification applications. The use of 2D or 3D ear images for human recognition varies from the use of ear prints. Many countries have commenced ear prints as physical proof in different criminal happenings. Ear prints are not widely adopted in court due to the absence of scientific consonance as to their identity. The usage of ear thermo grams helps to alleviate the matter of occlusion due to accessories and hair. In addition, the adjacent background of the ear is very predictable, as it is permanently present on the side of the head. The ear biometrics does not crave closeness to capture images. Ears have performed an important preface in forensic science.

Like all other biometrics ear recognition consists of four stages. The first step is ear detection which localizes the ear's position in an image. The ear recognition system generally uses a rectangular border to imply the ear's spatial border from the side profile of an image of face. Ear detection is a complex task and errors at this step can affect the perfection of the system. Second step is feature extraction which extracts the discriminating feature from ear. Third step is matching which assimilates the extracted features of the test ear image to the templates stored in the database to make the ear's identity. The matching stage yields scores introducing the matching proximity to other ear images. Last stage is matching which uses the match scores to generate a final decision.

Advantages

- Ear does not change over time during a human life.
- Colour distribution is more uniform in ear.
- Ear is not sensitive to expression or colour changes.
- Images of ear can be taken from a distance without knowing of a person.
- Ear is also works in low resolution.

Disadvantages

- Sometimes ear is partially covered by hair; in those cases capturing image of ear is very difficult.
- The uniqueness of ear is moderate.

Voice Recognition

Voice recognition or speech recognition (Kinnunen & Li, 2010; Campbell, 1997) is an automated system which used computer software program or hardware device having ability to decode the human voice. To handle a device or enact commands without using a mouse, keyboard, or any buttons, voice recognition is mainly used. Recently Automatic speech recognition (ASR) software requires the user to train the ASR software to identify the voice of a person.

A single digit spoken by a user is first recognized ASR device in 1952 and it was not computer driven. Nowadays, ASR programs are used in many fields like Healthcare, Military, Telecommunications and personal computing.

The main use of voice recognition system in healthcare organization is the generation of customized patient's room that will deliver unique services and technology. For example, researchers rely on the future hospitals that will launch using touch screen monitors having voice recognition software in every patient's room. These will support the health care personnel to access patient health records very quickly with simple voice commands. According to the National institute of Health (NIH), speech-recognition technology should be accepted by every health care organization due to its excellent benefits. The success of voice recognition technology in healthcare organization is depends on proper employee training and implementation.

Advantages

- Voice recognition increases productivity.
- It helps with menial computer tasks, such as scrolling and browsing.
- It can support people who have trouble using their hands or trouble in movements.
- It can help people who have cognitive disabilities.

Disadvantages

- Voice recognition can be hacked easily with prerecorded vocal messages.
- It needs some time to adjust to each user's voice.
- Its accuracy affected due to background noise.
- It can be problematic in a dense circumstance.
- Sometimes the monopoly of software companies causes a problem.

Signature Recognition_

Writing is human's behavioral characteristics as well as an acquired skill. Signature recognition (Faundez-Zanuy, 2005) is a simple means of person identification which is used in banks, offices, and different sectors. Signature recognition demands a person to show a specimen of text which induced as a basis of measure of their writing. The aim of the signature recognition system is to recognize the writer of a given signature. There are two different ways to recognize a signature.

The first technique is static which demands the individual to write their signature on paper. Then this signature will be digitized by a camera or an optical scanner. The digitized data then recognized by some software algorithm that checks text of the signature by testing its shape. This approach is known as an "off-line" mode or static mode of recognition.

The off-line mode of handwriting recognitions are a socially adopted identification process which are formerly used for bank, offices, credit card and different business transactions. Off-line signature processing can be used for healthcare organizations to identify patients, healthcare personnel, validate the contracts and past health record's documents.

Another approach for signature recognition is dynamic which used dynamic characteristics of the signature, like, acceleration, velocity, pressure, timing, and direction of the signature strokes which are very difficult to replicate. Static signature can be replicated by a trained human counterfeiter or manipulation by a computer, but dynamic characteristics are unique and very complicated in nature to the handwriting style of a person. These strengths of dynamic signature recognition make it suitable for true person authentication in healthcare.

An electronic signature system enables a healthcare organization to speed up admission, registration, and documentation processes and reduce errors while saving on printing and filing costs. Signature pads are the ideal choice for capturing electronic handwritten signatures at the admission desk or at the point of medical care. It seems like a conventional ink pen. Patients can read information regarding their healthcare services on the screen and see their signature in real-time when signing.

These reliable devices guarantee a low total cost of ownership (TCO) and ensure a Return on Investment in less than 1 year.

Advantages

- **Speed:** This reduces paper documents maintained by healthcare organizations. Within very short amount of time patients can get registered, healthcare personnel as well as patients can be identified.
- **Costs:** The digital signatures on electronic documents are cheap compared to paper documents
- **Security:** The risks of healthcare documents being intercepted, read, dilapidated, or deflect are reduced by digital signature.
- **Authenticity:** Like signed paper document, a digital signature which is present in a digital document can stand up in court.

Disadvantages

- **Expiry:** Technological products which use digital signatures are highly dependent on the technology used by it and technological productions have a low life span.
- **Certificates:** Digital certificates are required from trusted certification authorities to use digital signatures.
- **Software:** Some verification software at any cost is required to work with digital signature.
- **Law:** In some states and countries, laws regarding cyber and technology-based issues are powerless or even non-existent. In these cases digital signature is very risky.

Selection Criteria of Biometrics Technology

The selection criteria of biometrics for healthcare system will rely on several application-specific factors and the circumstances in which the recognition process is completed. The factors those are very important while selecting a biometrics technology are throughput, the user profile, the overall system cost, requirements for matching perfection, complexity, long term maintenance cost, abilities, and cultural and social issues that could affect the adoption of user.

The main criterion in the election of appropriate biometrics technology is its perfection. When the probe biometrics sample is assimilated to the templates stored in the database for verification process, a matching score is used to assure or refuse the identity of a person. A threshold for this matching score is decided by the system designers to provide the intended level of matching perfection for the system. The perfection of a biometrics based authentication system is generally standardized by the false acceptance rate (FAR) and false rejection rate (FRR). The system administrators can adjust sensitivity to FAR and FRR to get to the intended level of matching accuracy protecting the security need of system. The false acceptance rate means the possibility that an individual is verified incorrectly by a biometrics system or accept an imposter. The false rejection rate means the possibility that a correct person is rejected by a biometrics system.

A comparative study of some biometrics traits is given in Table 1.

Biometrics Identity and Identity Management

Biometrics identity refers to the biological characteristics of a person which help to identify a person successfully. Identity Management or Identity and Access Management (IAM) is mainly used to ensure that resources are accessed only by the right people, at the correct time and for the true causes in the

Table 1. Comparative study of some biometrics traits

Biometrics Technology	Accuracy	Cost	Equipment used	Social acceptance	Use	Interference
Fingerprint	High	Medium	Scanner	Medium	Patient, healthcare personnel identification and access control	Cut or wound on finger, dirtiness and roughness of finger.
Face Recognition	Medium	Medium	Camera	High		Illumination, facial expression, pose, occlusion
Retinal Scan	High	High	Camera	Low		Irritation
Iris Recognition	High	High	Camera	Medium		Glasses
Hand Geometry Recognition	Medium	Low	Scanner	High		Rheumatism, arthritis
DNA	High	High	Test equipment	Low		Multiple samples are required, complex
Signature Recognition	Low	Medium	Touch panel, optic pen	High		Changeable
Voice Recognition	Medium	Medium	Microphone	High		Cold, noise

field of computer security and business discipline. Biometrics authentication is used in computer science as a tool for access control and identification. Also, it is used to recognize individuals in crowd that are under surveillance.

Also, it comprises the management of explicative information of the user and how and by whom that information can be attained and changed. Identity management manages entities like users, hardware, network resources and software applications. Biometrics identity management gives the authority to set biometrics to that system as a substitute for users' credentials.

Today keeping the security of patient's data is a high priority for healthcare systems. As more organizations replace data by electronic medical records, healthcare identity management resolutions can aid to restrain unauthorized access. Biometrics identity management provides faster access to patient's data as well as securing patient's information. Users can simply interact with biometrics sensor and immediately access their remote session.

Biometrics identity management can be deployed in various environments like self-service kiosks to the emergency room, patient's identification for different healthcare services and access to their Electronic Health Records accurately. Biometrics identity management provides several facilities to healthcare users -

1. **Token Replacement:** Biometrics identity management eliminates token based identity management (token can be lost or stolen) and improves security using biometrics. It eliminates the task of using and managing tokens in healthcare facilities.
2. **Insurance Fraud Prevention:** To acquire valuable private data hackers is targeting healthcare databases. Mainly health records are stolen from doctor's offices or major health insurer. Biometrics identity management through biometrics authentication adds an extra layer of security that can significantly reduce insurance frauds by ensuring that every patient is exactly who he or she claims to be. Biometrics identity management reduces fraudulent services. It replaces insurance cards totally by biometrics or adds biometrics as a second factor. For this, patients can easily sign in at the doctor's office, hospital or pharmacy.
3. **Electronic Health Records (EHRs) Maintenance:** To protect the Electronic Health Records (EHRs) from unauthorized access is the biggest challenge in medical industry. Biometrics identity management enables very fast and secure access to the patient's records. It also controls who can access patient's medical records. It generates reports on any type of access request.
4. **Physical Access Control:** Medical facilities require high security to protect patients or their private data. Replacing token with a biometrics identity access management solution will provide a better security in accessing any building or a room in healthcare system. Biometrics identity access management keeps records of people entering any facilities (building or room). It secures any facility in healthcare system with biometrics authentication. It reduces the cost associated with the loss of tokens.

Multi-Biometrics Fusion for Healthcare Organization

Most of existing biometrics systems use a single biometrics trait for verification or identification. However, monomodal biometrics systems are unable to fulfill the needs of large-scale biometrics systems such as larger population coverage, varied deployment circumstance, and more demanding perfection requirements. To fulfill the demands of large-scale biometrics systems, integration of extra sources of information is required to tighten the decision process. A multi-biometrics system integrates useful data from multiple biometrics traits or from algorithms or from sensors, and other elements to create a recognition procedure more harden. The fusion of biometrics has various benefits such as growing population coverage, impeding spoofing activities and decreasing enrolment failure.

Although it is evident that the biometrics technology based applications are already received a considerable amount of attention and they also reached a particular level of maturity, however, any biometrics system is not free from challenges which are to be addressed to minimize their risks. Multi-biometrics system (Jain & Ross, 2004) is a way that takes advantage of uniqueness of various physiological and behavioral characteristics of a person and combines data at different fusion levels. This gives additional tools to healthcare for identifying patients in the cases of injury or trauma correctly.

To solve problems faced by the conventional uni-modal biometrics systems, multi-biometrics fusion provides more discriminating information by combining multiple biometrics evidences at different implementation levels to generate a more accurate result or description. Multi-biometrics systems are proposed with a motive to raise the perfection and robustness to intra-person changes and to noisy data. Fusion in multi-biometrics is mainly introduced to make an identification or verification decision based on the information combined from various biometrics traits.

Uni-Biometrics vs. Multi-Biometrics Systems

Uni-biometrics systems use a single biometrics trait while multi-biometrics systems use multiple biometrics traits of a person. Multi-biometrics system can enhance matching perfection, security and reliability of user authentication method. Multi-biometrics systems have efficiency to address inabilities of unibiometrics systems like, noise in data that is captured by a sensor, non-universality, impressibility to spoof attacks, and large intra-class changes.

Types of Multi-Biometrics Systems

Multi-biometrics systems use multiple samples captured from more than one biometrics trait or a single biometrics trait. This type of system uses one of the following integration cases to test biometrics features:

Multi-Sensor Systems

Multi-sensor systems utilize various sensors to generate more robust feature using a fusion methodology. For example, an optical and a capacitive sensor can be used to capture fingerprint samples of biometrics data, and then integrate it using sensor level fusion process.

Multi-Modal Systems

Multi-modal systems exploit multiple biometrics traits to recognize an individual. For example, a person provides his/ her fingerprint and iris scan for authentication on a multi-biometrics system.

Multi-Instance Systems

Multi-instance biometrics system integrates multiple instances of a single biometrics trait to recognize a person. For example: A multi-instance system may give fingerprints of more than one finger to identify a person. If this system uses single sensor to capture multiple instances, then cost of these systems will reduce.

Multi-Sample Systems

Multi-sample systems capture multiple samples of a same biometrics trait. For example, a multi-sample face recognition system may want left and right (both) profiles of the face along with frontal profile. Taking different samples of a biometrics trait is very useful to capture changes that can occur in a biometrics trait and at the same time get a more complete representation of a given trait.

Multi-Algorithm Systems

Multi-algorithm systems use multiple algorithms for feature extraction as well as matching on a single biometrics trait. These systems are very cheap and simple as they use single sensor to take biometrics samples.

Multi-Biometrics Fusion Techniques

In a multi-biometrics system, fusion (Ross &Jain, 2003) can be done on different levels to consolidate sampled data collected from different biometrics traits.

Multi-biometrics fusion techniques are categorized into two ways –

1. Fusion Prior to Matching
    a. **Sensor Level Fusion:** In this fusion, data collected at sensor level from multiple biometrics traits are combined.
    b. **Feature Level Fusion:** In this fusion, feature sets generated from multiple biometrics traits are combined to form a new feature set.
2. Fusion After Matching
    a. **Decision Level Fusion:** Decision level fusion works by taking the decisions generated by different biometrics matching modules. An ultimate decision is taken by getting individual decision of various biometrics traits.
    b. **Score Level Fusion:** The score level fusion technique fused match scores (similarity or difference between the probe sample and enrolled templates) to get the final decision.
    c. **Rank Level Fusion:** Rank level fusion integrates ranks assigned to each enrolled user by different sub modules to get a new rank for each user.

Challenges of Biometrics Technologies

Designing a robust biometrics authentication system is a challenging task. Two biometrics samples of same biometrics trait of a person may not be the same due to several factors. Sometimes two biometrics samples of same biometrics trait of two different persons may appear to be the same like face images of twins or look-alikes. There are several challenges of a biometrics authentication systems and they are summarized as follows:

- **Variation Within Persons:** Biometrics samples taken by biometrics authentication systems may be influenced by several factors like, changes in environment, age, stress, disease, occupational factors, socio-cultural aspects of the circumstance, changes in human interface with the system, and so on. Therefore, every time when a person interacts with the biometrics system he/she may provide different biometrics information. This type of situations leads uncertainty in biometrics authentication systems.
- **Sensors:** Performance of sensor plays a significant role in biometrics authentication system. Sensor age, the sensitivity of sensor performance, and interface of a sensor affect the perfection of a biometrics authentication system.
- **Feature Extraction and Matching Algorithms:** Biometrics traits cannot be directly associated for authentication. Therefore robust feature extraction algorithms are required to extract durable and distinguishable features. Designing robust feature extraction and matching algorithms are challenging tasks.
- **Data Integrity:** Information may be changed through transformation, manipulation or mismanagement, inappropriate compression, or some other means.

No any single biometrics trait is known to be fully durable and distinguishable across all groups. Biometrics traits have primary distinguishable statistical features and varied degrees of durability under ordinary physiological situations as well as challenging circumstances. The several aspects of biometrics traits are not well convinced, specifically at large scales. Complicating phenomenon, the fundamental biological features and distribution of biometrics traits in demography are usually seen only through filters arbitrate by assessment procedures, devices and subsequent biometrics feature extraction. Better understanding of biometrics traits in human beings can be possible by sufficient data collection and analysis. It is very important to understand fundamental discriminating features of a biometrics trait and also the durability, exactness, and implicit variableness of a given measure.

Another main feature of biometrics based authentication system is to make decision under ambiguous situations. A biometrics match yields not definite recognition but a probability of true recognition and a non-match yields a probability that an individual is unknown to the system.

Future Trends

It is very important to identify the tools and technologies which facilitate true user identification. Currently the papers which are available and information about biometrics disclose that the main contention for applying biometrics solutions is high security. This aspect is permitted by policy makers, governments for fighting against terrorism, help to identify criminals, or immigrants at border crossing etc. Therefore, the best solution for person identification is biometrics (using biometrics cues) through some automatic devices. Biometrics readers are now used in homes, offices, computers, machines, devices, etc to get security. In near future biometrics technology will be the largest technology in market which will be used in everywhere. However, for the most part, the use of these technologies will only replace existing user authentication and access control methods, giving increased security and amenities. There will be no need to take keys, identity cards, personal documents and don't even remember passwords, etc. This implementation of biometrics will assemble to the overall security solution: counteracting the possibility of theft or illegal use of equipments/technologies. Biometrics devices will grant new attributes and approaches towards security solutions.

CONCLUSION

Identity verification of human beings plays a significant role in today's linked and fast going world that can give entree to very secure facilities, perform transactions and also used for the tasks like border crossing, passport verification, banking transactions, etc. User authentication is very important for security measures. As biometrics is being used for physical as well as digital identity recognition, therefore it is essential to enhance the performance as well as use in new domain like healthcare of current biometrics technologies. Biometrics systems could be found useful in healthcare domain due to its irreplaceable properties towards successful person identification. As biometrics are used for personal identification for accessing high security facilities where security is susceptible to attack, it becomes necessary to increase performance and reliability of these systems more and more. To increase the accuracy and develop robust biometrics systems, multi-biometrics systems can be found competitive over monomodal biometrics. This chapter presents a comprehensive study on different biometrics systems including physiological and behavioral biometrics traits and their suitableness for healthcare domain where every live entity is important to be authenticated successfully. Besides, challenges and future scopes of biometrics systems are also discussed in this chapter. As healthcare organizations play a serious role in our daily life as well as the security is an important concern in healthcare system, therefore choosing appropriate unibiometrics or multi-biometrics systems and then integrate with healthcare systems would be an imperative aspect for keeping things secure.

REFERENCES

Bowyer, K. W., Hollingsworth, K., & Flynn, P. J. (2008). Image understanding for iris biometrics: A survey. *Computer Vision and Image Understanding*, *110*(2), 281–307. doi:10.1016/j.cviu.2007.08.005

Campbell, J. P. (1997). Speaker recognition: A tutorial. *Proceedings of the IEEE*, *85*(9), 1437–1462. doi:10.1109/5.628714

Chandra, A., Durand, R., & Weaver, S. (2008). The uses and potential of biometrics in health care: Are consumers and providers ready for it? *International Journal of Pharmaceutical and Healthcare Marketing*, *2*(1), 22–34. doi:10.1108/17506120810865406

Chellappa, R., Wilson, C. L., & Sirohey, S. (1995). Human and machine recognition of faces: A survey. *Proceedings of the IEEE*, *83*(5), 705–741. doi:10.1109/5.381842

Choras, M. (2007, June). Image feature extraction methods for ear biometrics--a survey. In Null (pp. 261-265). IEEE.

Diwakar, M., & Maharshi, M. (2013). An extraction and recognition of tongue-print images for biometrics authentication system. *International Journal of Computers and Applications*, *61*(3).

Faundez-Zanuy, M. (2005). Signature recognition state-of-the-art. *IEEE Aerospace and Electronic Systems Magazine*, *20*(7), 28–32. doi:10.1109/MAES.2005.1499249

Gehring, W. J., Yan, Q. Q., Billeter, M., Furukubo-Tokunaga, K., Schier, A. F., Resendez-Perez, D., . . . Wüthrich, K. (1995). Homeodomain-DNA recognition. In Nmr In Structural Biology: A Collection of Papers by Kurt Wüthrich (pp. 493-505). Academic Press. doi:10.1142/9789812795830_0042

Han, W. Y., & Lee, J. C. (2012). Palm vein recognition using adaptive Gabor filter. *Expert Systems with Applications*, *39*(18), 13225–13234. doi:10.1016/j.eswa.2012.05.079

Harb, A., Abbas, M., Cherry, A., Jaber, H., & Ayache, M. (2015, September). Palm print recognition. In *Advances in Biomedical Engineering (ICABME), 2015 International Conference on* (pp. 13-16). IEEE. 10.1109/ICABME.2015.7323239

Jahan, S., Chowdhury, M., & Islam, R. (2017, December). Robust fingerprint verification for enhancing security in healthcare system. In *2017 International Conference on Image and Vision Computing New Zealand (IVCNZ)* (pp. 1-5). IEEE.

Jain, A. K., & Ross, A. (2004). Multibiometrics systems. *Communications of the ACM*, *47*(1), 34–40. doi:10.1145/962081.962102

Jain, A. K., Ross, A., & Prabhakar, S. (2004). An introduction to biometrics recognition. *IEEE Transactions on Circuits and Systems for Video Technology*, *14*(1), 4–20. doi:10.1109/TCSVT.2003.818349

Jia, W., Cai, H. Y., Gui, J., Hu, R. X., Lei, Y. K., & Wang, X. F. (2012). Newborn footprint recognition using orientation feature. *Neural Computing & Applications*, *21*(8), 1855–1863. doi:10.100700521-011-0530-9

Kinnunen, T., & Li, H. (2010). An overview of text-independent speaker recognition: From features to supervectors. *Speech Communication*, *52*(1), 12–40. doi:10.1016/j.specom.2009.08.009

Kumar, T., Braeken, A., Liyanage, M., & Ylianttila, M. (2017, May). Identity privacy preserving biometrics based authentication scheme for Naked healthcare environment. In *Communications (ICC), 2017 IEEE International Conference on* (pp. 1-7). IEEE.

Kumar, V. A., & Ramakrishnan, M. (2010). Footprint recognition using modified sequential haar energy transform (MSHET). *International Journal of Computer Science Issues, 7*(3).

Leonard, D. C., Pons, A. P., & Asfour, S. S. (2009). Realization of a universal patient identifier for electronic medical records through biometrics technology. *IEEE Transactions on Information Technology in Biomedicine*, *13*(4), 494–500. doi:10.1109/TITB.2008.926438 PMID:19273015

Maltoni, D., Maio, D., Jain, A. K., & Prabhakar, S. (2009). *Handbook of fingerprint recognition*. Springer Science & Business Media. doi:10.1007/978-1-84882-254-2

Mirmohamadsadeghi, L., & Drygajlo, A. (2011, October). Palm vein recognition with local binary patterns and local derivative patterns. In *Biometrics (IJCB), 2011 International Joint Conference on* (pp. 1-6). IEEE.

Mohammedi, M., Omar, M., Aitabdelmalek, W., Mansouri, A., & Bouabdallah, A. (2018, April). Secure and lightweight biometrics-based remote patient authentication scheme for home healthcare systems. *In Programming and Systems (ISPS), 2018 International Symposium on* (pp. 1-6). IEEE.

Ng, R. Y. F., Tay, Y. H., & Mok, K. M. (2008, August). A review of iris recognition algorithms. In *Information Technology, 2008. ITSim 2008. International Symposium on* (Vol. 2, pp. 1-7). IEEE. 10.1109/ITSIM.2008.4631656

Nwosu, K. C., & Igbonagwam, O. (2014, October). BEPOC: A biometrics-enabled point-of-care patients and electronic medical records management system for medical missions. In *Healthcare Innovation Conference (HIC), 2014 IEEE* (pp. 223-226). IEEE.

Riera, A., Soria-Frisch, A., Caparrini, M., Grau, C., & Ruffini, G. (2008). Unobtrusive biometrics system based on electroencephalogram analysis. *EURASIP Journal on Advances in Signal Processing*, *2008*, 18.

Ross, A., & Jain, A. (2003). Information fusion in biometrics. *Pattern Recognition Letters*, *24*(13), 2115–2125. doi:10.1016/S0167-8655(03)00079-5

Sanchez-Reillo, R., Sanchez-Avila, C., & Gonzalez-Marcos, A. (2000). Biometrics identification through hand geometry measurements. *IEEE Transactions on Pattern Analysis and Machine Intelligence*, *22*(10), 1168–1171. doi:10.1109/34.879796

Shen, T. W. (2005). *Biometrics identity verification based on electrocardiogram (ECG)* (PhD dissertation). University of Wisconsin.

Sidlauskas, D. P., & Tamer, S. (2008). Hand geometry recognition. In *Handbook of Biometrics* (pp. 91–107). Boston, MA: Springer. doi:10.1007/978-0-387-71041-9_5

Sumathi, S., & Malini, R. R. (2010, April). Face recognition system to enhance E Health. In *E-Health Networking, Digital Ecosystems and Technologies (EDT), 2010 International Conference on* (Vol. 1, pp. 195-198). IEEE.

Tabatabaee, H., Milani-Fard, A., & Jafariani, H. (2006, April). A novel human identifier system using retina image and fuzzy clustering approach. In *Proceedings of the 2nd IEEE International Conference on Information and Communication Technologies* (pp. 1031-1036). IEEE.

Zhang, D., Liu, Z., & Yan, J. Q. (2010). Dynamic tongueprint: A novel biometrics identifier. *Pattern Recognition*, *43*(3), 1071–1082. doi:10.1016/j.patcog.2009.09.002

Zhang, D., Liu, Z., Yan, J. Q., & Shi, P. F. (2007, August). Tongue-print: a novel biometrics pattern. In *International Conference on Biometrics* (pp. 1174-1183). Springer.

Zhang, G. H., Poon, C. C., & Zhang, Y. T. (2009, November). A biometrics based security solution for encryption and authentication in tele-healthcare systems. In *Applied Sciences in Biomedical and Communication Technologies, 2009. ISABEL 2009. 2nd International Symposium on* (pp. 1-4). IEEE.

KEY TERMS AND DEFINITIONS

Biometrics Identity: Biometrics identity refers to biometrics traits which are used to identify a person.

Biometrics Identity Management: Identity management is also known as identity and access management (IAM) which is used to ensure that resources are accessed only by the right people, at the right time and for the right reasons in the field of computer security and business discipline.

Feature Extraction: Feature extraction transforms an input image into a set of features.

Matching: Matching compares the test sample with the training samples.

Multi-Biometrics System: A multi-biometrics system integrates information from multiple biometrics traits, algorithms, sensors, and other components to make a recognition procedure more secure.

Preprocessing: Preprocessing dispels artifacts from the sensor to enhance the quality of the input biometrics sample (e.g., removing background noise).

Similarity Score: Similarity score is the score generated at the matching stage of a biometrics system.

Verification: Verification is the act of verifying an identity what he/she claims (1:1 matching).

Chapter 2
Security, Reliability, and Performance Assessment for Healthcare Biometrics

Ramgopal Kashyap
Amity University Chhattisgarh, India

ABSTRACT

A health system comprises, all things considered, individuals and activities whose essential purpose is to advance, reestablish, or look after wellbeing. This incorporates endeavors to impact the determinants of wellbeing-enhancing exercises. Biometrics has reformed the human services industry; gadgets can take exceptional data about you from your eye, your imprint, or your thumbprint and utilize it to recognize you. This data is utilized to guarantee that you are who you say you are, and you have the authorization to work with the medicinal services data you are endeavoring to get to. It offers a thorough review of biometric innovation applications intending to electronic social insurance security issues. Information classification, information legitimacy, information respectability, and client/substance verification are discussed in this chapter along with security, which is viewed as a foundation for medicinal services data frameworks as they contain amazingly touchy data.

INTRODUCTION

The technique by which medical records are put away and recovered is experiencing a change in a framework that will better use current innovation. Security dangers required with this new arrangement of chronicling and recovering patient records have achieved the beginning of a few government directions on the insurance and security of therapeutic records which thusly has expanded the need for a dependable client validation conspire in this area. Medicinal services suppliers are organizations or people giving human services administrations people, counting wellbeing experts and united wellbeing callings can act naturally used or filling in as a representative in a doctor's facility, center, or other medicinal services foundation, regardless of whether government worked, private for-benefit, or private e.g. non-legislative association. They may likewise work outside of direct patient care, such as, in an

DOI: 10.4018/978-1-5225-7525-2.ch002

administration wellbeing office or other organization, medicinal lab, or wellbeing preparing foundation. Cases of well-being specialists are specialists, attendants, birthing specialists, dietician, paramedics, dental practitioners, restorative research center technologists, advisors, clinicians, drug specialists, chiropractor, optometrists, network wellbeing laborers, conventional solution professionals, and others (Kashyap & Piersson, 2018b). Healing facilities are open associations that give long haul remains to patients, and are staffed by proficient doctors, specialists, and medical caretakers. A plenty of data is exchanged, translated, made, what's more, put away ordinary inside these associations. The part of Information and Communications Technology (ICT) data security and hazard administration is particularly imperative in these associations that keep a tremendous measure of classified data, to guarantee for the conveyance of better, more secure and more proficient social insurance administrations. In this exploration venture, we will investigate how specialists of restorative participation in doctor's facilities take after rules of security and how security issues and hazard administration is taken care of by IT workforce. In the regular universe of social insurance suppliers and facilitator, they are progressively baffled by the need to use convoluted passwords, tokens, cards or other lumbering types of validation to sign in to their EHR (Electronic health records) framework or to prescribe a medication. Much of the time, they are required to confirm 50-100 times each day (Lo, Wu & Chuang, 2017). These strategies are repetitive, tedious and barely secure, as passwords cards and tokens can be hacked or overlooked, shared or stolen, et cetera. In this way, quality time with patients is yielded.

BIOMETRIC

Biometric innovation is a safe and helpful distinguishing proof technique and it doesn't have to keep in mind complex passwords, nor brilliant cards, keys, and so forth. Biometrics is the quantifiable attributes of people in view of their personal conduct standards or physiological highlights that can be utilized to confirm or perceive their character. Physical qualities join fingerprints, palm or hand geometry, iris, retina, and facial qualities. Conduct attributes incorporate mark, keystroke and voice design with the mix of biometric innovation items and current PC innovation, it is anything but difficult to do checking, administration, frameworks mix, computerized administration, and security applications. Biometric innovation has been set up in a few markets now and this innovation looks extremely encouraging. Some vertical markets utilizing biometrics incorporate 1. Government – driver's licenses, voter cards, and so on; 2. Transportation – airplane terminal security, tickets; 3. Social insurance – quiet/worker personality cards; 4. Budgetary – ATM cards, Master cards (which contain a photograph of the holder) 5. Security – staff gets to control and personality checks (which incorporates time and participation); 6. Open equity and security ID's 7. Instruction – understudy/educator personality check and access control. A large part of us know about or have used a significant number of the kinds of ID techniques said above. These are cases of the least difficult sorts of distinguishing proof strategies that we use regular (driver's let, worker identity cards). Biometrics is certifiable not another idea, we (omit) have been liable to it somehow or another and a few of us really want to give ID through biometrics. Others may have security concerns with regards to putting away their organic attribute data in a focal database some place or inspite of enabling that fact to be contained in a little chip on an ID or Master card. Architects of biometric frameworks must remember that person organic quality data is touchy and must be defended with the suitable security systems (Gold, 2012). No biometric arrangement will be 100% secure, yet when contrasted with a PIN or a secret key, biometrics may offer a more prominent level of security.

BIOMETRIC IN HEALTH CARE INDUSTRIES

Biometric innovation speaks to the future for positive human services distinguishing proof and will upgrade the secure use, stockpiling, and trade of personal wellbeing data. Biometric character arrangements dissuade and lessen misrepresentation by the provider's area, benefit programs, forestalling supplier charging for "ghost claims" or administrations when a patient is not at the supplier area on the administration date. Verifying oversaw mind "experience information" or administrations from suppliers so Medicare and Medicaid projects can depend on this revealed information for the setting of oversaw mind rates (Varol, Orhan, Tuncer & Akyuz, 2016). Verification is an essential part of the human connection with PCs and the distinguishing proof of people has turned out to be significant. An individual is exceptionally related to either Password, Stick, Smart card, token, unique finger impression, iris, discourse and so on (Kashyap & Tiwari, 2018). These highlights are classified into a physiological or social trademark. The biometric framework is an example acknowledgment framework that perceives a man either by physiological attributes or social portrayal. It looks at the current biometric information caught with put away, affirmed legitimate information in the database. Among the different biometric validation techniques, iris acknowledgment is more solid and exact technique in light of its high acknowledgment rate and uniqueness in acknowledgment (Kashyap & Piersson, 2018a). Kashyap proposed a fundamental differential administrator for the legitimate division of an iris. Iris acknowledgment calculation is executed utilizing half and half systems. A strategy proposed in utilized crossover system utilizing histogram and wavelets examination to section the iris pictures acquired from Phoenix database. They initially presented this approach on combination iris pictures. The amalgamation pictures build coarse iris pictures with the given coefficients at that point, combination iris pictures are improved utilizing super determination by controlling the coefficients, and they make numerous iris pictures with determined classes. Broad analyses of this strategy prove that the orchestrated iris pictures have an acceptable group for positioning the iris. (Kashyap & Tiwari, 2017) Introduced a segmentation method with a specific end goal to confirm the uniqueness of the person's iris and furthermore its execution as a biometric sign (Kashyap & Gautam, 2016). Created dynamic form based bi-level division of iris from the human eye pictures. In this chapter, the iris is perceived in view of some computational calculation. A technique to section the iris in the eye picture utilizing confined vitality based dynamic shape strategy is proposed in (Waoo, Kashyap & Jaiswal, 2010). Keeping in mind the end goal to decrease the usage time and increment the division exactness, another division, strategy in view of dynamic form technique. A part of the fundamental biometric recognizable proof advances and how they're being used as a part of clinical practice are depicted in the accompanying areas.

Electronic Record Keeping

Radiologists are the undisputed pioneers in receiving electronic record keeping in any case, the calling is failing to meet expectations with biometric recognizable proof, an innovation that is creating an ocean change in security and IT and making noteworthy picks up in doctor's facility use. Biometric recognizable proof dispenses with watchword manhandle, evacuates the need to recollect various passwords, rolls out 90-day secret key improvements straightforward, and keeps up security measures in a virtual work area medicinal services IT condition (Kashyap & Gautam, Tiwari 2018). At the point when connected to patients, it can drastically lessen information passage mistakes and streamline tolerant work process. Like discourse acknowledgment transcription frameworks, the innovation for biometric distinguishing proof has been popularized for a considerable length of time (Ogundipe, 2011). Its essential markets have been law authorization, the military and safeguard enterprises, what's more, organizations with stringent security necessities.

Bio Key

BIO-key's innovation is sent in driving healing facilities, centers, and private practices. Medicinal services suppliers use BIO-key's unique mark biometric innovation to upgrade security, increment accommodation and meet consistence necessities of the State Board of Pharmacy, Specialists, medical caretakers, and manager's use. BIO-key to get to EHR records and to give consistent two-factor confirmation to electronic recommending of controlled substances. Moreover, our innovation is utilized to give anchor access to a huge number of drugs administering cupboards in healing facilities all around the world. BIO-key's better one-than numerous coordinating capacities to enable any association to enlist a medicinal services supplier, or understanding and precisely recoup their records without requiring any further type of distinguishing proof (Klich, 2017). This is a progressive technique to oversee quiet data, care and installments. BIO-key innovation accomplishes high precision through its protected enhanced image technology in excess of 1,600 personal information, focuses are extricated from each unique mark.

Biometric Devices: Properties of Biometrics

The planned catching i.e. enlistment or validation of biometric test information and correlation i.e. coordinating with beforehand put away trademark or standardizing information requires the accompanying properties of biometric attributes: Invariance: The biometric trademark ought to be consistent over an extensive stretch of time. This would dispose of the need for consistent refreshing of the layouts that are put away in the framework. For instance, the iris is steady all through a man's lifetime when contrasted with facial qualities which may change because of maturing. Desirability and Timeliness: The person trademark must have the capacity to be consequently contrasted with a normal standard. The biometric test ought to be proper for catch without holding uptime, which is imperative for constant confirmation and different complexities since we need to use a system which will give close continuous distinguishing proof. The flight deck biometric verification framework should have the capacity to catch the biometric data from the true blue client with diminished framework result holding up time (Mansfield-Devine, 2011). Singularity: The biometric trademark ought to have adequate special properties keeping in mind the end goal to recognize one person from some other this is valid for all biometric qualities Educability: The caught information ought to be fit for being lessened to a size that is anything but difficult to deal with however difficult to copy. This property is imperative particularly when we are managing conveying the biometric information crosswise over secure channels i.e. from the confirming gadget to the controller of the outcomes which might be in a remote region. Reliability: The biometric system ought to guarantee high unwavering quality and uprightness. The flight deck biometric validation framework should be dependable on the grounds that it is exorbitant to have a framework that does not give reliable outcomes. Privacy: The biometric method ought to guarantee the protection of the person utilizing the framework with the goal that they are persuaded that their security isn't being disregarded in any capacity. These properties are essential to all biometric qualities e.g. iris, retina, unique mark, and facial qualities since we need to have the capacity to give a precise method for verifying approved faculty in the flight deck.

Biometric Performance Measurements

The execution of biometric frameworks is tried for the most part as far as false rejection rate (FRR), false acceptance rate (FAR), failure to enroll rate (FER), enlistment time, and confirmation time. The false acceptance rate is most imperative when security is a need though low false dismissal rates are favored when comfort is the need. The biometric framework utilized in the flight deck must have a low false acceptance rate since security is the need. On the off-chance that the false acknowledgment rate is as low as conceivable then we have a superior shot of not permitting unapproved subjects into the framework. The time when the FAR and FRR meet or hybrid is known as the equivalent mistake rate. This rate gives a more sensible measure of the execution of the biometric framework as opposed to utilizing either the FAR or FRR independently. The failure to enroll rate is the rate which a subject can't acquaint his or her biometric with the framework which is satisfactory to the framework (Kukula, Sutton & Elliott, 2010). For instance, if there is a unique mark filtering gadget which is extremely delicate to the pictures introduced to it and a subject can't give an obvious picture then he or she won't have the capacity to select into the framework. Normally, there are frameworks that will enable the subject a few endeavors to select biometric data into the framework. Both the enlistment and live introduction times are vital factors in deciding or testing framework execution. The enlistment time is that course of events in the middle of and including the catching of the biometric test and making the put away layout of that example. The confirmation time is an estimation of the procedure of live introduction. This procedure incorporates the catch of the crude information, live layout handling, examination of the put away format to the live format and the time it takes for the framework to give a choice i.e. coordinate or non-coordinate. To give the persistent validation component wanted for the flight deck, the check time must be close continuous for a fruitful flight deck biometric verification framework.

Multi-Biometrics

Biometric frameworks need to battle with loud information, confined degrees of opportunity, and inability to select issues, parody assaults, and unsuitable blunder rates. In a few circumstances, it might be achievable to convey a biometric framework that exploits in excess of one strategy for recognizable proof or confirmation to beat these issues. A biometric gadget can either be incorporated with non-biometric types of validation or with different types of biometric confirmation gadgets. At the point when a biometric gadget is incorporated with different types of biometric verification gadgets, it can be depicted as a multi-biometric framework. Multi-biometric frameworks might be more solid and give higher confirmation rates because of the nearness of various, free bits of proof. Multi-biometric frameworks address the issue of non-comprehensiveness, since different characteristics guarantee adequate populace scope, and give against mocking measures by making it troublesome for a gatecrasher to take many biometric attributes of a bona fide client (Neves, Moreno & Proença, 2017). On the off-chance that there is a shortcoming in one technique for biometrics, at that point joining it with a biometric strategy that is more grounded about that shortcoming will lighten that issue.

In this design, shows that figure 1, it is fundamental that both of the biometric strategies carry out a coordinating score equal to the acknowledgment score set for the framework which is set up at first. Multi-Biometric System "OR" Configuration a multi-biometric framework utilizing the OR arrangement, in this design as sown in figure 2, general acknowledgment by the framework can be accomplished either by both biometric techniques having a coordinating score equal to the acknowledgment score set for the

Figure 1. Portrays a multi-biometric framework utilizing and arrangement

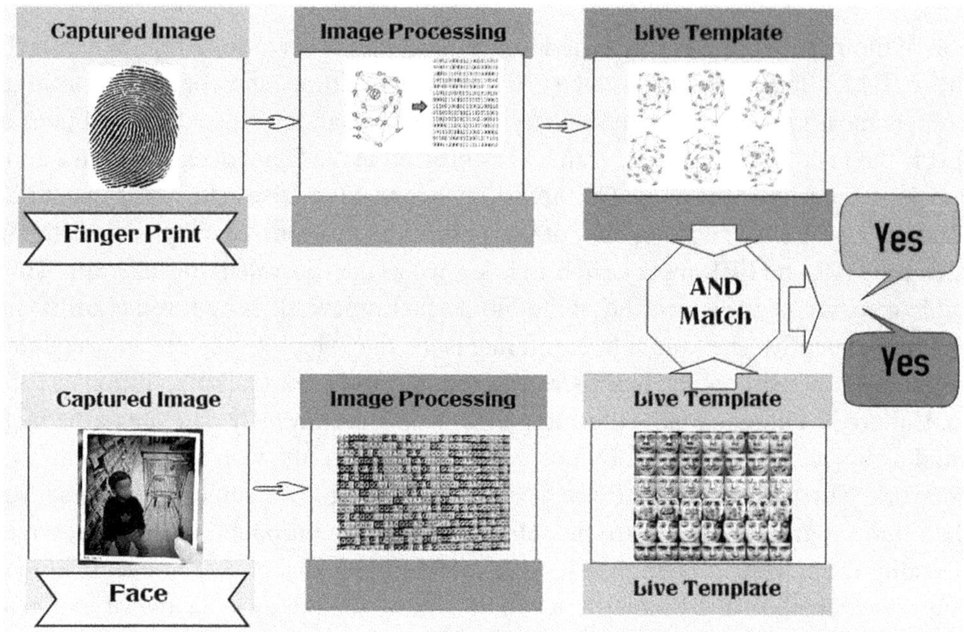

framework at first or by either biometric strategy having a coordinating score equal to the acknowledgment score set for the framework at first. This design does not give the certainty that the person is who they say they are and in addition, the AND arrangement does.

This design may diminish the false dismissal rate general on the grounds that the client will be acknowledged into the framework by for instance, either their unique mark format coordinating the already put away unique finger impression picture or by their facial layout coordinating the beforehand put away facial picture or both (Gawande, Zaveri & Kapur, 2011). Since utilizing this design may diminish the false dismissal rate, the false acknowledgment rate will expand, which is definitely not a smart thought for profoundly anchored zones.

BIOMETRIC SYSTEM PROCESS

All biometric frameworks fundamentally take after a similar arrangement of rules for biometric highlight coordinating spoke to in Figure 3. Biometric catch happens at the biometric gadget i.e. unique mark scanner. Biometric catch happens at the biometric gadget (i.e. unique mark scanner). The picture of the biometric is shown in figure 3 that is handled utilizing particular calculations custom-made for that biometric strategy to deliver a live format. The live layout of the biometric is a numerical portrayal now obtained biometric.

From the capacity gadget, the layout of the biometric which was put away as a major aspect of client enlistment is recovered and should coordinate the incentive from the live format. At the point when this happens, a biometric coordinate is obtained. Figure 3 speaks to the procedure that happens between biometric information enlistment and DVD stockpiling of the format. This procedure starts with the

Figure 2. Portrays a multi-biometric framework utilizing or arrangement

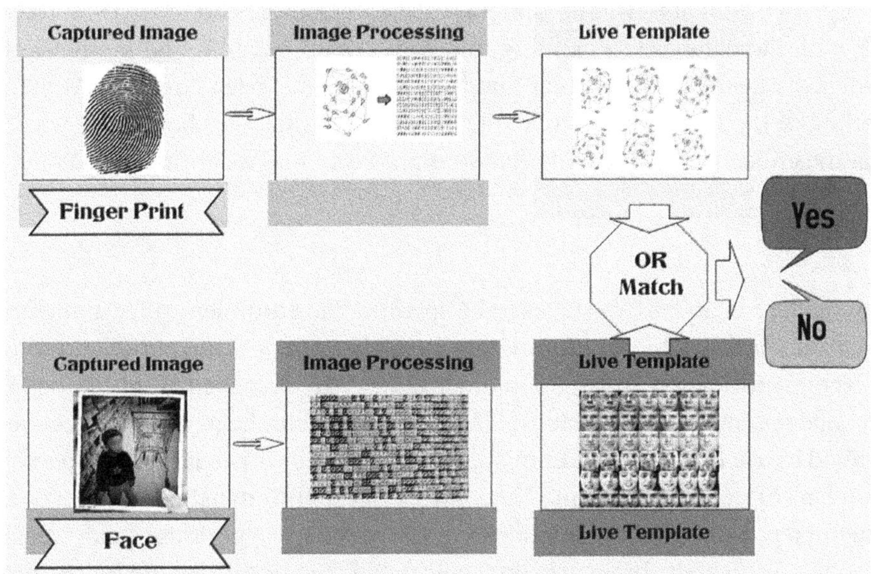

Figure 3. Standards biometric system process

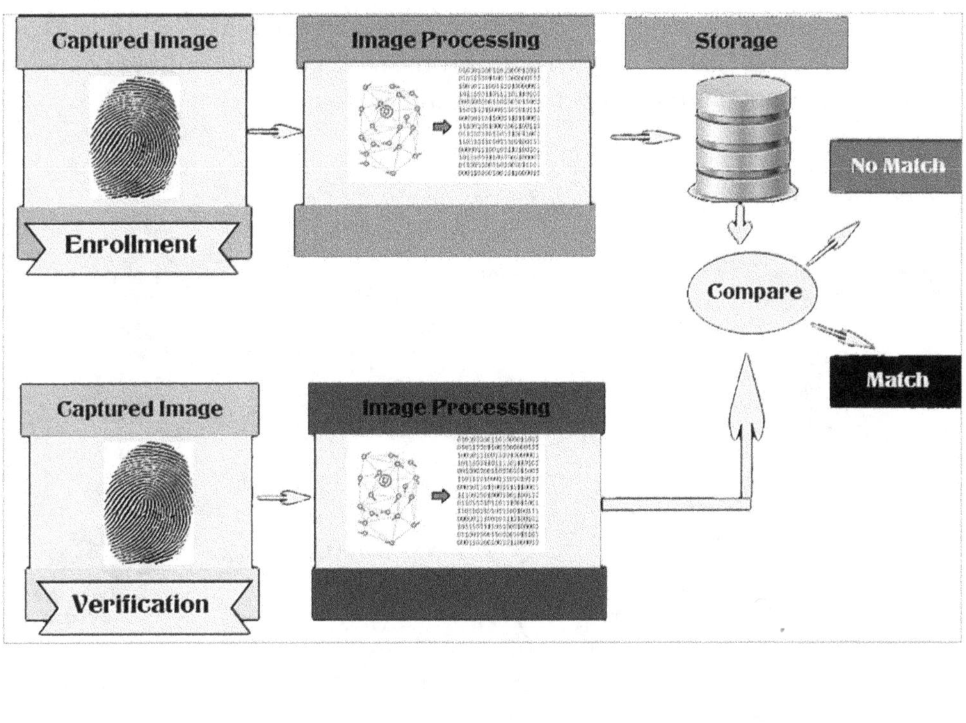

biometric information enlistment. Once the gadget has caught the biometric information, a format is produced. This format will be used later for correlation at the season of confirmation/coordinating. On the off-chance that the biometric information isn't effectively changed over into its proportionate layout then the biometric information must be re-elected into the framework. After the biometric information has been changed over into its proportional layout, it will be put away into the database, along the biometric formats of other approved clients (Biggio, fumera, Russu, Didaci & Roli, 2015). On the off-chance that there is an issue with the database stockpiling, the way toward catching the biometric from the client and capacity endeavors again until the point that achievement has been come to. At the point when the database stockpiling of the layout is effective, the biometric format will be put away into the reinforcement database at the enlistment office and afterward put away into a confided in DVD.

Operational Mode

The validation procedure includes two stages to be specific, the enrolment stage and recognizable proof stage. Amid the enrolment stage, biometric tests are caught from people and put away as biometric formats in databases. In the identification stage, the biometric tests are removed from the people and thought about against the one put away in the database. If there is a match, the person is effectively verified and the user is permitted to use the asset or administrations and the user is a legitimate person. It there is no match, the person isn't effectively confirmed and the user is kept from using the asset or administrations and the user is an illicit person. Figure 4 demonstrates the two distinct periods of individual confirmation.

Amid the enrolment stage, the biometric tests extricated from the client are put away in the database as biometric layouts. Layouts uncover data about a man and are inclined to assaults. Hence, it is urgent for any biometric framework to guarantee the security of the biometric formats. Biometric frameworks substantiate the personality of a man utilizing his organic or conduct attributes, such as, face, unique finger impression, iris, palm print, hand geometry, voice, mark, and stride and so on, biometric modalities or characteristics (Kashyap & Gautam, 2017a) probably the most generally utilized biometric attributes by the personality administration frameworks are appeared in Figure 4.

Figure 4. Biometric enrollments and storage process

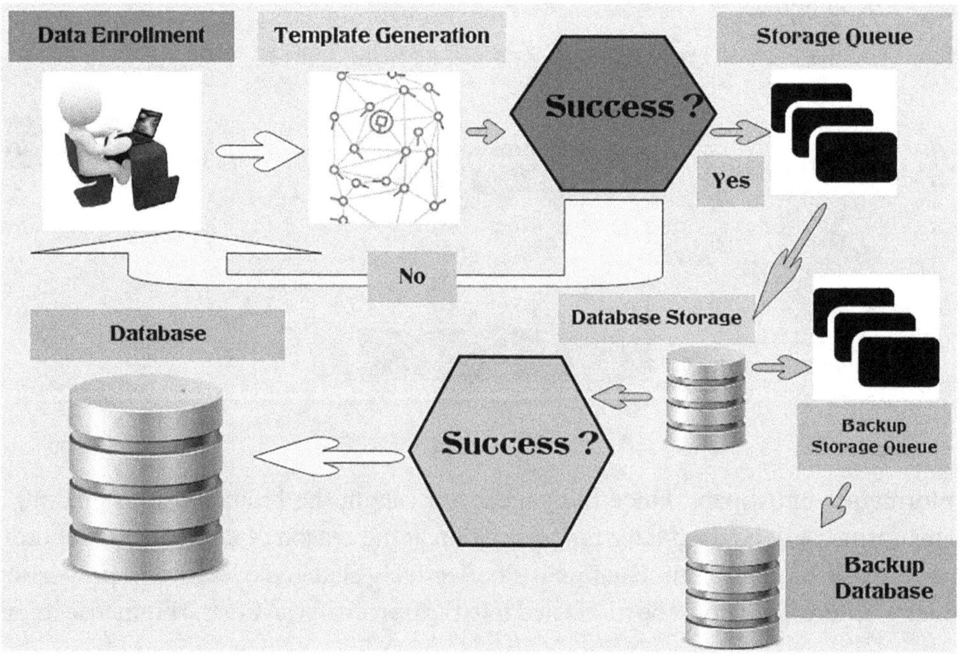

The prime preferred standpoint of biometric frameworks contrasted with ordinary recognizable proof techniques is supplanting what the individual conveys and what the individual recalls ideal models with which the person is, subsequently counteracting personality misrepresentation by utilizing biometrics designs that are asserted to be extraordinary, lasting and difficult to manufacture. Since, the person is required to be available during the season of the verification by the biometric framework, in this way additionally averting false denial claims (Pointner, 2017). So as to battle expanding security dangers in data time, governments, businesses, and institutes have to a great extent received and empowered research on biometric verification frameworks.

Confinements of Unimodal Biometric System

A portion of the fundamental variables influencing the exactness of the unimodal biometric frameworks are as per the following.

Clamor in Detected Information

Noise in the obtained biometric test may come about from faulty and despicably kept up sensors or ominous encompassing conditions (Gawande & Golhar, 2018). For example, aggregation of earth or the lingering stays on a unique mark sensor may bring about a boisterous unique mark picture as appeared in Figure 5. The uproarious biometric test may not be effectively coordinated, for in good faith clients, with their separate layouts in the database or may be erroneously coordinated with the impostors, in this way prompting a critical lessening in the execution of the framework.

Figure 5. Class variations of a biometric system

Intra-Class Varieties

Intra-class varieties in biometric tests are normally delivered by the client's unseemly collaboration with the sensor e.g., wrong facial posture sees figure 6, changes in the natural conditions e.g., light changes, use of various sensors amid enlistment and confirmation, or transient variety in the biometric qualities such as maturing. Substantial intra-class varieties more often than not diminish the genuine acceptance rate (GAR) of a biometric framework.

Class Likenesses

Inter-class similitude is characterized as the cover of the biometric tests, in the element space, on many classes or people. The absence of uniqueness in the biometric include set prompts an expansion in the false acceptance rate (FAR) of the framework (Dae Hyun Yum, Jin Seok Kim, Sung Je Hong & Pil Joong Lee, 2011). Henceforth, there is an upper bound on the measure of one of a kind people who can be obliged by the biometric framework.

Interoperability Issues

Most biometric frameworks are composed and worked under the supposition that the biometric test to be looked at is acquired utilizing a similar sensor and, so, are limited in their capacity to match or think about biometric tests starting from various sensors. Biometric parody assault is the think attempt to control ones biometric characteristics keeping in mind the end goal to stay away from the acknowledgment, or the creation of physical biometric ancient rarities keeping in mind the end goal to go up against the personality of another person. In this way, the unimodal biometric framework is not adequate to meet the assortment of necessities, including coordinating execution, forced by a few real scale confirmation frameworks. The restrictions of the unimodal biometric framework can be moderated by one of the standard arrangements known as multimodal biometric frameworks. The combination at the match score level has been broadly examined in the writing and is likewise the prevailing level of combination in multimodal biometric frameworks, because of straightforwardness in getting to and joining the match scores (Hariri, 2011). Thusly, in this proposition, we likewise embraced combination at the match score level. Figure 6 shows the engineering of a multimodal biometric framework, with reference to the one considered in this proposition, to be specific a multimodal framework made of a face and a unique mark matcher. Such frameworks work as takes after:

At the outline stage, honest to goodness clients are selected into the framework, by putting away their biometric qualities in a database together with the comparing characters. At the confirmation stage, the client gives his face what's more, unique mark to the particular sensors, and cases his character.

Attacks Against Biometric Systems

Foe assaults abuse the framework vulnerabilities for the most part at least one module or interfaces. Six conceivable distinctive focuses where security of biometric frameworks can be endangered have been distinguished in see figure 7, as depicted beneath:

Figure 6. Illustrates the architecture of a multimodal biometric system

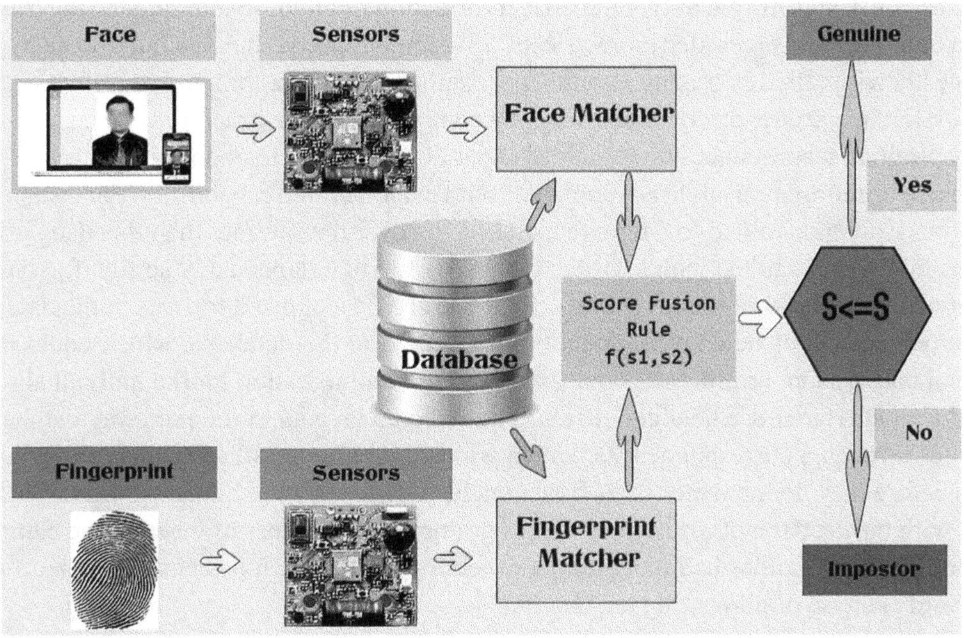

Figure 7. Biometric measurement components in human body

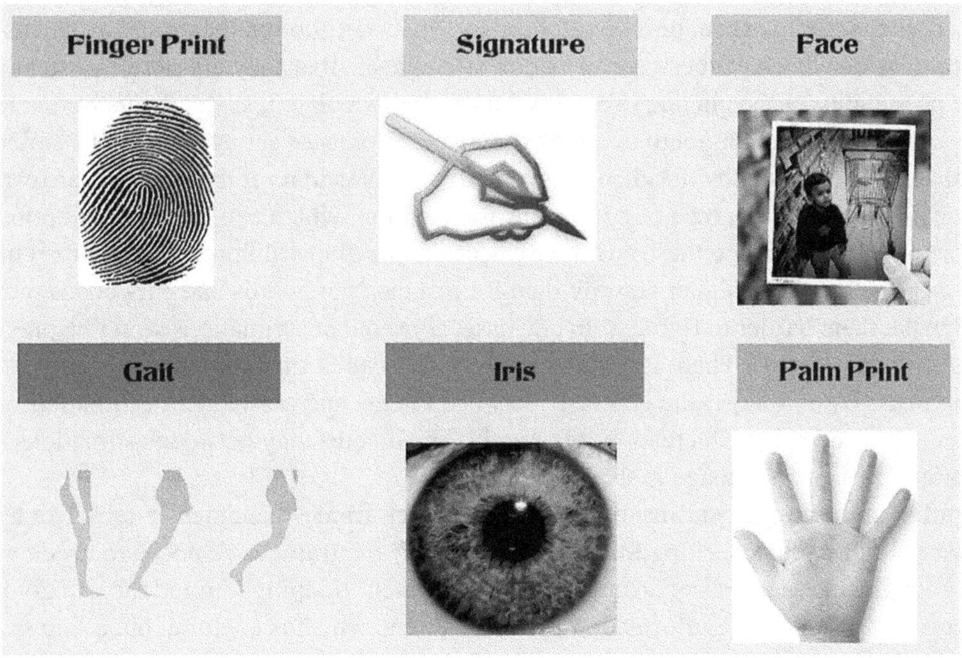

A phony biometric attribute might be displayed at the sensor, such as, a phony finger, a duplicate of a mark, or a face veil. Carefully put away biometric information might be resubmitted to the framework. In this sort of assault, formerly recorded biometric information is replayed into the framework bypassing the sensor, hence likewise called as "replay assault". For example, display a computerized duplicate of unique finger impression image or recorded sound flag of a person. The part extractor might be assaulted with a Trojan pony program that produces foreordained capabilities. Honest to goodness include sets separated from the biometric information might be supplanted with manufactured capabilities. For instance, if details of a unique mark are transmitted to a remote matched say over the Internet than this danger is genuine. The matches might be assaulted with a Trojan pony program that dependably straightforwardly creates a predetermined outcome - coordinate, no match, or a score. The enlisted formats in the database might be altered or evacuated, or new layouts might be presented in the database, which could bring about approval for a fake person, or if nothing else disavowal of administration for the individual related with the debased format (Hariri & Shokouhi, 2011). The selected layouts in the put away database are sent to the matches through a correspondence channel which could be assaulted to change the substance of the formats before they do the matches. A final conclusion yield by the biometric framework might be superseded with the decision of result from the programmer. Regardless of whether the element extraction and coordinating modules had magnificent execution attributes, it has been rendered futile by the straightforward exercise of abrogating the outcome.

Security With Noisy Data Biometric Authentication

Conventional confirmation techniques, for the most part, depend on something a client knows or has. The information can, for instance, be a secret word or PIN code, though the protest is normally a token or shrewd card. A drawback of these confirmation strategies is that they are not intrinsically bound to the genuine personality of the client. A secret word can be speculated, since clients have a tendency to pick basic passwords altogether not to overlook them, and a token or savvy card can be stolen. Besides, passwords that are overlooked by the client require extra systems to reset it and so present extra security vulnerabilities. Biometrics is a type of uproarious information, which can fill in as an option or can be used as a part of an expansion to these customary access control instruments. Highlights removed from uproarious information yield higher entropy than for instance passwords and PIN codes picked by clients, which make them harder to figure 8. In any case, consequent estimations won't be precisely equal (Wójtowicz & Ogiela, 2014). There is a high connection between them, in any case, in any case, there are little contrasts. By and large, the contrasts between clients and consequent estimations are alluded to as between-and inside class fluctuation, separately. The inconstancy between estimations is a test for the use of biometrics a noisy image is shown in figure 8.

A straightforward arrangement makes a biometric check framework intently look like a traditional grouping issue. Rather than checking for a "correct match" the framework needs to check whether the new info is enough close a put away layout, i.e., a close match. Keeping in mind the end goal to do this, a verifier requires free reference information about the client, which is anything but a safe arrangement, since it can be stolen. A stolen format can be used for, in addition to other things, cross coordinating, work crawl and personality misrepresentation. For instance encircling somebody, by abandoning counterfeit confirmation, such as, fingerprints on a wrongdoing scene, could be a difficult issue with stolen formats. As a rule, the reprocessing steps connected to the crude information to acquire the format can be switched with a specific end goal to make an ancient rarity that looks like the first biometric enough

Figure 8. Noisy fingerprints image

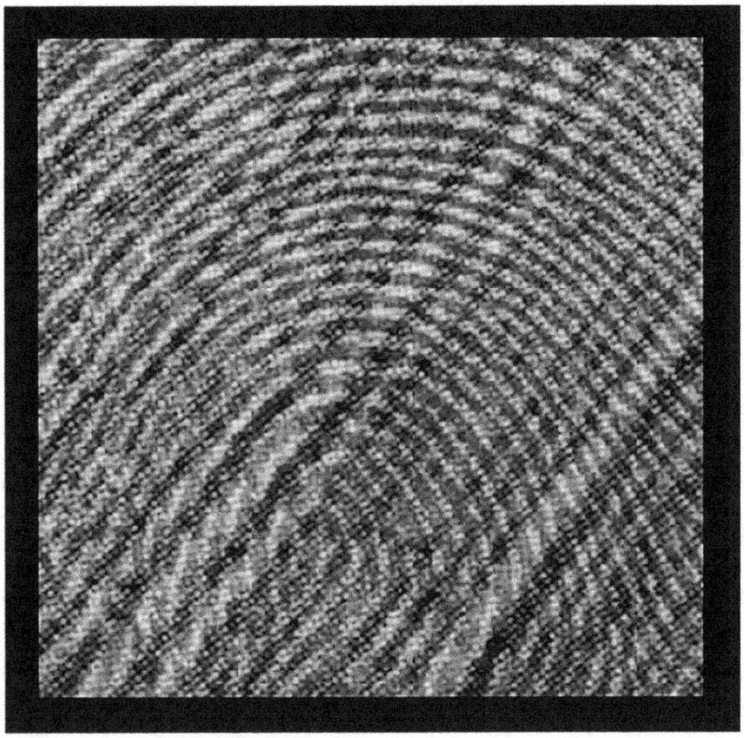

to breeze through the test (Chung, 2016). It has been illustrated that phony fingerprints could be made from layouts that were sufficient to mimic the proprietor. Besides, once traded off there is no answer for re-enlists the client. Toward the end, we do have just 10 fingers, 2 irises, and so forth. Other than security issues, there are protection dangers required for putting away biometric formats. It can happen that the put away reference information uncovers private data unnoticed. This can, for instance, be sexual orientation, age, and even certain maladies and clutters are known to show themselves in biometric information. Additionally, bargaining a database with many free formats is more engaging than investing a similar push to get biometric highlights of a solitary client. Regardless of whether all biometrics would be open verification can be founded on them. This requires a hands-on confirmation with for instance recognition to make sure that we are estimating biometrics of a real client (Pointner, 2017). Types of remote confirmation ought to be kept away from in such a circumstance since these are more helpless against misleading. Despite that security can be kept up with somewhat more exertion, format assurance is as yet required to make sure the security of the selected clients. A straightforward yet gullible way to deal with secure the reference information is to apply encryption, i.e., putting away cipher text CX of X together with an ID in an open database as delineated. Gatherings permitted to play out a validation, e.g., outskirt control, are given the decoding key K. This arrangement fills in as long as all gatherings can be trusted. Be that as it may, it doesn't forestall inside assaults, i.e., trusted verifiers taking the information for their own benefit. In some cases, it can even be undesired or otherwise difficult to confide in all gatherings which make this approach infeasible. Biometrics is not a perfect answer to all security issues they might enthusiasm for their benefit and inconspicuousness; however, the real security they can give is constrained. Other than the issue showed above, there is another issue. The variety between clients is adequate to separate high entropy keys, however, variety between resulting readings, i.e., clamor, from a solitary client's highlights can't be ignored. This seriously restrains the measure of data that can be dependably extricated. At last, it is this proportion among between and infraclass fluctuation that impacts the quality of the security arrangement worked with it. At last a biometric confirmation plan should attempt to carry out the accompanying necessities, which are more regularly an exchange off than something that can be accomplished at the same time.

Methodology Related Research

Each biometric framework depends on biometric modalities the decision of method is a key driver of how the framework is architected, how it is displayed to the client, and how coordinate versus no match choices are made. Understanding specific modalities and how best to use the modalities is basic to generally speaking framework viability. The investigation into a few interrelated zones will bring preceded with change: diminishing the cost of sensor equipment; enhancing the flag to-clamor proportion, the convenience and reasonableness, and the repeatability of measures; and broadening future. Enhancing the unwavering quality of distinguishing a locale of intrigue when the client introduces his or her biometric attributes to the framework, for instance, finding the faces in a picture or isolating discourse motion from encompassing commotion (Kashyap, 2018). Discovering better approaches to extricate invariant portrayal from the inalienable fluctuating biometric flag that is, the thing that sort of computerized portrayal ought to be utilized for a face to such an extent that the attribute can be perceived in spite of changes in posture, brightening, appearance, maturing. Enhancing the execution of the coordinating calculation within the sight of defective division, boisterous highlights, and intrinsic flag change. Creating approaches to refresh references with the goal that they can represent varieties and the maturing of reference information in enduring frameworks. Creating binding and dividing plans to accelerate seeks in real databases. Enhancing heartiness to assaults, including the introduction of misrepresented biometric attributes. Investigating the peculiarity of a specific biometric characteristic and its relationship to the coordinating execution. Does data about uniqueness serve to build comprehension of as far as possible on coordinating execution, for instance? Notwithstanding the general difficulties depicted above, there are likewise challenges particular to specific biometric modalities and characteristics. While the accompanying discourse does not depict every one of the difficulties for every methodology, it offers some possibly productive roads of examination for the most well-known ones.

BIOMETRIC SECURITY ISSUES

There are two kinds of shortcoming with biometrics validation/ID either client don't confirm/distinguish mistakenly or clients verify/recognize wrongly. In these cases, the trouble of recognizable proof example ought to be considered. Clients try not to confirm or distinguish wrongly when likeness examination is harsh while likeness examination nearly is simple when clients confirm or recognize wrongly. In alternate words, there must be a satisfactory adjusts between the principal issue which is False Negative and the second one which is a False Positive mistake that is a test for biometric frameworks. As indicated by past encounters one of the mistakes can have a little scale while the other blunder right has a high scale. Utilizing biometric innovations can be alright for a particular space if both blunders compose happen rarely. Another regular test identified with doing of validation/recognizable proof biometric innovation. The information gathered from the client ought to be kept private. Hence, the information will be anchored enough for the specific biometric framework. It is clear information accumulation from clients ought to be at present and any reiteration of the information gathering for confirmation designs isn't normal. This is essential since unique finger impression databases encourage finger duplicate that could be misused against another person at the wrongdoing scene. Another case is outskirt control framework can't be a la mode rapidly like independent machines to build fingers duplicate. The second biometric security issue is taking body part which put in danger physical uprightness of people. Be that as it may; hijacking and extorting can occur of the body parts taking. The last issue is needed various personalities could be showed also. For example, a few nations may characterize a man for their biometric databases in any event for remote nationals or shelters which reveal delegate for mystery administrations (Mahnken, 2014). There must be extraordinary thought and consideration on picking a biometric trademark to use in a specific place or for a characterized aim not just in light of the requirements and necessities of the objective association yet in addition because of biometrics innovations shortcomings, such as, fabrication and cheat. For example, slope climbing assault is a well-known danger for biometrics that will happen when an aggressor or unapproved client has accessed a biometric framework format or database counting people computerized mark and attempt to change edge even include new format or change accessible layouts. Advanced mark could be imperiled when marking key removed from advanced mark totally relies upon the measure of biometrics selected by the general population. The answer for taking out the assault against computerized signature which can be used for electronic mail, pull back from the account, contracts and so forth is utilizing encryption systems. Keeping in mind the end goal to make it more secure, mystery scratch for the advanced mark can be put away onto a smart card. Unique mark could be deceived by call uses by residual focuses and stamp ink on the per-user gadget, hand cream or oily hand. Facial acknowledgment is another illustration that can be endangered by imitates and camouflage when condition light is not appropriate. In any case; some biometrics organizations battle to perform measurable strategies to kill the progressions in biometric gadgets yield as a result of different ecological conditions, such as, lighting at the season of taking the test. Iris acknowledgment can have diminishment on quality or inspite of expanding danger of misrepresentation when individuals use focal points and due to "watery eyes".

Biometrics Privacy Issues

In investigating the lawful part of utilizing biometric innovation, we ought to think about information security close to an individual's protection. In alternate words, information experts must apply all conceivable rules to counteract coincidental occasion and composed misrepresentation furthermore, assaults to secure private and data. Sudden occasion incorporates misfortune furthermore, demolition of information because of human mix-up or mechanical issues and crash, additionally characteristic emergency. The cases of sorted out assault are abused of information for a unique structure, information exposure, and defilement and so on. We should consider the security issue especially when biometric information is executing. This procedure normally includes a few stages: putting away information, exchanging, separating design and assessing similitude's. Also, there must be exceptional defend for instance by applying solid encryption calculations when biometric information ought to be disregarded a hazardous all-inclusive system like the web or systems with poor security. Legitimate contemplations ought to be considered being used of biometric information since information security is an unavoidable segment of security assurance law. Different arrangements ought to be prescribed to heighten biometric information security (Khalaf & Sulaiman, 2015). Information must get anonymous, mysterious or utilizing accepted names and erasing superfluous data after enrolment process completed, biometric information changed to layouts and putting away information. False Rejection Error, blunder type, and False Acceptance Error or blunder must be at the most minimal scale these two present biometrics framework precision which clearly impacts on security. The last consequences of these kinds of a mistake are basic and indispensable for information proprietor. For instance, one may use an anchored building or a database that ought not to be permitted to use to the regions. Indeed, even worth one could pass outskirt of a nation wrongly because of uncalled for recognizing travelers.

Biometrics advances can be a danger for person's security in some ways since it joins pivotal private data. For example, retina checks uncovered use of liquor; from the past two days. Another illustration that is under discourse, however, is researching weather topsy-turvy fingerprints show data on homosexuality life. More data is in some biometric rules may gather information without learning of people who called aloof biometrics acknowledgment, such as, confront acknowledgment and stride verification. That is another case against people security. Likewise, utilizing at least two biometrics includes all the while building security of a framework, making the protection issues twofold. An essential subject is the means by which biometric ought to be used and how it ought not to be used by any means. Above all else, we clarify how biometrics ought to be utilized. Biometrics mythologies can be used as a part of the hardware that has been shared just by one person since there is no mutual database between vast quantities of clients to happen extortion. Besides, biometrics can be utilized with ownership validation like key, attractive card and savvy card and validation by learning like a secret word in a request to build security of validation technique. What's more, it can be used at the point when there is no lessening being used of great legal advances, such as, unique finger impression. The biometric highlights have been saved money on the database could be uncovered to obscure gadgets by the individual who approaches the biometric database (Kashyap & Gautam, 2017b). Moreover, biometric advances can be utilized when there is no protection issues got from biometrics in light of the fact that every individual controls their own types of gear. On the off-chance that conceivable turning biometric framework off when confirmation has been finished (Rabuzin, Bača & Maleković, 2007). Thusly, the wellbeing and physical honesty of clients won't be in danger regardless of whether the client is occupied with working with the aggressor together. Cases that biometrics ought not to be used at all incorporate the latent use biometrics by gadgets accommodating for confronting acknowledgment at open spots since individual does not mindful of the account. While individual on dynamic biometrics catching mindful of the procedure, such as, travel papers control and can disregard dynamic biometrics thus, utilizing latent biometric techniques in mystery ought to be averted by law.

Security and Privacy Enhancement in Biometric Databases

A few strategies have conveyed to improve security and ensuring biometric databases security by thinking about check or confirmation as a key point. Conventional strategy to control protection and security called zero-total worldview stresses some place protection must be a bargain for security and someplace security for security. While positive-whole model "Untraceable Biometric" approach incorporates two strategies: biometric encryption and cancelable biometric. Positive-aggregate model raises protection and security at the same time. Biometric Encryption, there is no database including tests for instance for the unique mark with a specific end goal to use in biometric encryption rather than putting away biometric information in a database, Biometric Encryption makes a computerized key from the biometric information, at that point use it to scramble or encode other data, such as, PIN code, account number, social number and so forth. Not just the key yet additionally the biometric highlight can't be recuperated from the spared data. At last, the Biometric encryption method encodes the code and store it not biometric highlight. The aftereffects of Biometric encryption confirmation would be either a computerized key or otherwise a mistake message in light of the fact that the key will be created again if proper biometric include has been displayed by the client. While the second system, cancelable biometric, apply a few changes to the essential biometric highlight and putting away the new layout. The comparative changes will be connected to biometric highlight at the season of validation, and if remedy coordinating done between the biometric highlight and the format, the answer will be "Yes" generally "No". Biometric encryption expands security in three different ways (Wechsler, 2012). To start with, there is no compelling reason to safeguard biometric formats then the likelihood of missing, defilement and abusing diminish. Second, People can control and point of confinement the use of their biometric data for a particular plan. Third, Security will rise since confirmation strategy and information broadcasting have been enhanced. Multimodal biometric framework, multi-biometric or biometric combination is another strategy to give security, protection, and better solace. See figure 9. As an issue of actuality, multimodal biometric expands exactness and accessibility of a biometric framework.

Multiple Sensors

It implies utilizing different sensors to separate assortment of acknowledgment includes as yields for one particular biometric highlight. For case, utilizing a blend of developments in two-dimensional, such as, development to send and sides and three-dimensional developments in step verification. Movements to advances and sides in addition to vertical movements can be used to enhance add up to acknowledgment precision. Employing various biometrics highlights to remove set of acknowledgment yields to build exactness in the confirmation/check cases and speed in recognizable proof circumstance. For instance, stride verification and facial acknowledgment are two quick validation techniques while unique finger impression and iris acknowledgment are moderate however more exact (Tan & Perkowski, 2017). Applying many units of biometric trademark for illustration, gathering biometric information for more than one finger, or checking the retina of the two eyes. It implies gathering, For example, iris biometric information for a few times or a few pictures of ear or face. As such, unique calculations or otherwise highlight extraction techniques must give the same outcomes.

Biometric Template Protection

As the world turns out to be carefully exceptional, each coherently initiated gadget can be gotten to and adjusted locally or remotely; genuinely or misguided. The wide use of biometric security frameworks requires geological dispersion of a huge amount layouts procured amid the enlistment stage. Creation and improvement of unbreakable calculations and cryptography have turned into an essential advance to be grasped for the headway of security of such formats to insure and guarantee the aversion of assaults of any kind or encroachment for false employments (Isobe, Ohki & Komatsu, 2013).

Multi-Biometrics Template Protection Schemes

Perfect biometric format insurance ought to have properties, given as underneath: Diversity: the safe layout must not let cross-coordinating crosswise over databases, along these lines guaranteeing the client's security. Educability: it ought to be direct to deny a bargained layout and reissue another one in view of the same biometric information. Security: it must be computationally difficult to get the first biometric layout from the protected format. This property keeps an enemy from making a physical satire of the biometric characteristic from a stolen layout. Performance: the biometric format security plan ought not to corrupt the acknowledgment execution (FAR and FRR) of the biometric framework. The interpretation, bending and turn to cause an event of biometric intro class manageability because of this fluctuation, it isn't conceivable or safe to encode any biometric formats utilizing standard encryption strategies, such as, AES, RSA and so on. Utilizing this standard encryption will make the format be uncovered amid the validation procedure, which increments the risk to enemy access and takes (Brindha, 2012).

Feature Transformation

As put by Anil K. Jain et al., "In Feature change approach a transformation work (F) is connected to the biometric template (T) and just the changed format (F (T; K)) is stored in the database. The limits of the transformation work are regularly gotten from a random key (K) or secret key. A similar change function is connected to inquiry highlights (Q) and the transformed query (F (Q; K)) is directly matched against the transformed template (F (T; K))." An element change part used to make sure the format. Confirmation system when the biometric format is makes sure utilizing an element change approach. The element change can be additionally ordered into two orders: Salting (Salt) is a cryptography calculation intended to shield against assaults, such as, lexicon assaults and rainbow table assaults which are pre-processed. Salting is additionally alluded to as rehashing in biometrics; it is a way to deal with ensure layout in which the biometric highlights are changed utilizing a capacity characterized by a client particular key or secret word. Here, amid the verification stage, the key or the password is required to be recalled by the client or it should be put away safely due to the invertible change. The entropy of the biometric format is expanded due to the extra necessity of such keys and passwords, which so builds the trouble for any enemy to figure the layout. Convertible change invertible change uses a convertible change work which is a restricted capacity. Here, of beast compels multifaceted nature, regardless of whether the key (which must be exhibited amid the verification stage) or the changed biometric layout is acquired by an enemy; it is computationally relatively difficult to recoup the first biometric format (Valenzuela, Schwartz & Pedrini, 2014).

Biometric Cryptosystems

Biometric formats and the crude information once lost have an unfortunate result; they can be effectively used to cross-coordinate through the diverse database for the wanted outcomes by a foe. Since this crude information is beside difficult to drop when they are lost or stolen, it is essential to fall back on a more secure framework named as biometric cryptosystem or biometric encryption, it is likewise named as aid information strategies. In a key-awaiting cryptosystem, the biometric format is anchored by solidly restricting it with a key inside a cryptographic structure. The assistant information put away in a database which is a solitary element installing both the layout and the key makes it computationally hard to translate the format. Here, the intra user varieties in biometric information can be ensured which is controlled by the ability of the code-word for rectifying the mistake. Enter creating biometric cryptosystem with regards to biometrics key age, (Mai, Lim & Yuen, 2017) the idea of Secure portray and Fuzzy extractor, where an aide information is a protected draw which releases constrained measure of information estimated as far as entropy misfortune. Still, it can remake the correct layout if gave an inquiry which coordinates exceptionally to the format. The fluffy extractor is cryptographic crude which can produce a cryptographic key from the biometric highlights. In any case, it is a troublesome errand to create such keys with high entropy and strength figure 9 beneath demonstrates the simple template extraction process.

Remote Authentication

The world has been digitized ceaselessly with such advancement, it has turned out to be practically open to the majority of the general population from anyplace to anyplace. Remote validation is a type of confirming an individual through the Internet or any system association where a client presents her/his qualifications as a proof of character to check the inventiveness and the authenticity of herself which she/he can claim to be. Since the conventional username and secret word approach for confirmation have turned out to be defenseless against various blemishes and assaults which have been disentangled and generally spread recently; advancement of thorough security convention against disguises must be started. Minimal couple of creative work has been accomplished in the ramifications of multi-biometrics frameworks for remote validation. This convention permits a remote client for single enlistment and much validation with a certain level of security (Thanh & Hải, 2012). The significant advances incorporated into their convention are i) Registration, ii) Authentication and iii) Ticket conceding.

Figure 9. Biometric template extraction process

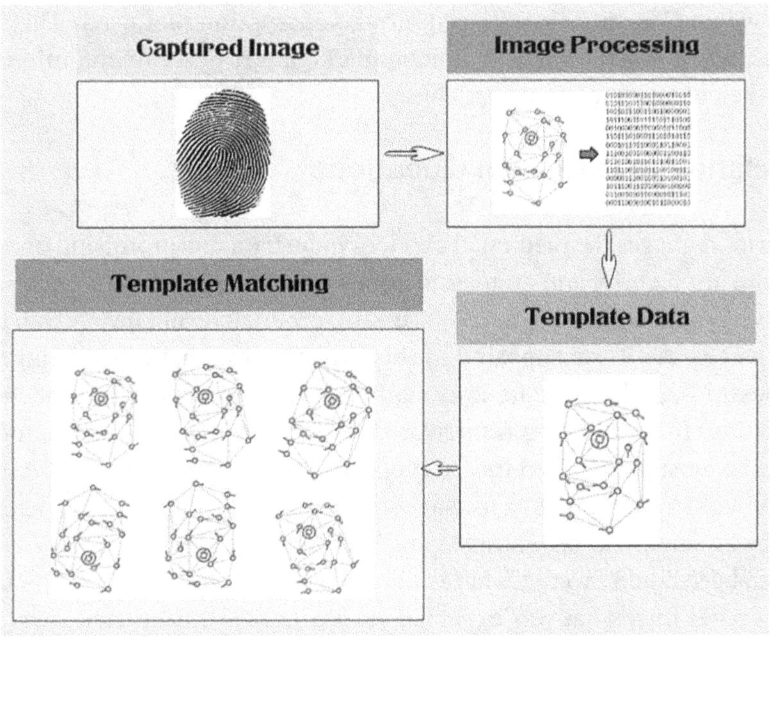

Figure 10. Essential thoughts of remote authentication stages

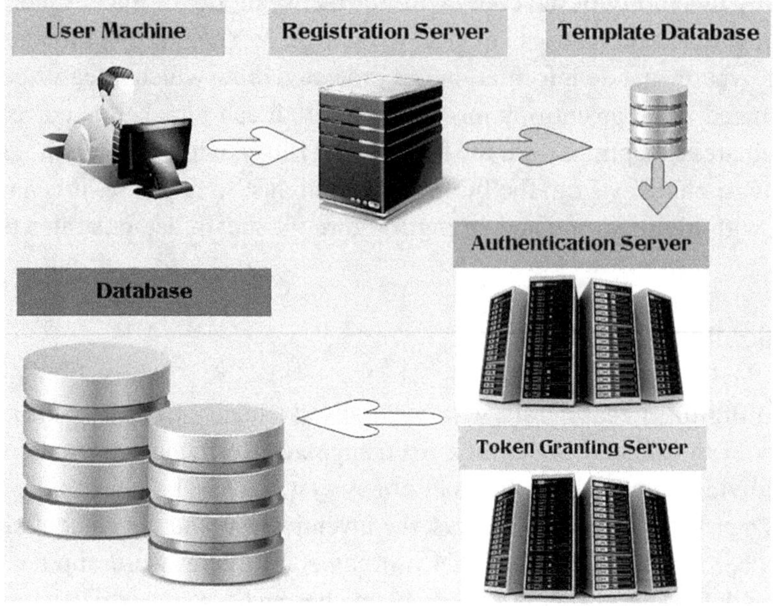

Alluding to Figure 10, accepting a remote customer 'Alice' has all the equipment necessities, she at first registers/enlists herself to the Registration Server which likewise effectively spares the biometric layouts got; second, she confirms herself with the validation server which drives her to be conceded with the tickets and session keys. These tickets and session keys have expiry time allotment which permits higher security. In any case, while getting to Bob, Alice does not really need to experience all the three stages specified before. Since the tickets can be reused for certain timeframe, she can simply conjure and confirm to reach to Bob safely. All the modulo-activities in the encryption area uses the articulation (M task N) mod P, where P is specifically needy to the encryption plot used. This should be possible with a biometric catching gadget, such as, a unique mark gadget or a camera, info is caught, removed, and contrasted and the put away reference layout.

Exchange in Relation to Research Questions

It is obvious biometrics manages the medicinal services industry a few promising openings as innovation keeps on propelling so are security and protection issues which keep on being a noteworthy worry in the industry. Biometrics has appeared to be more successful, productive and hearty in handling the security and protection challenges. As it has appeared in this investigation that e-wellbeing data ought to have certain security necessities, such as, secure information privacy, information respectability, information readiness and validation. In e-wellbeing frameworks, these customary biometric models display highlights which are for the most part abused for encryption and validation purposes. Assaults on biometric layouts could prompt vulnerabilities, such as, substitution of layouts by a sham to get entrance; formats could be parodied by an enemy to get entrance and can enable a foe to do replay assaults. Once more formats from the database could prompt what is known as 'work crawl' where the biometric identifier is used for purposes other than what was expected (Hassan, 2003). Biometric layouts when bargained are difficult to be supplanted since they are for all time connected to the client. These issues influence open certainty and client acknowledgment through biometric cryptosystems also, calculable biometrics, a large part of these issues could be tended to cancelable biometrics tries to explain the biometric format unalterable quality by contorting biometric layout or signs in view of picked change. Most patients are troubled about their protection with regards to the use of biometric innovation. The incongruity of the circumstance is that, in the attempt to use biometrics to safeguard the security of patients or people by limiting access to their own data; its application could prompt protection worries since biometric data is thought to be close to home and crucial data in view of its substance. Plainly, protection concerns encompassing biometrics are for the most part connected with personal acknowledgment. Interruption is viewed as one of the greatest dangers to protection. The greater part of the security issues demonstrated over those reason vulnerabilities in the biometric framework can prompt interruption. A part of these protection concerns includes Fear of gathering extra data from the biometric data taken from people. The worry here is that biometric data could be connected to different purposes aside from the expected reason. For instance, with the progression of the human genome, there is the dread that data could be gotten from organic estimations. Loss of obscurity there is a worry that the utilization of biometrics may encroach on the privilege of people to stay unknown. This in light of the fact that most biometric attributes are not a mystery and are freely recognizable which could be gotten without the individual's learning and utilized as a part of biometric acknowledgment. Fear of utilizing it to track people. The worry here is that the gathering of biometric data could be used to track the development of people. This breeds "The government" impact on people having the dread that their biometric data would be used to track or screen them without their insight. Fear biometric data would be shared without authorization or enough clarification, people express worry over the control they have over their own data about degree use.

CONCLUSION

Electronic medicinal services influences on Information and Communications Technology (ICT) to enhance the effectiveness of wellbeing administration lessen the cost guarantees security, and help contact a huge number of patients or social insurance buyers outside the bounds of human services foundations. The incapability of ICT makes it imperative in social insurance particularly even with advancements in web advances, PC, and media transmission frameworks. These improvements present protection and security worries in social insurance frameworks. Keeping up security and protection in medicinal services frameworks in this way turns out to be critical as when patients' delicate data get uncovered could have a genuine harming impact. Data security as for social insurance information security and patients' security is a fundamental prerequisite in the human services division. The electronic medicinal services like security necessities; information privacy, information realness, information respectability, client/element validation are very important. Data privacy: It is fundamental that entrance to patients' wellbeing data is confined to those approved as it guarantees the protection of the patient's information. This applies to information away and on transmission which must be kept confined just between the imparting parties. Many articles propose the use of biometrics to carry out the secrecy of wellbeing information for the most part through biometric encryption. Data readiness: with a specific end goal to avert fraud especially in information correspondence, guarantee that the cause of the wellbeing information isn't traded off. Data uprightness: Patient's well-being data being conveyed between two gatherings ought to be free from unapproved changes, cancellations, and substitutions. User/substance confirmation: Effective and helpful client/element validation is accomplished through the utilization of biometrics. It is watched that every one of the examination papers addresses the issue of validation which makes it extremely vital for the security of wellbeing data. Calculations: The audit demonstrates that various calculations are used as a part of tending to the security prerequisites of electronic social insurance through the utilization of biometrics. These calculations or then again techniques can extensively be assembled under; encryption and verification. Key age/appropriation: Keyage and conveyance assume an essential part in any cryptographic framework. Through encryption methods access to delicate wellbeing, data is limited in this way giving security and assurance to the wellbeing information. A critical part of biometric cryptography is the utilization of biometrics to create cryptographic keys and furthermore ensure cryptographic keys for secure dissemination. Authentication: Authentication techniques are utilized to acknowledge or dismiss the personality of a client of a biometric framework guaranteed by a person. In the remote sensor organizes the techniques used help to produce validation keys for biosensors. Key age/dispersion: The execution assessment of key age/dispersion plans shows how well the plans accomplish better execution as far as key understanding. Authentication: The execution of a biometric framework in connection to of client validation measures the adequacy of the framework as far as FAR and FRR. The viability portrays the capacity of the biometric framework to select and check people. From the articles evaluated up until now, it turns out to be clear that keeping in mind the end goal to guarantee electronic human services security, at that point steps must be taken to discuss the security prerequisites successfully. Biometrics has appeared to accomplish that in two different ways; through biometric encryption and through biometric confirmation. Biometric encryption innovation connected to wellbeing information expands security of the encoded information. The information scrambled winds up available just to the person whose biometric highlight was used to encode the information in this way guaranteeing information protection. Be that as it may, because of the advances in versatile wellbeing and telemedicine, the majority of the exploration papers concentrated on the use of biometric encryption rules in remote sensor systems to anchor information correspondence. In remote sensor correspondence, guaranteeing the security of the biosensors includes having a compelling encryption and key age/appropriation framework. Biometrics accomplishes that practical and advantageously since the biometric signals are now being caught by the sensors. It must be specified that, before any encryption, conveying parties must a secure channel for the appropriation of the shared key to avoid man-in-the-center assaults. Dispersion plans utilizing on biometrics could be used to secure the channel for compelling key circulation. Biometric additionally offers a compelling and secure technique of validation than conventional methodologies. Despite that it is sensible to use one of a kind quality/organ as methods for her/his ID and check in the present computerized world, it is of no use if the gadgets and sensors used as a part of digitizing such characteristics are not reliable and tried and true. The use of multi-biometrics sounds conceivable being referred to, yet the functional thought of its use holds generally the complex development of such hello there tech cryptosystems and calculations which have not yet been consummated to its coveted zenith. Addition in revelation and confirmation of uniqueness of various biometric layouts won't influence and help much in check and security if the ease of use and responsibility can't be dependable and be acknowledged in an aggregate field. Notwithstanding, because of the expanded improvement of complex PC frameworks and cryptography to split the passwords and safety efforts of the present security structure, it is somewhat trustworthy to actualized biometrics and multi-biometrics for validation purposes. The everyday acknowledgment of weakness while privatizing one's having a place has raised, ensuring characterized matters and flexibility to keep one's mystery a mystery to oneself has gradually moved toward becoming imperiled. This requests a fast development being developed of abnormal state safety efforts. In view of authentic and present research and discoveries, biometric security frameworks have turned into the rising and promising answer for such uncertainties. The present security framework is made and administered by human-created methods and gadgets; the future security means will join human itself.

REFERENCES

Biggio, B., fumera, g.,, Russu, P., Didaci, L., & Roli, F. (2015). Adversarial Biometric Recognition: A review on biometric system security from the adversarial machine-learning perspective. *IEEE Signal Processing Magazine, 32*(5), 31–41. doi:10.1109/MSP.2015.2426728

Brindha, V. (2012). Multi-Modal Biometric Template Security: Fingerprint and Palmprint Based Fuzzy Vault. *Journal of Biometrics & Biostatistics, 03*(06). doi:10.4172/2155-6180.1000150

Chung, W. (2016). Social media analytics: Security and privacy issues. *Journal Of Information Privacy And Security, 12*(3), 105–106. doi:10.1080/15536548.2016.1213994

Gawande, U., & Golhar, Y. (2018). Biometric security system: A rigorous review of unimodal and multimodal biometrics techniques. *International Journal of Biometrics, 10*(2), 142. doi:10.1504/IJBM.2018.091629

Gawande, U., Zaveri, M., & Kapur, A. (2011). Multi-algorithmic approaches to iris recognition. *Biometric Technology Today, 2011*(4), 8-10. doi:10.10160969-4765(11)70075-3

Gold, S. (2012). Biometrics fit to secure the London Olympics. *Biometric Technology Today, 2012*(3), 5-8. doi:10.10160969-4765(12)70072-3

Hariri, M. (2011). Possibility of spoof attack against robustness of multibiometric authentication systems. *Optical Engineering (Redondo Beach, Calif.), 50*(7), 079001. doi:10.1117/1.3599874

Hariri, M., & Shokouhi, S. (2011). Robustness of Multi Biometric Authentication Systems against Spoofing. *Computer And Information Science, 5*(1). doi:10.5539/cis.v5n1p77

Hassan, B. (2003). Examining data accuracy and authenticity with leading digit frequency analysis. *Industrial Management & Data Systems, 103*(2), 121–125. doi:10.1108/02635570310463447

Isobe, Y., Ohki, T., & Komatsu, N. (2013). Security performance evaluation for biometric template protection techniques. *International Journal of Biometrics, 5*(1), 53. doi:10.1504/IJBM.2013.050733

Kashyap, R. (2018). Object boundary detection through robust active contour based method with global information. *International Journal of Image Mining, 3*(1), 22. doi:10.1504/IJIM.2018.093008

Kashyap, R., & Gautam, P. (2016). Fast level set method for segmentation of medical images. In *Proceedings of the International Conference on Informatics and Analytics (ICIA-16)*. ACM. 10.1145/2980258.2980302

Kashyap, R., & Gautam, P. (2017a). 'Fast Medical Image Segmentation Using Energy-Based Method', Biometrics. *Concepts, Methodologies, Tools, and Applications*, *3*(1), 1017–1042. doi:10.4018/978-1-5225-0983-7.ch040

Kashyap, R., & Gautam, P. (2017b). *Fast Medical Image Segmentation Using Energy-Based Method*. In *Pattern and Data Analysis in Healthcare Settings*. Medical Information Science Reference.

Kashyap, R., Gautam, P., & Tiwari, V. (2018). Management and Monitoring Patterns and Future Scope. In Handbook of Research on Pattern Engineering System Development for Big Data Analytics. IGI Global.

Kashyap, R., & Piersson, A. (2018a). Big Data Challenges and Solutions in the Medical Industries. In Handbook of Research on Pattern Engineering System Development for Big Data Analytics. IGI Global.

Kashyap, R., & Piersson, A. (2018b). *Impact of Big Data on Security. In Handbook of Research on Network Forensics and Analysis Techniques* (pp. 283–299). IGI Global. doi:10.4018/978-1-5225-4100-4.ch015

Kashyap, R., & Tiwari, V. (2017). Energy-based active contour method for image segmentation. *International Journal of Electronic Healthcare*, *9*(2–3), 210–225. doi:10.1504/IJEH.2017.083165

Kashyap, R., & Tiwari, V. (2018). Active contours using global models for medical image segmentation. *International Journal of Computational Systems Engineering*, *4*(2/3), 195. doi:10.1504/IJCSYSE.2018.091404

Khalaf, E., & Sulaiman, N. (2015). A New Secure Storing System for Biometric Templates Based Encryption and Concealment. *Journal of Applied Sciences (Faisalabad)*, *15*(5), 773–782. doi:10.3923/jas.2015.773.782

Klich, A. (2017). Electronic health records (ehr)- selected legal issues. *Ekonomiczne Problemy Usług*, *126*, 351–360. doi:10.18276/epu.2017.126/2-35

Kukula, E., Sutton, M., & Elliott, S. (2010). The Human–Biometric-Sensor Interaction Evaluation Method: Biometric Performance and Usability Measurements. *IEEE Transactions on Instrumentation and Measurement*, *59*(4), 784–791. doi:10.1109/TIM.2009.2037878

Lo, N., Wu, C., & Chuang, Y. (2017). An authentication and authorization mechanism for long-term electronic health records management. *Procedia Computer Science*, *111*, 145–153. doi:10.1016/j.procs.2017.06.021

Mahnken, S. (2014). Today's authentication options: the need for adaptive multifactor authentication. *Biometric Technology Today*, *2014*(7), 8-10. doi:10.10160969-4765(14)70126-2

Mai, G., Lim, M., & Yuen, P. (2017). Binary feature fusion for discriminative and secure multi-biometric cryptosystems. *Image and Vision Computing*, *58*, 254–265. doi:10.1016/j.imavis.2016.11.011

Mansfield-Devine, S. (2011). Biometrics for mobile devices struggle to go mainstream. *Biometric Technology Today*, *2011*(9), 10-11. doi:10.10160969-4765(11)70172-2

Neves, J., Moreno, J., & Proença, H. (2017). *QUIS-CAMPI: an annotated multi-biometrics data feed from surveillance scenarios.* IET Biometrics; doi:10.1049/iet-bmt.2016.0178

Ogundipe, O. (2011). Electronic Medical Record Keeping: Eleven Years Experience at Life Support Eye Clinic, Lagos, Nigeria. *Nigerian Journal of Ophthalmology*, *19*(1). doi:10.4314/njo.v19i1.70771

Pointner, P. (2017). Hybrid tech: the future of biometric ID verification? *Biometric Technology Today, 2017*(8), 8-10. doi:10.10160969-4765(17)30183-2

Pointner, P. (2017). Hybrid tech: the future of biometric ID verification? *Biometric Technology Today, 2017*(8), 8-10. doi:10.10160969-4765(17)30183-2

Rabuzin, K., Bača, M., & Maleković, M. (2007). A Multimodal Biometric System Implemented within an Active Database Management System. *Journal of Software*, *2*(4). doi:10.4304/jsw.2.4.24-31

Tan, R., & Perkowski, M. (2017). Toward Improving Electrocardiogram (ECG) Biometric Verification using Mobile Sensors: A Two-Stage Classifier Approach. *Sensors (Basel)*, *17*(2), 410. doi:10.339017020410 PMID:28230745

Thanh, L., & Hải, N. (2012). Developping Kerberos-role authentication protocol for resource management system. *Journal Of Computer Science And Cybernetics*, *20*(4). doi:10.15625/1813-9663/20/4/1899

Valenzuela, R., Schwartz, W., & Pedrini, H. (2014). Linear dimensionality reduction applied to scale invariant feature transformation and speeded up robust feature descriptors. *Journal of Electronic Imaging*, *23*(3), 033017. doi:10.1117/1.JEI.23.3.033017

Varol, S., Orhan, F., Tuncer, S., & Akyuz, S. (2016). Biometric systems in the context of information security in health institutions. *Health Care Academician Journal*, *3*(4), 155. doi:10.5455ad.13-1483706096

Waoo, N., Kashyap, R., & Jaiswal, A. (2010). DNA nano array analysis using hierarchical quality threshold clustering. In *2010 2nd IEEE International Conference on Information Management and Engineering*. IEEE.

Wechsler, H. (2012). Biometric Security and Privacy Using Smart Identity Management and Interoperability: Validation and Vulnerabilities of Various Techniques. *The Review of Policy Research*, *29*(1), 63–89. doi:10.1111/j.1541-1338.2011.00538.x

Wójtowicz, W., & Ogiela, M. (2014). Biometric watermarks based on face recognition methods for authentication of digital images. *Security and Communication Networks*, *8*(9), 1672–1687. doi:10.1002ec.1114

Yum, D. H., Kim, J. S., Hong, S. J., & Lee, P. J. (2011). Distance Bounding Protocol with Adjustable False Acceptance Rate. *IEEE Communications Letters*, *15*(4), 434–436. doi:10.1109/LCOMM.2011.030911.110066

KEY TERMS AND DEFINITIONS

EHR: Electronic health record (EHR) is an advanced adaptation of a patient's paper diagram. Ehrs are constant, tolerant focused records that make data accessible immediately and safely to approved clients. While an EHR contains the therapeutic and treatment narratives of patients, an EHR framework is worked to go past standard clinical information gathered in a supplier's office and can be comprehensive of a more extensive perspective of a patient's care. EHRs can: contain a patient's therapeutic history, analyze, prescriptions, treatment designs, vaccination dates, hypersensitivities, radiology pictures, and research center and test outcomes. Enable access to confirm based devices that suppliers can use to settle on choices about a patient's care k process. One of the key highlights of an EHR is that wellbeing data can be made and overseen by approved suppliers in a computerized arranges fit for being imparted to different suppliers crosswise over in excess of one medicinal services association. EHRs are worked to impart data to other human services suppliers and associations, for example, research facilities, masters, restorative imaging offices, drug stores, crisis offices, and school and working environment centers so they contain data from all clinicians engaged with a patient's care.

ICT: Information and communications technology (ICT) alludes to all the innovation used to deal with broadcast communications, communicate media, wise building administration frameworks, varying media handling and transmission frameworks, and system based control and observing capacities. Despite the fact that ICT is regularly viewed as an expanded equivalent word for data innovation (IT), its degree is wider. ICT has all the more as of late been utilized to portray the union of a few advances and the utilization of basic transmission lines conveying extremely differing information and correspondence writes and designs. Uniting innovations that epitomize ICT incorporate the converging of varying media, phone and PC organizes through a typical cabling framework. Network access suppliers ordinarily give web, telephone and TV administrations to homes and organizations through a solitary optical link. The disposal of the phone systems has given tremendous monetary motivating forces to actualize this merging, which kills huge numbers of the expenses related to cabling, flag circulation, client establishment, adjusting, and upkeep costs.

Chapter 3
Indexing for Healthcare Biometric Databases

Kamlesh Tiwari
Birla Institute of Technology and Science, India

Geetika Arora
Birla Institute of Technology and Science, India

Phalguni Gupta
National Institute of Technical Teachers Training and Research, India

ABSTRACT

The healthcare industry offers highly personalized services to its patients. It is necessary to correctly identify the patient and efficiently link his medical records as and when needed. It is important to note that the need of identification arises in many desperate scenarios when the patient may not be able to tell anything about himself. Biometrics can help in this scenario by using physiological or behavioral characteristics. Some of the biometrics traits could be acquired without direct participation of the person, and therefore, the patient need not provide any pin, password, or token for his identification. Biometrics can handle challenges of duplicate medical records and identity theft. However, there is an important issue that may arise when a large number of patients get registered to the system. Increase in the size of the biometric database gradually escalates the time required for identification. This calls for the need of an efficient indexing approach that can confine the search space and decrease the response time. This chapter highlights biometric indexing approaches suitable for the healthcare industry.

INTRODUCTION

With the availability of large volume of data originating from low-cost sensors, it is now possible to offer useful and personalized services to users. Healthcare can greatly be benefited by this as it offers highly personalized and specific services its patients. The healthcare industry comprises of the companies that offer medical services, develop pharmaceuticals, manufacture medical equipment, and offer medical insurance to its clients. There exist many challenges in healthcare industry that are in terms of existence of

DOI: 10.4018/978-1-5225-7525-2.ch003

duplicate medical records and identity theft (Mogli, 2012). Also, with an accelerated growth in patients' data, this industry has been experiencing many faults due to mixing up of patients' data, confusion of medical charts among different patients and many more. This sometimes results in serious flaws such as giving medication to a wrong patient. In order to improve the quality of patient care, the healthcare industry has been migrating towards the digitization of health records, popularly known as Electronic Health Records (EHR) (Hoerbst & Ammenwerth, 2010). EHR comprises of a person's multiple electronic medical records which are generated by the physicians. It also contains personal health records generated by the patient himself (Ambinder, 2005). With this advancement, all the information related to a patient can easily be accessed at any place and at any point of time. It has also resulted in elimination of medical errors caused due to wrongly entered or redundant paperwork. With the presence of EHRs, physicians can easily trace a patient's complete medical history and the patient can enjoy higher quality of care. Patients also have an advantage in terms of less needed maintenance of medical charts and records.

An efficient healthcare system needs to deliver the right information to the right stack holder (such as patient or doctor) which in term depends on various factors such as accuracy of the records, their update state, data consistency *etc.* With everything going digital these days, security of a patient's information has become an important challenge in healthcare. Data privacy is a major concern. To address the issue, it is needed to restrict the access of patients' information only to the authorized persons. Authentication is the first step in access control, which is depends on accurately recognizing genuine users of the system. Traditional modes of authentication involve the use of the personal identification number (PIN), passwords, access cards, tokens *etc.* all these methods suffers with the loss or leakage of information. Biometrics offer better authentication mechanism by using physiological and/or biological characteristics of the user. Physiological characteristics involve the use of physical measurements of the body such as fingerprint, knuckle, ear, iris, palmprint, hand geometry *etc.* Behavioral characteristics refer to certain behavior associated with a person in performing specific task such as, voice pitch, signature, gait, keystroke, typing style *etc.* It has been established that biometrics should be preferred over the traditional methods for authentication as it overcomes various shortcomings of the later mode of authentication (Maltoni, Maio, Jain, & Prabhakar, 2009). Biometric is hard to spoof or forge and it is always available with the user. There is no need to explicitly carry/memorize it. Sometime biometrics could even be acquired covertly without the direct involvement of the user. Biometric authentication has widely been used in various areas such as financial sector, government sector, military establishments *etc.* for higher security.

A biometric system can operate in two modes *viz.*, enrollment and authentication. In enrollment mode, a new user is registered with the system and his/her biometric samples are stored in the database along with a unique identifier. The stored biometric sample is also known as a biometric template of the user. Biometric pertains highly binding information of the user. Since it is not possible for the user to change his biometric trait therefore the responsibility to protect template from malicious use and to maintain the privacy of user lies with the system itself. For authentication, a newly acquired biometric sample is presented to the system as input query. The system has to find another biometric template from the stored templates that matches the most with the query sample. The authentication can further be subdivided into verification and identification based on the type of applications biometric security is applied to. In verification, user provides his identity to the system along with the query sample and the system has to verify if the query sample matches the claimed identity. While, in case of identification, the system matches the query sample with all the templates stored in the database to establish its identity (Tiwari, & Gupta, 2015). Identification is the most needed application in the healthcare industry because the patient may not be able to claim his identity due to his/her sickness. A biometric template can directly be linked to an electronic health record and can be used during identification. Since identification involves matching the query or input sample with all the templates stored in the database. This one-to-many matchings process becomes very compute intensive as the number of comparison increases. Therefore, as the size of database increases, the time taken in identification also increases. This calls for a need of efficient retrieval approach to reduce the search space and response time of the identification process. This can be achieved by exploring the possibility of biometric indexing. The key concept is to find an appropriate representation of the biometric sample that helps to make similarity based clusters. The retrieval approach aims to produce small sized candidate list from the clusters that is used for matching the query sample instead of the complete database.

An indexing approach involves two phases namely, 1) indexing and 2) retrieval. In indexing phase features are extracted from the biometric sample to construct feature vector. Extracted feature vector contains important and distinguishing characteristics of the trait. During indexing, each feature vector that corresponds to an index is stored in the index table. The aim of the retrieval strategy is to use index table to select templates from the database that have similar feature vectors. Extracted templates form the candidate list for the matching. This list is supposed to be small and therefore, reduces the number of candidates for matching (Khodadoust, & Khodadoust, 2017). A block diagram depicting this process is shown in Figure 1.

However, designing an indexing approach for a biometric database is not the same as that for just another database. There are certain challenges that are observed while designing the indexing approach for a biometric database. Some of the challenges are: (1) features extracted from a biometric sample are large and they vary in number, (2) query sample may be translated or rotated as compared to the stored database templates, (3) quality of sample images may be poor which results in appearance of false feature

Figure 1. Block Diagram showing indexing/retrieval process

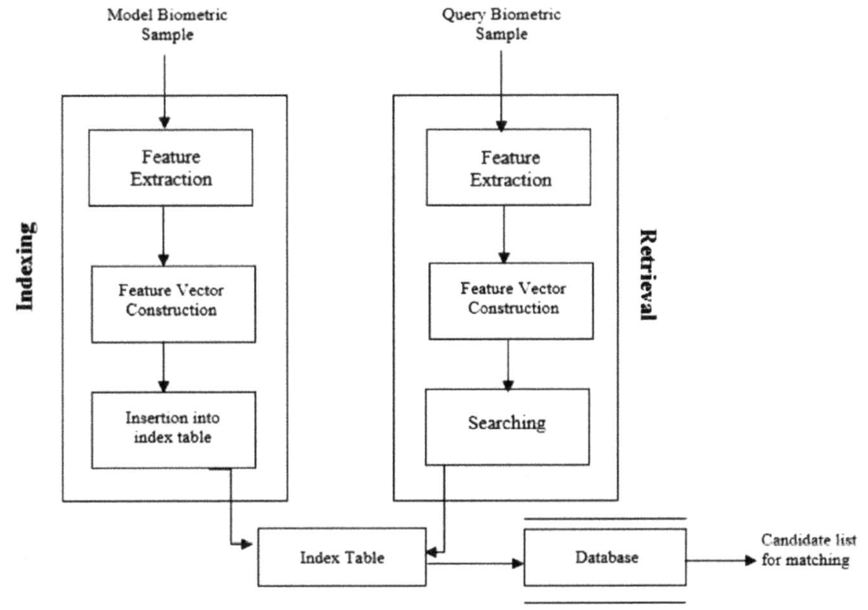

points and absence of true feature points from the sample, (4) features extracted from a sample at a given point of time may not be in same order when extracted again from the sample taken at a different time.

To evaluate the performance of an indexing approach one can use parameters such as *hit-rate, penetration-rate, false-accept-rate,* and *false-reject-rate*. *Hit-Rate* refers to the correct number of matchings that have been obtained among the top *t* candidates retrieved from the total number of queries made. *Penetration-Rate* refers to the number of candidates that have to retrieved from the total number of templates from the database for a successful retrieval. *False-Acceptance-Rate* (FAR) implies the number of imposter samples that have been categorized as genuine matches by the system and *False-Reject- Rate* (FRR) is the number of rejections of the genuine samples. The relation between FAR and FRR is depicted in Figure 2. As the FAR decreases, FRR increases and vice versa. That is because with the increase in security level, the system will become less user friendly and the number of rejections of genuine user would increase. The point at which both the entities meet is termed as *Equal-Error-Rate* (EER). It is used to determine the operating value of threshold.

FINGERPRINT

Fingerprint is an impression that gets developed automatically on nearly all the surfaces coming in direct contact to the human finger. Digital fingerprint sensors can very effectively capture this print however, they may vary in size and quality. Fingerprint is one of the most acceptable biometric traits because it has good accuracy and there is no social stigma in providing fingerprint samples. As compared to other biometric traits, it offers many advantages such as easy collectability, acceptability, uniqueness and permanence. Fingerprint has been used widely for access control, forensic applications and government services. Structural pattern of ridge and valley present on the fingerprint is found to make a unique pattern

Figure 2. A graph depicting the relationship between FAR and FRR

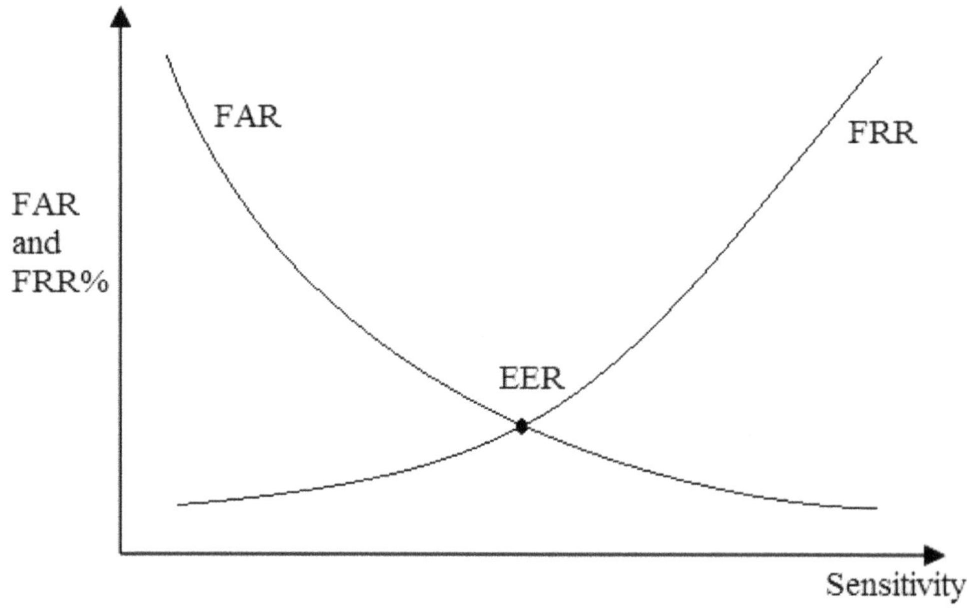

for every finger and therefore, are used for authentication. Irregularities in the ridge structure, called as minutiae points, are used as features during fingerprint recognition. Presence of discriminating patterns such as loops, whorl and arch makes fingerprint highly expressible biometric trait. Loop and whorl are called fingerprint singularities that can be treated as global features. In healthcare domain, if the patient cannot participate in fingerprint acquisition, one can put his finger on the scanner to get the fingerprint. We would focus a specific problem related to identification of fingerprints that arises when a large number of fingerprints are registered in the system and the task is to identity of a newly given fingerprint.

Fingerprint identification work efficiently for small databases but the performance starts degrading as the size of database increases. There is a need to reduce the search space to limit the number of comparisons (Bai, Wang, Zhao, & Li, 2018). To do so, implementation of a pre-selection techniques is required which aims at filtering the search space for comparison with respect to the query sample. This can happen through either 1) Exclusive Classification or 2) Continuous Classification. Exclusive classification is based on five classes of fingerprints namely, the whorl, the tented arch, plane arch, ulnar loop and radial loop (Yager, & Amin, 2004). By using them, fingerprints could be divided into five mutually exclusive sets. During identification, the class of the query fingerprint is found out and the templates belonging to its class are extracted from the stored database as a candidate list for matching. This approach faces a serious limitation due to non-uniform class distribution. Majority of the fingerprints (more than 93%) belongs to only three classes the loops (radial and ulnar) and the whorl (Wilson, Candela, & Watson, 1994). Due to this reason, the reduction in search space is sometime found to be insignificant. Another problem is that the class of the query fingerprint is not known apriori and it may be ambiguous. Therefore, in such cases where the class of the query sample is not clear, it is not possible to find the candidates in which the matching could be done. The other technique, continuous classification or indexing has come as a rescue to overcome the former's limitations. In this approach, features of fingerprints are extracted and stored in the index table (Khodadoust, & Khodadoust, 2017). During retrieval, the feature vector corresponding to query fingerprint and all those samples that have similar feature vectors are extracted from the database. The extracted templates are then used as candidates for matching with the query fingerprint. The retrieval strategy can be an incremental approach such as clustering or hashing or a simple filtering approach.

Indexing Approaches for Fingerprint Database

Many fingerprint indexing approaches have been proposed in literature. These approaches can be categorized into four broad categories based on the feature used for index construction, 1) texture based, 2) minutiae based, 3) hybrid and 4) deep neural network based.

Texture Based Fingerprint Indexing Approaches

These approaches uses texture or global features of the fingerprint pattern for indexing. An approach based on orientation and frequency of ridge pattern of fingerprint has been proposed by Cappelli et.al. in (Cappelli, 2011). It has been found that the proposed approach could search for a query fingerprint from a database containing one million fingerprints in less than a second. Another rotation invariant indexing approach has been proposed by Lee *et.al.* (Lee, Kim, & Park, 2005) that uses polar complex moments (PCMs). The approach on FVC2002 DB1a and NIST DB4 databases has achieved perfect matching at penetration rate of 7.9% and 5.34% respectively. However, since the derived features have less correlation

Figure 3. Sample fingerprint impressions taken from FVC2004 DB1 (Maio et.al.)

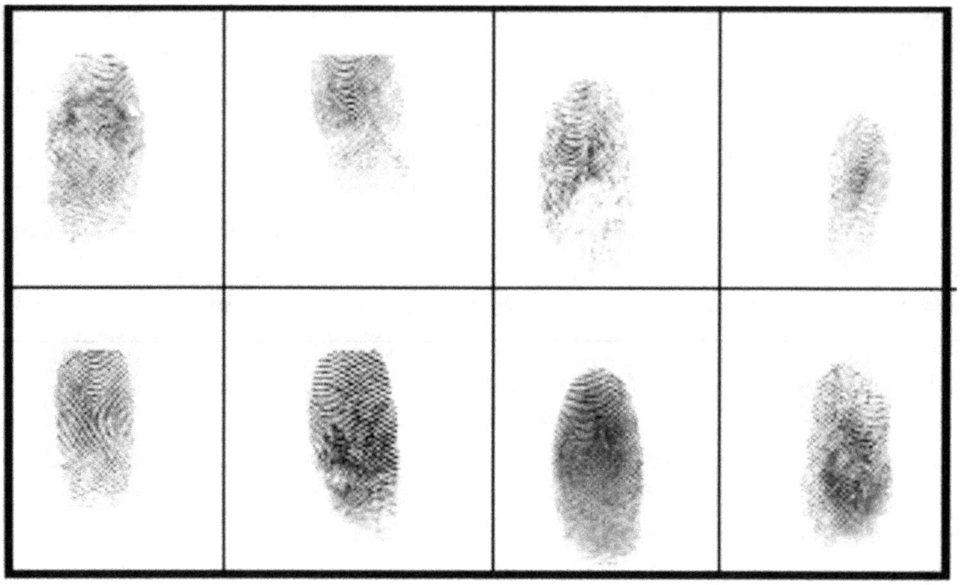

with ridge frequency and minutiae, these can be integrated with them to improve accuracy. Anand *et.al.* proposed an indexing approach based on pores found on fingerprint (Anand and V. Kanhangad, 2017). The features have been extracted using Delaunay triangulation (Lee, & Schachter, 1980) and they are clustered using unsupervised k-means clustering algorithm (Hartigan & Wong, 1979). During retrieval stage, some clusters are selected on the basis of their Euclidean distance (Danielsson, 1980) from the query print. These approaches suffer from limitations that the global features are not robust to distortions such as translation, sheer, rotation, occlusion, scale *etc.* Also, since these approaches make use of location of the singular points, it is expected that the fingerprints are already aligned in the database.

Minutiae Based Fingerprint Indexing Approaches

Minutiae based indexing approaches use local features *i.e.* minutiae points to construct feature vector. A minutia is represented by a 3-tuple (x_m, y_m, θ_m) where x_m and y_m coordinates give the location and θ_m is the angular direction or orientation of the minutiae whose value lies in range $[0, 2\pi]$. Single Minutia based approaches uses a lone minutia to construct feature vector/index. A geometric hashing based approach has been proposed in (Jayaraman, Gupta, & Gupta, 2014). It has two stages, namely, indexing and searching and both of the operations have been performed with linear time complexity. Minutia Binary Pattern which is a fixed length feature vector is constructed from a minutia point. Each minutia and its corresponding feature vector is inserted into the hash table exactly once. This reduces computational cost and memory requirement. The approach has been experimentally tested on FVC2004 databases (Maio, Maltoni, Cappelli, Wayman, & Jain, 2004), DB1, DB2 and DB3 and has achieved a hit rate of 96%, 95% and 94% at a penetration of 2.61%, 2.33% and 1.95% respectively. An indexing approach based on Coaxial Gaussian Track Code (CGTC) vector has been proposed by Tiwari *et.al.* (Tiwari, & Gupta, 2015). CGTC is a fixed length feature vector constructed for every minutia to place in a Quantized

Lookup Table (QLT). Insertion is done only once to reduce computational and memory cost. Another advantage of this approach is that rehashing is not required because of uniform distribution of minutia points in QLT. This approach has achieved a hit rate of 100%, 98.6% and 100% with penetration rates of 1.90%, 3.86% and 1.62% respectively, when tested on FVC2002-Db 2a, FVC2004 DB1 and IITK-Sel500FP respectively. Both of the discussed indexing approaches are robust to global distortions like rotation and translation and they also have linear time complexity. But a single minutia is not robust to elastic deformation.

Minutiae pair based approaches take into consideration of two minutiae at a time for feature vector construction. This is so because minutiae pairs are invariant to rigid transformations and exploiting redundant combination of minutiae pair provides immunity against noise (Jin., Teoh, Ong, & Tee, 2010). It has proposed a fingerprint indexing approach that uses two minutiae or a minutiae pair. This approach extracts the difference between their angles, distance between these points and other two angles formed between the line joining them and their orientation. It achieved 3% Equal Error Rate on FVC2004 DB1 database. For these approaches, the computation complexity is very high because many features need to be computed and that too, for every pair of minutiae.

Minutiae triplet based approaches use triangles formed from a group of three minutiae points for making index for the index table. The concept of minutiae triangles has been first introduced by Germain *et.al.* (Germain, Califano, & Colville, 1997). This approach has considered all possible minutiae triangles and pairwise matching has been among them. It has used transformation parameter clustering to improve the accuracy. An effort has been made by Bhanu *et.al.* to improve the former approach by incorporating some additional features to the minutiae triplets (Bhanu, & Tan, 2003). This approach has used triplet features such as angle, headedness, greatest side, type, direction and geometric constraints in order to eliminate false correspondences. These approaches suffer from high computational complexity *i.e.* $O(n^3)$ because n^3 triangles are possible for n minutiae points and therefore, these many triangles need to be considered during identification. To reduce the number of minutiae triangles for matching, a Delaunay Triangles based approach has been proposed (Bebis, Deaconu, & Georgiopoulos, 1999). Suppose, there is a set of Q number of points; for this, a Delaunay Triangulation is generated by using its Voronoi Diagram. The Delaunay Triangles can be efficiently computed and are unique. The proposed approach has been successful in reducing the number of minutiae triangles to linear *i.e.* from $O(n^3)$ to $O(n)$. An indexing approach that is hierarchical in nature and uses barycenter has been proposed by Elmouhtadi *et.al.* (Elmouhtadi, Aboutajdine, & El Fkihi, 2015) to account for false identification. To ensure that the matching is robust to variation in orientation, it is carried out by considering all the angles of the triangle extracted from both the samples. The issue with this approach is that some similar triangles may be found which do not belong to same minutiae points.

Minutiae Cylinder Code (MCC) based approaches use a three dimensional data structure which represents spatial relationship of distance and orientation among minutiae lying in neighborhood. A large number of approaches has been proposed that takes into account minutia orientation with other features such as ridge curvature, singularities *etc.* However, in case of large databases, it is favorable to extract features from minutiae because information about minutiae is most commonly used and is robust (Li, Yang, & Busch, 2015). A Locality Sensitive Hashing (LSH) (Slaney, & Casey, 2008) based on minutia cylinder code (MCC), has been proposed for indexing fingerprint database (Cappelli, Ferrara, & Maltoni, 2011). An MCC based approach that uses pose estimation has been proposed by Su *et.al.* (Su, Feng, & Zhou, 2016). It is assumed that the poses remain invariant among different images of same fingerprint. Therefore, fingerprints can be aligned in an efficient manner. This paper also introduced a new version of LSH algorithm with the purpose of improving the the accuracy. This has been tested on FVC and NIST databases. The MCC representation is high dimensional and redundant. To address this, an indexing approach using a feature mapping matrix to map real-valued high dimensional version of MCC into a low dimensional binary code has been proposed by Bai *et.al.* (Bai, Wang, Zhao, Li, 2018). The main objective of using this representation is that redundant information can be removed by minimizing inter-bit variation and maximizing intra-bit variation. It has used multi-index hashing to find the exact k-nearest neighbors on the binary codes to obtain a candidate list for matching.

Hybrid Fingerprint Indexing Schemes

These approaches make use of both local and global features for indexing. An indexing approach that uses FOMFE coefficients along with minutiae triplets to generate two different feature vectors has been proposed by Zhou *et.al.* (Zhou, Hu, Wang, Petersen, & Bennamoun, 2016). Two candidate lists are generated using both the features and they are integrated into a single list using fuzzy rules. An indexing approach using both minutiae neighborhood and minutiae cylinder code for extracting features has been proposed (Parmar, & Degadwala, 2016). The constructed feature vectors are put into a category from the five different classes by using a support vector machine (SVM).

Deep Neural Network based Fingerprint Indexing Approaches

Deep Convolutional Neural Network (DCNN) architecture can suitably filtering an image in a parallel way. DCNN has been used in image processing and has made significant contribution in tasks such as image classification, object detection, image segmentation *etc.* A Convolution Neural Network (CNN) based indexing approach that generates a fixed length feature vector without explicitly extracting the minutiae points has been proposed by Cao *et.al.* (Cao, & Jain, 2017). For aligning the fingerprints in a combined coordinate system, it has made use of an orientation field dictionary. The training of the CNN has been done on a longitudinal fingerprint database and last fully connected layer of the network outputs a fixed length feature vector that forms the index in the index table. The feature vector can also be used for template protection due to its fixed length. An experiment has been conducted on two rolled fingerprint databases, NIST SD4 and NIST SD14 and it has shown an error rate of 0.40% and 0.26% respectively at penetration rate of 10% that is quite impressive.

IRIS

Iris is outer circular strip of black portion of human eye. It is considered to be one of the most accurate biometric traits. It has relatively many advantages in authentication over other biometric traits such as iris can be imaged from a distance of less than a meter *i.e.* without actually making a contact with the sensor. It is not sensitive to illumination angle because of being a planar object. Being an internal organ, it is secure from the environment and remains invariant over time. Sample of iris images are shown in Figure 4. Easy localization of eyes from the face and distinguishing shape of the iris makes extraction of this feature reliable and precise and a feature vector can be constructed which is size invariant (Wang & Tan, 2002). Moreover, even though the iris expands and contracts to adjust the size of the pupil, its

Figure 4. Sample iris image taken from UBIRIS database (Proença, & Alexandre, 2005)

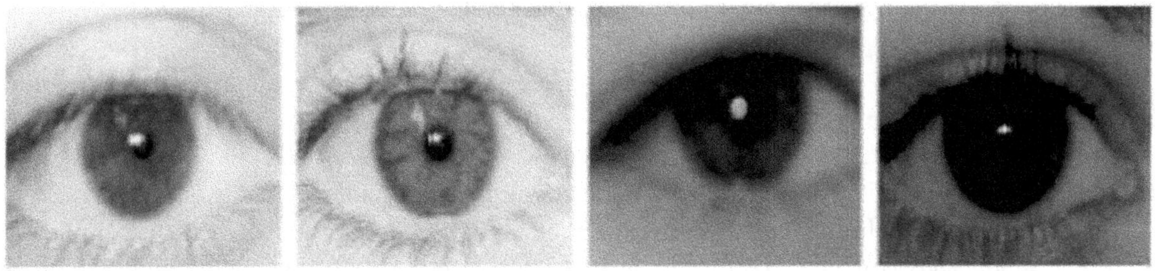

complex texture remains intact. Such alterations in the texture due to expansion and contraction can be inverted mathematically while analyzing an iris image, to extract and encode iris features.

Indexing Approaches for Iris Database

Mukherjee *et.al.* have proposed two different indexing approaches for iris database. One of these approaches is based on iris codes, known as post-encoding indexing and the other one is based on the features mined from iris texture. The second one is known as pre-encoding indexing (Mukherjee & Ross, 2008). For the first approach, it generates iris code which corresponds to binary code and has dimensionality equal to that of unwrapped iris structure. This involves three steps, 1) iris segmentation, 2) geometric normalization and 3) feature extraction. But, there are two problems associated with iris codes: 1) they are high dimensional and 2) there may exist correlation between dimensions. To overcome this problem, dimension is reduced by using one of the following three methods. In the first approach, each row of the iris code is made single entity by taking average of the row entries. Second approach involves averaging column entries to make it a single entry. The last approach uses Principal Component Analysis (PCA) (Holland, 2008) to transform iris code into low dimension. After generating lower dimensionality iris code, k-means clustering (Hartigan & Wong, 1979) is applied to divide them in multiple groups. The second approach, texture based indexing, used pixel intensities and local group of pixels (block) in the iris texture. It assigns same index value to the blocks which has similar texture characteristics. The experimental test has been conducted on CASIA-IrisV3-Interval dataset which is a subset of CASIA iris image database V3.0. The first approach has achieved a hit rate of 80% for average penetration rate of 17% when PCA based method has been applied and a hit rate of 80% for an average penetration rate of 21% when column based method has been used. The hit rate of 80% has been achieved at average penetration rate of 8% when block based statistics has been used. However, the second approach based on iris texture achieved 84% hit rate at an average penetration rate of 30%.

Another indexing approach based on iris color has been proposed in (Puhan, & Sudha, 2011). The proposed approach has divided iris database into several groups based on color indices of the iris and it could be applied to iris databases containing either iris features for recognition or raw colored iris images. During enrolment, a new iris is added to the appropriate group based on its color index. This proposed approach has been tested on UBIRIS and UPOL database and achieved a hit rate of 98% with a penetration rate of less than 25%. An indexing approach for poor quality iris images, also known as degraded images, has been proposed by Proenca *et.al.* (Proenca, 2013). A bloom-filter based iris indexing technique has been proposed by Rathgeb *et.al.* (Rathgeb, Breitinger, Baier, & Busch, 2015). Bloom filters are extracted from iris codes which are basically binary iris templates. These extracted bloom filters have been used to construct binary tree which has reduced the search space to O(log *n*).

FACE

Face is one of the most intuitive biometric traits for recognition as human mostly rely on face to recognize a person. Face recognition has gathered a lot of attention in the recent times due to its major applications in both government and commercial sectors; for instance, face tagging, security systems, human-computer interface, surveillance, biometric authentication, video indexing etc. It aims at automatically authenticating a person's identity from a digital video sequence or an image taken in a controlled or uncontrolled environment. Face offers several advantages over other biometric traits. First, it can be acquired passively *i.e.* without any explicit involvement of the user. Secondly, preprocessing after face recognition make it robust to noise, scale, orientation and illumination. Since face recognition is non-intrusive, it does not involve health risks (Jafri, & Arabnia, 2009). Some of the face images taken from a public database, FERET (Phillips, Wechsler, Huang, & Rauss, 1998) are shown in Figure 5.

Indexing Approaches for Face Database

Am indexing approach for facial images proposed by Wu *et.al.* makes use of both local (such as nose, eyes, mouth *etc.*) and global features (whole face image) (Wu, Ke, Sun, & Shum, 2011). Local features are capable of encoding geometric constraints and are more robust to variation in expression and pose. After locating the five local features *i.e.* two eyes, nose tip and two mouth corners, a hamming signature of 40 byte length is computed for every face image that represents the high dimensional global feature. This approach takes into account the fact that the local and global features are complementary. Local features help in the efficient traversal of the index table and returns top candidates for matching. The hamming signature is used to re-rank the retrieved candidates. The proposed approach has been tested on Labeled Face in Wild (LFW) dataset (Huang *et.al.*, 2007) and it has been shown that the approach does not have any false positives in top-10 results. An indexing approach for face database using modified geometric hashing has been proposed by Kaushik *et.al.* (Kaushik, Umarani, Gupta, A. K., Gupta, A. K., & Gupta, P., 2013). The features of the face that form the control points have been extracted using Scale Invariant Feature Transform (SIFT) and a descriptor or a control point vector has been constructed

Figure 5. Sample facial images taken from FERET database (Phillips, 2004)

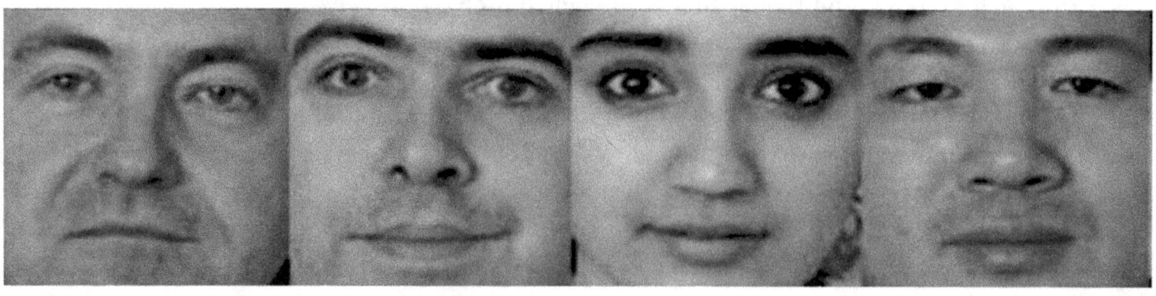

around each control point in order to represent its uniqueness in its neighborhood. The SURF descriptors for the control points for every face image has been used to map face images into hash table that is built with modified geometric hashing for face database. The proposed approach has been tested on FERET database (Phillips *et.al.*, 1998) and compared in terms of cost that includes computational cost and memory requirements with two other well known indexing approaches namely, Delaunay Triangulation (DT) based and All Possible Triangulation (APT) indexing approach. It has been recorded that the proposed approach has made least number of entries in the hash table and the matching time is found to be less as compared to the other two. The proposed approach has achieved 100% hit rate in case top-4 best matches.

Many indexing approaches use approximate k-NN search methods to examine only the a small subset of the biometric database for matching during identification. The candidate list for matching is selected by indexing with the help of either hash codes or vector quantization. But the disadvantage of these approaches is that hyperparameters tuning using enrolled database and training is required for the indices. The hyperparameters need to be tuned whenever there is an update in the database and this leads to high maintenance cost for the target database. To overcome the problem, a novel k-NN search method has been proposed (Ishii, Imaoka, & Sato, 2017). The proposed approach sums the similarity scores sequentially with their dimensions. The target samples are then rejected based on their aggregated score. This approach is more suitable because preliminary computation of the feature-wise score is not required and hence, the sorted access. Moreover, the maintenance cost of the enrolled database reduces as the computation of the aggregated score and bounds of the residual score do not need any data structure or training. This approach has been trained on LFW dataset (Huang *et.al.*, 2007) and the trained model has been used on WDRef database has for performance evaluation of identification. It has achieved an identification rate of 88.5% when $k=10$.

OTHER BIOMETRIC TRAITS

Apart from fingerprint, iris and face that have been widely used for authentication. There are several other biometric traits that could also be considered for the healthcare industry. Some of them such as palmprint, ear and vein pattern (shown in Figure 6) are discussed in this section.

Figure 6. Biometric traits (starting from left)- ear, palmprint, finger vein pattern and hand vein pattern

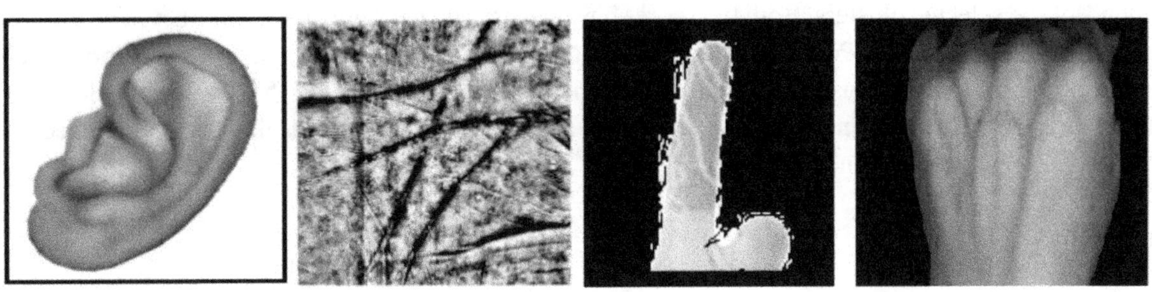

Palmprint

A palmprint is the inner skin area between the fingers and thumb. It consists of distinctive features such as wrinkles, minutiae points and ridges that form unique patterns for authentication (Jain, Ross, & Prabhakar, 2004). One of the palmprint image is shown in *Figure 6.* Many indexing approaches for palmprint database have been proposed in literature. A palmprint indexing approach that utilizes hierarchical matching criteria by using four different features at different levels of matching has been proposed in (You, Kong, Zhang, & Cheung, 2004). These features are, 1) distance between key points, 2) global texture energy, 3) fuzzy interest line and 4) local directional texture energy. The proposed approach has been tested on an in-house database containing 7752 palmprint images of 386 subjects. The first three level features have been successful in reducing the number of samples by 9.6%, 7.8% and 60.6% respectively. The last level feature is then used for further matching. A texture based palmprint indexing approach has been proposed in (Li, You, & Zhang, 2005). It has used global and local texture energy that have been used for construction of the feature vectors. The searching is first done using the global features. The reduced list of candidates are further searched using the local features. The best and worst response time achieved for searching a query sample has been reported as 0.3 and 3.7 seconds respectively. Another indexing approach for the palmprint database has been proposed by (Paliwal, Jayaraman, & Gupta, 2010). The construction of index has been done using the match score vector and palmprint texture has been used for retrieval of the top t candidates for matching. This approach has been tested on PolyU database that consists of 7752 palmprint images. The test revealed a hit rate of 98.01%. A palmprint indexing approach that has used geometric hashing to index SURF keypoints for generation of the index has been proposed in (Badrinath, Gupta, & Mehrotra, 2013). When a query palmprint is presented to the system, its corresponding index is extracted from the hash table and voting is casted for all the templates that correspond to the extracted index. All those templates whose voting score is greater than a specified threshold serve as candidates for matching. SURF matching is performed between the query sample and the candidates. Finally, the SURF score and voting score are fused together to attain better accuracy. The proposed approach has been tested on three publicly available datasets, IITK, CASIA and PolyU and the reported accuracy at the top best match is 98.94%, 98.75% and 97.76% respectively. A clustering based indexing scheme for palmprint database has been proposed in (Kavati, Prasad, & Bhagvati, 2014). It computes an index code for every palmprint image by calculating its similarity with a predefined set of sample images. These indices are stored in an index table and clustering is applied on the samples. When an input query palmprint image is given to the system, clusters that have similarity with the query sample are selected. Then, the voting strategy is utilized to compute similarity between the the query sample and elements of the selected clusters. Both the scores have been fused to attain better accuracy. The proposed approach has been tested on PolyU palmprint dataset. A minutia based palmprint indexing approach has been proposed in (Muñoz-Briseño, Hernández-Palancar, & Gago-Alonso, 2015). The approach has used minutiae triplets for construction of feature vectors and the experimental results achieved an accuracy of 99.5% for any value of penetration rate.

Ear

Human ear as a biometric has several advantages over other traits such as it is stable and has rich structure which is invariant with time. An ear is expression neutral as compared to face that could have various expressions. The ear does not suffer from variation in background as it is located at the middle side of the head therefore, it does not need to be captured in a controlled environment. The collection of ear sample does not involve hygiene issues and not much cooperation is required by the user (Hurley, Arbab-Zavar, & Nixon, 2008).

An indexing approach based on segmentation and classification of ear shape has been proposed in (Maity, & Abdel-Mottaleb, 2015). The paper consists of two parts, 1) automated 3D ear segmentation and 2) classification of the ear database. The database has been categorized hierarchically, first on the basis of shape with the categories being oval, rectangular, round and triangular. Then, for each shape category a tree has been constructed based on the depth information. Both k-dimensional (KD tree) and pyramid tree have been implemented to gauge their efficiency. During retrieval, the images that belong the hyper rectangle are included in the candidate list for matching. The experimental study conducted on University of Notre Dame Collection J2 dataset have shown a reduction of 10% - 50% in the search space while declining accuracy but increasing search speed. Another ear indexing approach on the basis of ear shape has been proposed in (Zhu, & Mu, 2018). The index is calculated on the basis of the category in which the shape of the ear lie. The shape feature is computed by using certain features of the ear such as, triangularity, rectangularity, circularity and ellipticity. The shape of the query ear sample is also computed and the ear images that are similar in the database are selected as the candidates for matching.

An indexing approach on ear database that is based on Iterative Closest Point (ICP) has been proposed in (Zhang, Mu, Yuan, Zeng, & Chen, 2017). It has used Local Surface Variation (LSV) to extract key features. LSV is responsible for eliminating redundant data. It only responds to concave and convex regions of the ear, thus ignoring flat skin surface to get the normalized ear image. The matching between the probe and database images is done by using the modified ICP-LSV algorithm, that is responsible for making system computationally more efficient. The proposed approach was tested on two databases, CASIA-3D FaceV1 and UND-J2 wherein it has achieved an accuracy of 100% and 98.55% respectively.

Vein Pattern

Vascular pattern on the human hand are relatively new trait for human recognition. It has found to have enough discriminative power and is very difficult to forge. The vein pattern also does not get affected with any disease or environmental changes (Wang, & Leedham, 2006). As a result, vein pattern offers unique, stable and repeatable features for authentication process.

A vein pattern indexing approach that uses the texture information and minutiae points has been proposed by Prasad *et.al.* (Kavati, Prasad, & Bhagvati, August). Firstly, a binary code is generated for every captured image using the texture information to eliminate wrong correspondences from the list of candidates. Then, delaunay triangulation is used along with the binary code to further narrow down the candidate list. A locality sensitive hashing (LSH) has been proposed in (Tang, Huang, Li, & Li, 2010). In this approach, the captured image is divided into 32x24 grids and the number of white pixels in every grid is counted. If the number of white pixels is larger than a certain threshold value then that grid is given a value equal to1. In this way, a 768 bit vector is formed that depicts the vein pattern. The same vector is obtained for a query image as well. Searching is done by using LSH based algorithm. Another palm vein pattern indexing approach has been proposed in (Zhou, Liu, Feng, Yang, Huang, & Nie, 2014). In this approach, Gaussian Radon transform has been employed during registration to obtain the orientation matrix and principal direction of the image. The images have been classified into six categories on the basis of the principal direction. The principal direction of the query sample is also extracted and the category is determined. All the samples contained in the determined category serve as the candidates for matching. The experiment has been conducted on PolyU and CASIA dataset and the proposed approach achieved an accuracy of 96.67% and 96.00% respectively.

CONCLUSION

Wide adaptability of biometrics in healthcare industry is needed for accurate and efficient access of medical records. However, large scale adaptation leads to a large size of patients' database which in term makes the process of identification very time consuming. One cannot afford high response time for an identification systems specially in an industry like healthcare where life can be at risk with a delay of few seconds. In order to make the system fast, it is required to lower the number of comparisons during the identification. To do so, efficient biometric database indexing schemes need to be applied that can produce small size candidate list against a query template presented for matching. This chapter reviews the state-of-the-art indexing approaches for fingerprint, iris and face. Other suitable biometric traits such as palmprint, ear and vein pattern have also been investigated. However, there exists other biometric traits such as retina, voice, signature, gait, keystroke *etc.* but they could not be used in healthcare industry for patient authentication as the user may not be in a suitable condition to give his/her corresponding biometric sample.

The discussed indexing approaches are found to be a strong candidate to improve the performance of the identification process. Among these traits, fingerprint recognition has been found to be the fastest followed by iris and face. However, if accuracy is a concern then iris is preferred followed by face and fingerprint. Nevertheless, it may be noted that none of these indexing approaches has been tested on real databases. Therefore, by implementing these on healthcare databases, their efficiency could be tested in real time.

REFERENCES

Ambinder, E. P. (2005). Electronic health records. *Journal of Oncology Practice / American Society of Clinical Oncology*, *1*(2), 57–63. doi:10.1200/jop.2005.1.2.57 PMID:20871681

Anand, V., & Kanhangad, V. (2017). Pore based indexing for High-Resolution Fingerprints. In *IEEE International Conference on Identity, Security and Behavior Analysis (ISBA)* (pp. 1-6). IEEE.

Badrinath, G. S., Gupta, P., & Mehrotra, H. (2013). Score level fusion of voting strategy of geometric hashing and SURF for an efficient palmprint-based identification. *Journal of Real-Time Image Processing*, *8*(3), 265–284. doi:10.100711554-011-0229-2

Bai, C., Wang, W., Zhao, T., & Li, M. (2018). Fast exact fingerprint indexing based on Compact Binary Minutia Cylinder Codes. *Neurocomputing*, *275*, 1711–1724. doi:10.1016/j.neucom.2017.10.027

Bebis, G., Deaconu, T., & Georgiopoulos, M. (1999). Fingerprint identification using delaunay triangulation. International Conference on Information Intelligence and Systems. *IEEE Proceedings*, 452–459.

Bhanu, B., & Tan, X. (2003). Fingerprint indexing based on novel features of minutiae triplets. *IEEE Transactions on Pattern Analysis and Machine Intelligence*, *25*(5), 616–622. doi:10.1109/TPAMI.2003.1195995

Cao, K., & Jain, A. K. (2017). Fingerprint indexing and matching: An integrated approach. In *IEEE International Joint Conference on Biometrics (IJCB)* (pp. 437-445). IEEE. 10.1109/BTAS.2017.8272728

Cappelli, R. (2011). Fast and accurate fingerprint indexing based on ridge orientation and frequency. *IEEE Transactions on Systems, Man, and Cybernetics. Part B, Cybernetics*, *41*(6), 1511–1521. doi:10.1109/TSMCB.2011.2155648 PMID:21708504

Cappelli, R., Ferrara, M., & Maltoni, D. (2011). Fingerprint indexing based on minutia cylinder-code. *IEEE Transactions on Pattern Analysis and Machine Intelligence*, *33*(5), 1051–1057. doi:10.1109/TPAMI.2010.228 PMID:21173446

Elmouhtadi, M., Aboutajdine, D., & El Fkihi, S. (2015). Fingerprint indexing based barycenter triangulation. In *Third World Conference on Complex Systems (WCCS)* (pp. 1-6). IEEE.

Germain, R. S., Califano, A., & Colville, S. (1997). Fingerprint matching using transformation parameter clustering. *IEEE Computational Science & Engineering*, *4*(4), 42–49. doi:10.1109/99.641608

Hartigan, J. A., & Wong, M. A. (1979). Algorithm AS 136: A k-means clustering algorithm. *Journal of the Royal Statistical Society. Series C, Applied Statistics*, *28*(1), 100–108.

Hoerbst, A., & Ammenwerth, E. (2010). Electronic health records. *Methods of Information in Medicine*, *49*(04), 320–336. doi:10.3414/ME10-01-0038 PMID:20603687

Holland, S. M. (2008). *Principal components analysis (PCA)*. Athens, GA: Department of Geology, University of Georgia.

Huang, G. B., Ramesh, M., Berg, T., & Learned-Miller, E. (2007). *Labeled faces in the wild: A database for studying face recognition in unconstrained environments*. Technical Report 07-49. University of Massachusetts, Amherst.

Hurley, D. J., Arbab-Zavar, B., & Nixon, M. S. (2008). The ear as a biometric. In *Handbook of biometrics* (pp. 131-150). Springer.

Ishii, M., Imaoka, H., & Sato, A. (2017). Fast k-Nearest Neighbor Search for Face Identification Using Bounds of Residual Score. In *Automatic Face & Gesture Recognition (FG 2017), 2017 12th IEEE International Conference on* (pp. 194-199). IEEE.

Jafri, R., & Arabnia, H. R. (2009). A survey of face recognition techniques. *JIPS*, *5*(2), 41–68.

Jain, A. K., Ross, A., & Prabhakar, S. (2004). An introduction to biometric recognition. *IEEE Transactions on Circuits and Systems for Video Technology*, *14*(1), 4–20. doi:10.1109/TCSVT.2003.818349

Jayaraman, U., Gupta, A. K., & Gupta, P. (2014). An efficient minutiae based geometric hashing for fingerprint database. *Neurocomputing*, *137*, 115–126. doi:10.1016/j.neucom.2013.02.059

Jayaraman, U., Prakash, S., & Gupta, P. (2010). An iris retrieval technique based on color and texture. In *Proceedings of the Seventh Indian Conference on Computer Vision, Graphics and Image Processing* (pp. 93-100). ACM.

Jin, Z., Teoh, A. B. J., Ong, T. S., & Tee, C. (2010). A Revocable Fingerprint Template for Security and Privacy Preserving. *Transactions on Internet and Information Systems (Seoul)*, *4*(6).

Karu, K., & Jain, A. K. (1996). Fingerprint classification. *Pattern Recognition, 29*(3), 389–404. doi:10.1016/0031-3203(95)00106-9

Kaushik, V. D., Umarani, J., Gupta, A. K., Gupta, A. K., & Gupta, P. (2013). An efficient indexing scheme for face database using modified geometric hashing. *Neurocomputing, 116*, 208–221. doi:10.1016/j.neucom.2011.12.056

Kavati, I., Prasad, M. V., & Bhagvati, C. (2013, August). Vein pattern indexing using texture and hierarchical decomposition of delaunay triangulation. In *International Symposium on Security in Computing and Communication* (pp. 213-222). Springer. 10.1007/978-3-642-40576-1_21

Kavati, I., Prasad, M. V., & Bhagvati, C. (2014). A new indexing method for biometric databases using match scores and decision level fusion. In *Advanced Computing, Networking and Informatics-Volume 1* (pp. 493–500). Cham: Springer. doi:10.1007/978-3-319-07353-8_57

Khodadoust, J., & Khodadoust, A. M. (2017). Fingerprint indexing based on expanded Delaunay triangulation. *Expert Systems with Applications, 81*, 251–267. doi:10.1016/j.eswa.2017.03.048

Lee, D. T., & Schachter, B. J. (1980). Two algorithms for constructing a Delaunay triangulation. *International Journal of Computer & Information Sciences, 9*(3), 219–242. doi:10.1007/BF00977785

Lee, S. O., Kim, Y. G., & Park, G. T. (2005). A feature map consisting of orientation and inter-ridge spacing for fingerprint retrieval. In *International Conference on Audio-and Video-Based Biometric Person Authentication* (pp. 184-190). Springer. 10.1007/11527923_19

Li, G., Yang, B., & Busch, C. (2015). A novel fingerprint indexing approach focusing on minutiae location and direction. In *Identity, IEEE International Conference on Security and Behavior Analysis (ISBA), 2015* (pp. 1-6). IEEE.

Li, W., You, J., & Zhang, D. (2005). Texture-based palmprint retrieval using a layered search scheme for personal identification. *IEEE Transactions on Multimedia, 7*(5), 891–898. doi:10.1109/TMM.2005.854380

Ma, L., Wang, Y., & Tan, T. (2002). Iris recognition based on multichannel Gabor filtering. In *Proc. Fifth Asian Conf. Computer Vision* (*Vol. 1*, pp. 279-283). Academic Press.

Maio, D., Maltoni, D., Cappelli, R., Wayman, J. L., & Jain, A. K. (2004). FVC2004: Third fingerprint verification competition. In *Biometric Authentication* (pp. 1–7). Berlin: Springer. doi:10.1007/978-3-540-25948-0_1

Maity, S., & Abdel-Mottaleb, M. (2015). 3D ear segmentation and classification through indexing. *IEEE Transactions on Information Forensics and Security, 10*(2), 423–435. doi:10.1109/TIFS.2014.2379437

Maltoni, D., Maio, D., Jain, A. K., & Prabhakar, S. (2009). *Handbook of fingerprint recognition*. Springer Science & Business Media. doi:10.1007/978-1-84882-254-2

Mogli, G. (2012). Role of Biometrics in healthcare privacy and security management system. *Sri Lanka Journal of Bio-Medical Informatics, 2*(4), 156. doi:10.4038ljbmi.v2i4.2245

Mukherjee, R., & Ross, A. (2008). Indexing iris images. In *Pattern Recognition, 2008. ICPR 2008. 19th International Conference on* (pp. 1-4). Academic Press. 10.1109/ICPR.2008.4761880

Muñoz-Briseño, A., Hernández-Palancar, J., & Gago-Alonso, A. (2015). Minutiae based palmprint indexing. In *Computational Forensics* (pp. 10–19). Cham: Springer. doi:10.1007/978-3-319-20125-2_2

Paliwal, A., Jayaraman, U., & Gupta, P. (2010, September). A score based indexing scheme for palmprint databases. In *Image Processing (ICIP), 2010 17th IEEE International Conference on* (pp. 2377-2380). IEEE.

Parmar, P. A., & Degadwala, S. D. (2016). A feature level fusion fingerprint indexing approach based on MV and MCC using SVM classifier. In *International Conference on Communication and Signal Processing (ICCSP), 2016* (pp. 1024-1028). IEEE.

Phillips, P. J. (2004). *The facial recognition technology (FERET) database*. Academic Press.

Phillips, P. J., Wechsler, H., Huang, J., & Rauss, P. J. (1998). The FERET database and evaluation procedure for face-recognition algorithms. *Image and Vision Computing*, *16*(5), 295–306. doi:10.1016/S0262-8856(97)00070-X

Proenca, H. (2013). Iris biometrics: Indexing and retrieving heavily degraded data. *IEEE Transactions on Information Forensics and Security*, *8*(12), 1975–1985. doi:10.1109/TIFS.2013.2283458

Proença, H., & Alexandre, L. A. (2005, September). UBIRIS: A noisy iris image database. In *International Conference on Image Analysis and Processing* (pp. 970-977). Springer.

Puhan, N. B., & Sudha, N. (2011). Coarse indexing of iris database based on iris colour. *International Journal of Biometrics*, *3*(4), 353–375. doi:10.1504/IJBM.2011.042817

Rathgeb, C., Breitinger, F., Baier, H., & Busch, C. (2015). Towards bloom filter-based indexing of iris biometric data. In *International Conference on Biometrics (ICB), 2015* (pp. 422-429). IEEE.

Slaney, M., & Casey, M. (2008). Locality-sensitive hashing for finding nearest neighbors. *IEEE Signal Processing Magazine*, *25*(2), 128–131. doi:10.1109/MSP.2007.914237

Su, Y., Feng, J., & Zhou, J. (2016). Fingerprint indexing with pose constraint. *Pattern Recognition*, *54*, 1–13. doi:10.1016/j.patcog.2016.01.006

Tang, D., Huang, B., Li, R., & Li, W. (2010, August). A person retrieval solution using finger vein patterns. In *Pattern Recognition (ICPR), 2010 20th International Conference on* (pp. 1306-1309). IEEE.

Tiwari, K., & Gupta, P. (2015). Indexing fingerprint database with minutiae based coaxial Gaussian track code and quantized lookup table. In *IEEE International Conference on Image Processing (ICIP)* (pp. 4773-4777). IEEE.

Wang, L., & Leedham, G. (2006). Near- and Far- Infrared Imaging for Vein Pattern Biometrics. *2006 IEEE International Conference on Video and Signal Based Surveillance*, 52-52. 10.1109/AVSS.2006.80

Wilson, C. L., Candela, G. T., & Watson, C. I. (1994). Neural network fingerprint classification. *Journal of Artificial Neural Networks*, *1*(2), 203–228.

Wu, Z., Ke, Q., Sun, J., & Shum, H. Y. (2011). Scalable face image retrieval with identity-based quantization and multireference reranking. *IEEE Transactions on Pattern Analysis and Machine Intelligence*, *33*(10), 1991–2001. doi:10.1109/TPAMI.2011.111 PMID:21646678

Yager, N., & Amin, A. (2004). Fingerprint classification: A review. *Pattern Analysis & Applications*, *7*(1), 77–93. doi:10.100710044-004-0204-7

You, J., Kong, W. K., Zhang, D., & Cheung, K. H. (2004). On hierarchical palmprint coding with multiple features for personal identification in large databases. *IEEE Transactions on Circuits and Systems for Video Technology*, *14*(2), 234–243. doi:10.1109/TCSVT.2003.821978

Zhang, Y., Mu, Z., Yuan, L., Zeng, H., & Chen, L. (2017). 3D Ear Normalization and Recognition Based on Local Surface Variation. *Applied Sciences*, *7*(1), 104. doi:10.3390/app7010104

Zhou, W., Hu, J., Wang, S., Petersen, I., & Bennamoun, M. (2016). Partial fingerprint indexing: A combination of local and reconstructed global features. *Concurrency and Computation*, *28*(10), 2940–2957. doi:10.1002/cpe.3600

Zhou, Y., Liu, Y., Feng, Q., Yang, F., Huang, J., & Nie, Y. (2014). Palm-vein classification based on principal orientation features. *PLoS One*, *9*(11), e112429. doi:10.1371/journal.pone.0112429 PMID:25383715

Zhu, Q., & Mu, Z. (2018, August). An Efficient 3D Ear Recognition System Based on Indexing. In *Chinese Conference on Biometric Recognition* (pp. 507-516). Springer. 10.1007/978-3-319-97909-0_54

Chapter 4
Deep Learning for Healthcare Biometrics

Upendra Kumar
Institute of Engineering and Technology Lucknow, India

Esha Tripathi
Institute of Engineering and Technology Lucknow, India

Surya Prakash Tripathi
Institute of Engineering and Technology Lucknow, India

Kapil Kumar Gupta
SRM University Lucknow, India

ABSTRACT

Mistakes in healthcare systems such as a mix-up of records or confusing medical charts lead to the wrong medications to patients. Major tasks such as administrative costs, legal expenses, and liabilities incur high cost to the healthcare industry using traditional, inaccurate patient identification processes. This can be resolved by biometric technology. Only physiological features can be used for patient identification to eliminate need of SSN, insurance card, or date of birth during registration. A biometric template can be directly mapped to an electronic health record to accurately authenticate individuals on subsequent visits. This technology ensures no medical records can be mimicked and the right care is provided to the right patient. Deep learning provides a platform to solve identification and diagnostic problems arising in medicine and can be used in healthcare biometrics to analyze clinical parameters and their combinations for disease prognosis (e.g., prediction of disease, extracting medical knowledge, therapy planning, and support).

DOI: 10.4018/978-1-5225-7525-2.ch004

INTRODUCTION

Health care is coming to a new phase in which large amount of available biomedical data are playing important roles. In this context, for example, the objective of precision medicine is to provide assurance that the right treatment is delivered to the right patient based on real time patient data at the right time by considering the variety of patient's data such as variability in environment, molecular traits, electronic health records (EHRs) and lifestyle (Lyman et al., 2016; Collins et al., 2015).

The availability of biomedical data brings tremendous opportunities and challenges in the field of health care research. Especially, exploring the associations among all the different pieces of information in these data sets is a fundamental problem for developing efficient medical tools based on data-driven methods and machine learning techniques. For fulfilling this objective, previous works tried to associate multiple data sources to construct joint knowledge databases which could be useful for predictive analysis and discovery (Xu et al., 2014; Wang et al., 2014). Even though some pre-define models exhibit great promises (e.g. (Tatonetti NP et al., 2012; Wang et al., 2014)), predictive tools mainly based on machine learning techniques have not been widely implemented in medical field (Bellazzi et al., 2008). As, there remain lots of challenges in making full utilization of the biomedical data, owing their high-dimension, dependency on temporal data, sparsity, heterogeneity and irregularity (Hripcsak et al., 2013; Luo et al., 2016). Further these challenges are complicated by several medical ontologies used to generalized the data (e.g. Unified Medical Language System (UMLS) (UMLS, 2016), Systematized Nomenclature of Medicine-Clinical Terms (SNOMED CT, 2016), International Classification of Disease-9th version (ICD-9) (18)), which contain many inconsistency and conflicts(Mohan et al.). Same clinical composition is also expressed in different manner across the data. Like in the EHRs, diagnosis of a patient with 'type 2 diabetes mellitus' can be identified by values of hemoglobin A1C >7.0 taken in laboratory, presence of 250.00 ICD-9 code. Accordingly, for making a higher-level semantic structure and better understand their correlations, it is nontrivial solution to harmonize these medical concepts (Wang et al., 2014; Gottlieb et al., 2013).

A general method in healthcare biometric research is having domain experts that identify the phenotypes for utilize in an ad hoc manner. Although, supervised definition of the feature space is inappropriate to determine new patterns sometimes. On the other hand, representations of learning technique allow to automatically detecting the representations required for prediction using the raw data (Bengio et al., 2013; Farhan et al., 2016). Deep learning approach are generally representation of various learning algorithms which contain multiple levels of representation that is obtained after composing simple but non-linear modules and each module transform the representation at one level (which start from the raw data) into a representation at a higher level (LeCun et al., 2015). Deep learning models demonstrated great performance in the field of speech recognition, computer vision and natural language processing area (Abdel-Hamid et al., 2014 ; Hannun et al., 2014).

Deep learning techniques introduce novel opportunities for healthcare biometric when its demonstrated performance in various fields and the rapid improvement of methodological progress is given. Efforts that are applied in deep learning methods to health care are already designed or in progress. Like Google DeepMind tool has announced various plans in which it apply its expertise to health care (Clark, 2016) whereas Enlitic tool is using deep learning concept to point out problems related to health care on X-rays and Computed Tomography (CT) scans (Bartlett, 2017).

Recent applications of deep learning in medical field basically emphasize the significant key aspect of health care. This will focus on biometric data only, especially the data which is originated from clinical imaging, genomes, wearable devices and EHRs. The healthcare corporate using electronic health records, by raising the question with how digitizing health records will create significantly and efficiently increase the quality of patient healthcare. As gradually all hospitals management and healthcare systems migrate to computerized entry and electronic health records, more and more exchange of health information are builds efficient networks. Many healthcare systems are conscious about data, how well manage data integrity to ensure for keeping free from any modification, corruption and even from unauthorized access.

Thirty years ago, majority of patients were not able to leave their house for treatment. Today Biometrics are being utilized as an authentic mechanism by several medical services. Even Electronic health records (EHRs) and Personal health records (PHRs) more frequently used. Although most of the developed countries, kept patient's record electronically whenever the patient's data is needed, it must be retrieved from records. By the use of a biometric identifier, patient's health data can be automatically stored and digitally recorded every time when record is opened. Some biometric health tools producers and service providers recommend turnkey application that maintain and track access to EHRs.

Deep learning concept basically works to categorize various trends in data sets, like patient's examine files, which may assist a doctor to be aware of the early symptoms of cancer or a possible risk based on another patient health data, habits, family history etc. Millions of people's lives can be saved by just awareness of electronically saved record and early diagnosis. There is some software which provides remote monitoring and care of patient. Patient must be outfitted with wearable technology which measures essential health parameters and provides as a feedback report. These devices could save a database storing millions of predictive health data points and predict a daily routine plan to follow: what to eat, when to exercise and how to dress. Additionally, these types of devices may connect from human body's nervous system and send neurons signals to transform blood flow, reduce stress and even slow down aging process. Basically, there is no limit on technology which is frequently learning from itself and making enhancement.

Deep learning technique can pay attention to wearable devices on particular conditions, such as remotely cardiac monitoring, to represent how to personalize algorithms according to one's particular biometric and patient data at an individual level. These devices can help in the interpretations of the vast analysis of the unstructured data through wearable medical devices. Initially analyze the ordinary mathematical algorithms which are used to trained data to detect anomalies in data. When deep learning technology combined with artificial intelligence, it would then seek to perform an interpretation of doctor's prescription report about patient, in order to save physician time. Such type of capabilities reduces time for physician effectively that enables them to focus on the serious patients and simplify the care process.

Biometric systems used in the field of medical incorporate with several technologies from the fields of medicine, biology, statistics, consumer's electronics and pervasive computing to construct systems of computer-aided therapy and diagnosis. Earlier such systems were expensive and limited to medical facilities, but increasingly, such systems are cheaper and integrated into wearable technologies, such as watches, bracelets and headbands. Such devices will be capable to provide medical diagnostic data through various cloud-based applications.

Blood banks initially use fingerprint scans of donors to assist them to obey federal set of laws that need perfect identification of every donor. The blood bank in this case study says it less frequently has to turn away donors without proper identification(ID). It has also reduced the likelihood of duplicate donor report and has been able to stop storing Social Security numbers, a practice that consumers gradually more object to. The benefits of biometric in healthcare is shown in Figure 1.

Other potential benefits of identifying patients through biometrics include:

Figure 1. Biometric based patient enrollment and authentication system

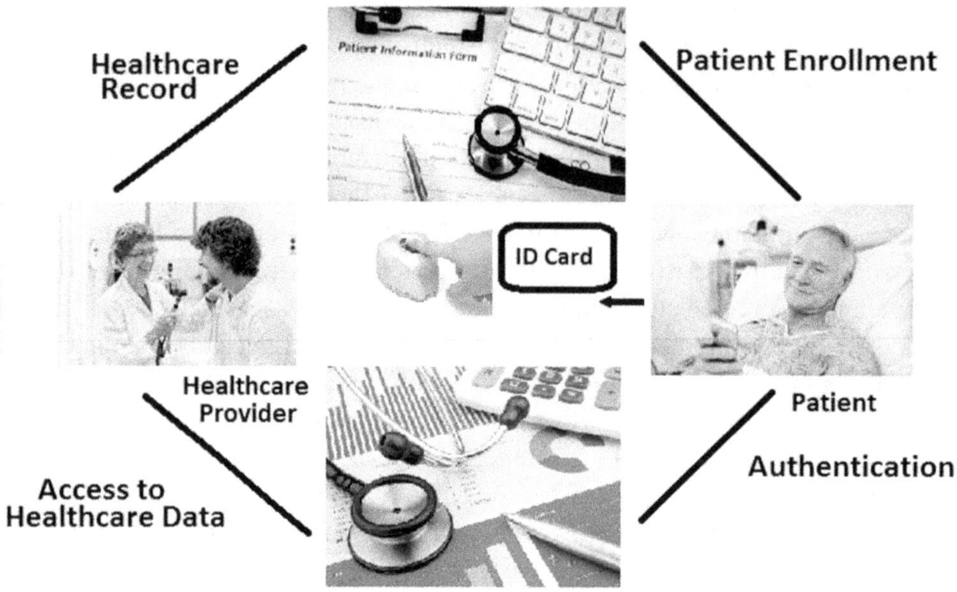

- **Preventing Medication Errors:** Biometrics can potentially replace patient wristbands and barcodes as identifiers when nurses are administering medication at the bedside.
- **Reducing Billing Fraud:** Biometric identification can be used instead of paper insurance cards, to prevent a common type of fraud that occurs when one patient poses as another to obtain insurance benefits. It can also deter unscrupulous providers from filing phantom claims, because a biometric scan offers proof of a patient's physical presence at a health care facility.
- **Improving the Capacity to React to Medical Emergencies:** With a fingerprint or palm scan, patients who are unconscious or unable to talk can still be quickly identified – along with pertinent medical history like drug allergies and current medications.
- **Furthering the State of *Health Information Exchange* (HIE)**: When digitized patient information is shared regionally and across providers, accurate patient identification is crucial. When multiple records exist for the same patient, there's a strong likelihood that one or all will be incomplete – potentially putting the patient at risk for drug interactions or inappropriate services. Biometric patient identification can eliminate the need for patients to fill out multiple forms at each provider visit, and ensure that all data are saved into the same patient record.

HEALTHCARE BIOMETRICS: OPPORTUNITY OF FUTURE GROWTH IN HEALTHCARE

Healthcare sector is one of the most crucial field which incurs high expenditure of any country and plays a significant role including whole surroundings at large. As per the survey done by World Health Organization(WHO) (published in La Times, by Etehad & Kim, 2017), the expenditure of United States per person on healthcare is more than any other country. With the huge economy and widespread healthcare services, this healthcare industry is always sensitive to threat from the people with malicious intents. Therefore, the security concerns are growing day by day and a need arises to integrate it with current age Information Technology (IT) based applications in order to provide easy and secure healthcare services. These services usually have wise spread involvement of the growth of biometrics based global healthcare market. A report of survey published by Coherent Market Insights (2016) says that the cost and effort estimation of Global Healthcare Biometrics market was projected to USD 2,848.3 Million by 2021 from USD 1,182.6 Million by 2016 and compound annual growth rate (CAGR) of 19.2% during the forecast period. This survey also says that around $80 billion costs of taxpayers of America usually lost in fraudulent activities with malicious intent in a year. In addition to these problems, the world can also face some other kinds of healthcare frauds such as applying fake bills, duplicate claims, identity theft threats etc. In 2017, the healthcare sector of America came across a malware attack and badly affected the healthcare centres which were equipped with large number of computers and devices having encrypted functionality. This may be considered a good enough example why biometrics with information technology has been incorporated with healthcare systems, because more and more dependency on information technology often leads to open the doors for cyber-criminals to steal or corrupt the patient's confidential records and other valuable asset information. These factors have encouraged the growth of healthcare biometric based market at a rapid rate. Widespread application of biometric technology in healthcare sector leads to the organization for medical cost reduction, secured storage of healthcare records, patient tracking (through online status), remote patient monitoring, workforce management (tracking in/out and online work-status), and granting access to the only authorized person into the hospitals. In this way, this technology facilitates the right treatment to the right person using invariant biometric features and their corresponding online medical records. Figure 2 discusses about the use of biometric in healthcare sector.

Framework for Deep Learning in Healthcare Biometric

Machine learning is a subset methodology of artificial intelligence that can learn relationships from the data while not to necessary to define them a priori (Murphy, 2012). Machine Learning provides ability to develop predictive models which does not require a need for strong assumptions regarding underlying mechanisms that are normally unknown (Bishop, 2006). Basically, machine learning process contains four steps: data harmonization process, representation of learning process, model fitting process and evaluation process (Jordan et al., 2015). Many years ago, construction of a machine learning system needs careful engineering and domain expertise that transform the raw information into an appropriate internal representation. The learning subsystem i.e. a classifier which could detect patterns in the data set. Conventional techniques are collection of a single, linear transformation of the input data and these techniques are limited in their ability to process natural data in their raw form (Bengio et al., 2013).

The technology used in biometric system usually relies on the variety of inputs acquired from a certain number of fields, starting with different kinds of sensors being used to create sample data for biometrics. Traditional signal or image processing and current age pattern recognition methods are obviously relevant to apply, as the captured raw data need to be pre-processed and prepared to get accurate and robust biometric based decisions for identification. Consequently at final stage of its processing, the biometric system produces outputs as a decision, which in turn links the acquired and processed biometric features to a distinct identity. Now a days, algorithms and mathematical formalism developed by the machine and deep learning community are frequently used in developing biometric systems by

Figure 2. Use of biometric technology in healthcare

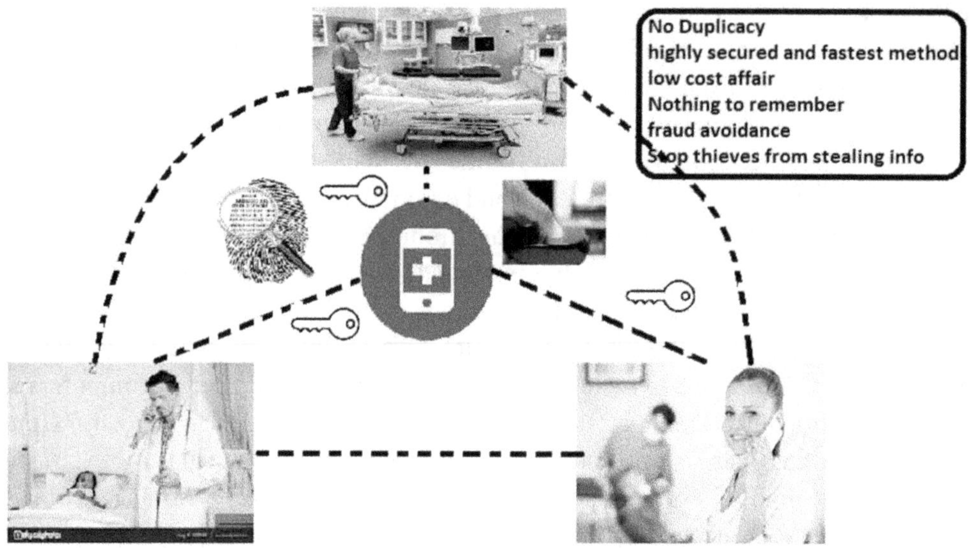

constructing decision function itself, but this is surely does not mean the only contribution done by the aforementioned technology worth mentioning. It has been tried in this chapter that how machine and deep learning methods are useful in selecting and extracting appropriate feature representations from the raw dataset (biometric traits) that will ease and facilitate the job of the employed decision functions, while dealing with exploring the temporal information of the data, and thereafter in fusing multi-modal information to build robust biometric based identification system especially in healthcare sector. The objective of study of this chapter is to familiarize the experts of machine and deep learning with the existing problems of biometrics especially in healthcare sector, to make them understand which techniques are currently employed to solve them, and what kinds of challenges are still open in the healthcare based biometric applications that may benefit from forthcoming advanced machine and deep learning methodologies. It is also intended in this study to familiarize the biometric technology oriented researcher to the methodologies and ways of deep learning and its more effective research methodology, and to develop advanced biometric system by providing a basic toolbox of machine and deep learning. In this chapter, exemplary and influential deep learning approaches have been introduced in the context of specific biometrics applications especially in healthcare sector. The author (Salah, 2010) has discussed many machine learning approaches and hopes to provide the researchers with the means to create novel deep learning solutions in order to tackle with the biometric problems based challenges.

The authors (Wu et al., 2018) discussed the implementation of biometric system with current age technologies. They showed that the maturity of different techniques i.e. sensor chip integration, artificial intelligence, and machine and deep learning algorithms in combination provide us with an effective solution for biometric based identification and authentication. The hardware component i.e. miniature sensors can be inconspicuously attached to the human body (without disturbing the subject) to discern and attest a user's identity card. In this work (Wu et al., 2018), the authors proposed an widely applicable biometric tool for continuous authentication based on an own-built fingertip which was employed on multi-sensor device. In order to get a high-precision with better accuracy, continuous authentication, in this work, a device was used to collect human's motion data and physiological data (physical traits) recognizing the state of motion. After recognizing the correct movement, they have applied one-class machine and deep learning algorithm for user authentication. In further experiments, they have also applied three light-weight deep learning classifiers for activity recognition and three one-class classification algorithms to achieve the authentication task. The incorporation of deep learning techniques with biometric system has outperformed the classification results.

Deep learning is different from traditional machine learning in how representations are learned from the raw data. In fact, deep learning allows computational models that are composed of multiple processing layers based on neural networks to learn representations of data with multiple levels of abstraction (LeCun et al., 2015). The major differences between deep learning and traditional artificial neural networks (ANNs) are the number of hidden layers, their connections and the capability to learn meaningful abstractions of the inputs. In fact, traditional ANNs are usually limited to three layers and are trained to obtain supervised representations that are optimized only for the specific task and are usually not generalizable (Rumelhart et al., 1986). Differently, every layer of a deep learning system produces a representation of the observed patterns based on the data it receives as inputs from the layer below, by optimizing a local unsupervised criterion (Bengio et al., 2009). The significant feature of deep learning is that these layers of features are learned from dataset using a normal learning method but they are not designed by human. Deep neural networks process the input data in a non-linear, layer-wise fashion to pre-train the nodes in successive hidden layers to learn representation of deep structure that are generalizable. These representations are then followed by a supervised layer to provide perfection of the entire network by applying the back propagation algorithm toward several representations. Framework for deap learning in healthcare biometrics is discussed in Figure 3.

Figure 3. Framework for deep learning in healthcare biometrics

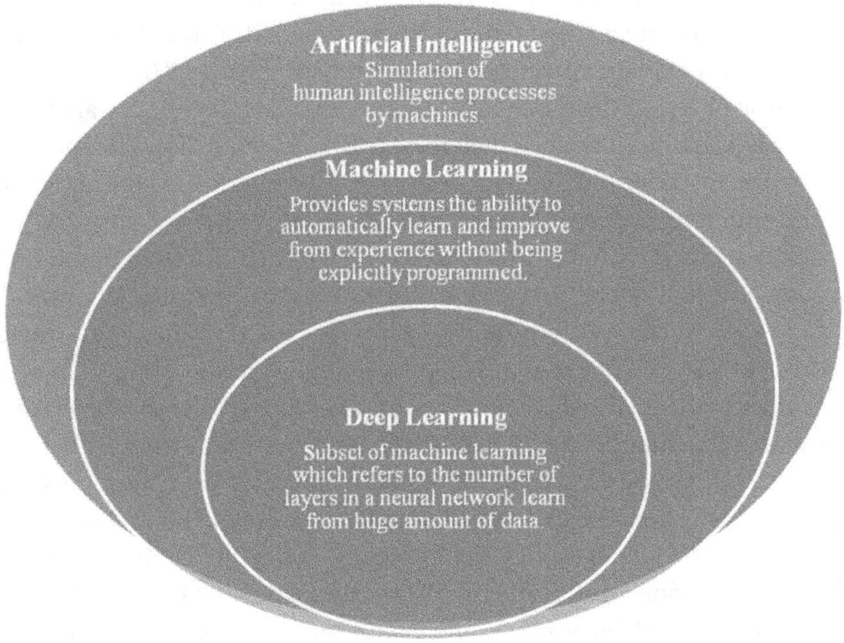

With the advent of new technology and population growth has revolutionized healthcare industry and different kinds of advancements have given rise to security threats to confidentiality of most of our personal and healthcare data. Usually, the records available in healthcare data management system (HDMS) consists of different multimedia formats such as text, numeric, audio, pictures and videos leading to such data which is large in amount and heavily unstructured. Also, most of the hospitals have many sub-branches at different geographical locations throughout a state, country and overseas, that leads to big data which may be highly unstructured. In view of these challenges and requirements a cloud computing technology based healthcare management system can be a possible effective solution for efficient and secured management of health care data (Shakil et. al., 2017). One of the major issues of a cloud computing based healthcare management system is the security aspect. These issues may include theft to identity, insurance frauds, tax fraudulence, medical frauds (crucial medical information may be intentionally changed for personal revenge) and defamation of patients with high and renowned personality. Therefore, a highly secured data access and information retrieval method is needed in order to secure of critical medical records stored and processed by current age health care management system. Biometric technology based authentication mechanism would be most suitable and appropriate in this scenario because it tries to overcome the limitations and problems of token theft (ATM card PIN being stolen) and forgetting passwords in traditional token id-password based mechanism which are usually used for providing security. It has been observed that it also provides high accuracy rate in order to secure access and retrieval of medical data. (Shakil et. al., 2017), proposed a biometric based healthcare tool i.e. BAMHealthCloud which is based on cloud computing technology to achieve secured and proper management of healthcare data, also provides security of data through biometric based authentication system. It has been developed by efforts of many researchers after performing a variety of case studies related to healthcare sector especially in developing countries. In this work, training of the classification model was performed using signature samples in parallel on Hadoop, MapReduce framework for authentication purpose. They also used a deep learning method based resilient back-propagation neural network. They performed rigorous experiments on it and concluded that their model can achieve a 9 times speedup of operations, the classification results with sensitivity of 0.98 and specificity of 0.95. Performance comparison charts of the proposed system with other state-of-art-algorithms showed that the proposed system has better performance than the existing systems discussed in various literatures.

The study and experiments done by (Lie et. al., 2017) showed how to get high authentication accuracy by systematically incorporating variety of design alternatives in type of sensors and their feature selection, application of deep learning techniques, context detection and utilization of multiple devices. They claimed that their authentication system performed with 98.1% accuracy though posses almost negligible system overhead and very low battery consumption. Fierrez-Aguilar et al., (2004) implemented a quality based multi-biometric systems (utilized multiple physiological traits) using support vector machines (SVM) methods in order to analyze and authenticate an individual to access the crucial data based on previously training data. The author (Toh et al., 2004) implemented a biometric authentication system for fingerprint and voice based physiological traits using multivariate polynomials regression. They claimed that their results showed better performance than optimal weighting methods. The author (Kryszczuk et al., 2005) worked on biometric fusion using a deep learning technique, i.e. bayesian belief network and obtained significant classification results. The benefit of bayesian belief based statistical model is that it allows to make decisions even with incomplete information and has most important ability to re-evaluate the existing decisions based on new information.

Existing Healthcare Models Following Healthcare Biometrics

This section discusses about the application of deep learning models in healthcare sector, hopefully which can be applied to implement models in the field of healthcare biometrics also. The role of deep learning concept for medical field is recent and not totally explored. Table 1 shows all the literature review related to healthcare, in this specially highlighting the various form of networks and the medical data considered. To the best of our knowledge, there are not any studies using deep learning to mix neither of these data sources, nor a part of them e.g. only clinical images and EHRs, only EHRs and genomics in a joint representation for medical analysis and prediction. Some preliminary studies evaluated the joined use of EHRs and genomics (e.g. see (Li et al., 2015; Gerstung et al., 2017)), not applying deep learning technique though, therefore, they weren't considered appropriate to this review. Mostly the deep learning architectures that are applied to the health care field have been based on convolutional neural networks (CNNs) (Lecun et al., 1998), recurrent neural networks (RNNs) (Williams et al., 1989), Restricted Boltzmann Machines (RBMs) (83) and Autoencoders (AEs) (Hinton et al., 2006).

Deep Learning in Clinical Imaging

The primary applications of deep learning to clinical data were on image processing, particularly prediction of Alzheimer disease depend on the analysis of brain Magnetic Resonance Imaging (MRI) scans and its variations (Liu et al., 2014; Brosch et al., 2013). In another medical domain, CNNs were utilize to infer a hierarchical representation of low-field knee MRI scans to automatically divide cartilage and predict the risk of osteoarthritis (Prasoon et al., 2013). Instead of using two-dimension images, this method obtained good results than a state-of-the-art method using manually selected three-dimension multi-scale features. Deep learning was also applied to segment multiple sclerosis lesions in multi-channel 3D MRI (Yoo et al., 2014) and for the differential diagnosis of benign and malignant breast nodules from ultrasound images (Cheng et al., 2016). More recently, (Gulshan et al., 2016) used CNNs to identify diabetic retinopathy in retinal fundus photographs, obtaining high sensitivity and specificity over about 10000 test images with respect to certified ophthalmologist annotations. CNNs also obtained performances on par with 21 board-certified dermatologists on classifying biopsy-proven clinical images of different types of skin cancer (keratinocyte carcinomas versus benign seborrheic keratoses and malignant melanomas versus benign nevi) over a large data set of 130000 images (1942 biopsy-labeled test images) (Esteva et al., 2017)

Deep Learning in Cardiac Diseases Using Electronic Health Records (EHRs) and ECG Based Biometrics

Recently deep learning technique has been applied to process aggregated EHRs, including both structured (e.g. diagnosis, medications, laboratory tests) and unstructured (e.g. free-text clinical notes) data. The greatest part of this literature study processed the EHRs of a health care system with a deep learning structural design for a specific reason, generally supervised, predictive clinical task. Specially, a common approach is to show that deep learning obtains better outcome than conventional machine learning models with respect to certain metrics, like area Under the Receiver Operating Characteristic Curve, accuracy and F-score (Manning et al., 2008). In this stage, while almost papers present end-to-end supervised

Table 1. Summary of the articles described in the literature review with highlighted the deep learning concept applied and the medical domain considered

Data	Author	Application	Model
Clinical imaging	Brosch et al.(2013)	Manifold learning of brain MRIs by deep learning	RBM
	Prasoon et al.(2013)	Deep feature learning for knee cartilage segmentation using a tri-planar Convolutional neural network	CNN
	Liu et al.(2014)	Early diagnosis of Alzheimer's disease with deep learning	Stacked Sparse AE
	Yao et al.(2014)	Deep learning of image features from unlabeled data for multiple sclerosis lesion segmentation	RBM
	Cheng et al.(2016)	Computer-aided diagnosis with deep learning architecture: applications to breast lesions in US images and pulmonary nodules in CT scans	Stacked Denoising Auto Encoder
	Gulshan et al.(2016)	Development and validation of a deep learning algorithm for detection of diabetic retinopathy in retinal fundus photographs	CNN
	Esteva et al.(2017)	Dermatologist-level classification of skin cancer with deep neural networks.	CNN
Electronic Health Record	Lasko et al.(2013)	Computational phenotype discovery using unsupervised feature learning over noisy, sparse, and irregular clinical data.	Stacked AE
	Liang et al.(2014)	Deep learning for healthcare decision making with EMRs	RBM
	Liu et al.(2015)	Risk prediction with electronic health records: a deep learning approach.	CNN
	Lipton et al.(2015)	Learning to diagnose with LSTM recurrent neural networks	LSTM RNN
	Che et al.(2015)	Deep computational phenotyping	Stacked AE
	Tran et al.(2015)	Learning vector representation of medical objects via EMR-driven nonnegative restricted Boltzmann machines (eNRBM)	RBN
	Choi et al.(2015)	Doctor AI: predicting clinical events via recurrent neural networks	GRU/RNN
	Miotto et al.(2016)	Deep patient: an unsupervised representation to predict the future of patients from the electronic health records	Stack denoising AE
	Miotto et al.(2016)	Deep learning to predict patient future diseases from the electronic health records	Stack denoising AE
	Pham et al.(2016)	DeepCare: a deep dynamic memory model for predictive medicine	LSTM RMM
	Razavian et al.(2016)	Multi-task prediction of disease onsets from longitudinal laboratory tests	LSTM RMM
	Dernoncourt et al.(2016)	De-identification of patient notes with recurrent neural networks	LSTM RMM
	Nguyen et al.(2017)	Deepr: a Convolutional Net for Medical Records	CNN
Genomics	Koh et al.(2013)	Denoising genome-wide histone ChIP-seq with convolutional neural networks	CNN
	Fakoor et al.(2014)	Using deep learning to enhance cancer diagnosis and classification	Stack Sparsed AE
	Lyons et al. (2014)	Predicting backbone Ca angles and dihedrals from protein sequences by stacked sparse auto-encoder deep neural network	Stack Sparsed AE
	Zhou et al.(2015)	Predicting effects of noncoding variants with deep learning-based sequence model	CNN
	Alipanahi et al.(2015)	Predicting the sequence specificities of DNA- and RNA-binding proteins by deep learning	CNN
	Kelley et al.(2016)	Basset: learning the regulatory code of the accessible genome with deep convolutional neural networks	CNN
Mobile	Zhu et al.(2015)	Using deep learning for energy expenditure estimation with wearable sensors	CNN
	Hammerla et al.(2016)	Deep, convolutional, and recurrent models for human activity recognition using wearables	CNN/RNN
	Jindal et al.(2016)	An adaptive deep learning approach for PPG-based identification	RBN
	Nurse et al.(2016)	Decoding EEG and LFP signals using deep learning: heading TrueNorth	CNN
	Sathyanarayana et al.(2016)	Correction of: sleep quality prediction from wearable data using deep learning	CNN

networks, some literature also propose unsupervised models to derive latent patient representations, which are then calculated using shallow classifiers for example logistic regression, random forest.

A number of works applied deep learning for prediction of diseases from the patient clinical state. (Liu et al., 2014) used a four-layer CNN to predict congestive heart failure and chronic obstructive pulmonary disease and showed significant benefit over the baselines. RNNs with long short-term memory (LSTM) hidden units, pooling and word embedding were utilize in DeepCare (Pham et al., 2016), an end-to-end deep dynamic network which infers current illness status and predicts medical outcomes for future scope. Some authors also proposed to moderate the LSTM unit with a decay effect to handle irregular timed events (which are typical in longitudinal EHRs). Moreover, they incorporated medical interventions in the model to dynamically shape the predictions. DeepCare was evaluated for disease progression modeling, intervention recommendation and future risk prediction on diabetes and mental health patient cohorts. (Recurrent Neural Networks) RNNs with gated recurrent unit (GRU) were used by (Choi et al., 2015) to develop Doctor AI, an end-to-end model that uses patient history to predict diagnoses and medications for subsequent encounters. The evaluation showed significantly higher recall than shallow baselines and good generalizability by adapting the resulting model from one institution to another without losing substantial accuracy. Differently, (Miotto et al., 2016) proposed to learn deep patient representations from the EHRs using a three-layer Stacked Denoising Autoencoder (SDA). They applied this novel representation on disease risk prediction using random forest as classifiers. The evaluation was performed on 76,214 patients comprising 78 diseases from diverse clinical domains and temporal windows (up to a 1 year). The results showed that the deep representation leads to significantly better predictions than using raw EHRs or conventional representation learning algorithms (e.g. Principal Component Analysis (PCA), k-means). Moreover, they also showed that results significantly improve when adding a logistic regression layer on top of the last AE to fine-tune the entire supervised network (Miotto et al., 2016). Similarly, (Liang et al., 2014) used RBMs to learn representations from EHRs that revealed novel concepts and demonstrated better prediction accuracy on a number of diseases.

Researchers found that sophisticated technologies (deep learning) can be successfully applied for implementation of many biometric authentication tools such as to provide protection of confidential documents and reliable and secured access to restricted areas. Recent advances observed in biomedical and clinical engineering fields not only require toimprove the recognition performance (improved efficiency) of biometric authentication systems but also provide support for its feasible implementations. In recent biometric research, electrocardiogram (ECG) signals collected from different individuals have been suggested to use as a biometric trait for identity verification (Chen et al., 2011). The authors (Safie et al., 2011) presented a novel deep learning-based framework using individual's electrocardiogram (ECG) as a biometric system for human authentication. In this work they devised pulse active ratio (PAR) as a new feature extraction technique and further it was utilized to generate novel feature vectors for ECG. They performed experiments on 112 different subjects and 9800 ECG comparisons were done by the proposed methods. They got around 10% improvement in the authentication accuracy when compared to traditional temporal and amplitude-based feature extraction methods.

Deep Learning in Healthcare Monitoring Using Mobile Based Dataset Followed by Healthcare Biometrics

Sensor-equipped smartphones and wearable are transforming a variety of mobile apps, including health monitoring (Shameer et al., 2017). As the difference between consumer health wearable and medical devices begins to soften, it is now possible for a single wearable device to monitor a range of medical risk factors. Potentially, these devices could give patients direct access to personal analytics that can contribute to their health, facilitate preventive care and aid in the management of ongoing illness (Piwek et al., 2016). Deep learning is considered to be a key element in analyzing this new type of data. However, only a few recent works used deep models within the health care sensing domain, mostly owing to hardware limitations. In fact, running an efficient and reliable deep architecture on a mobile device to process noisy and complex sensor data is still a challenging task that is likely to drain the device resources (Ravi et al., 2017). Several studies investigated solutions to overcome such hardware limitations. As an example, (Lane et al., 2015) proposed a low-power deep neural network inference engine that exploited both Central Processing Unit (CPU) and Digital Signal Processor (DSP) of the mobile device, without leading to any major overburden of the hardware. They also proposed DeepX, a software accelerator capable of lowering the device resources required by deep learning that currently act as a severe bottleneck to mobile adoption. This architecture enabled large-scale deep learning to execute efficiently on mobile devices and significantly outperformed cloud-based off-loading solutions (Lane et al., 2016).

We did not find any relevant study applying deep learning on commercial wearable devices for health monitoring. However, a few works processed data from phones and medical monitor devices. In particular, relevant studies based on deep learning were done on Human Activity Recognition (HAR). While not directly exploring a medical application, many studies argue that the accurate predictions obtained by deep models on HAR can leverage clinical applications as well. In the health care domain, (Hammerla et al., 2016) evaluated CNNs and RNNs with LSTM to predict the freezing of gait in Parkinson disease (PD) patients. Freezing is a common motor complication in PD, where affected individuals struggle to initiate movements such as walking. Results based on accelerometer data from above the ankle, above the knee and on the trunk of 10 patients showed that RNNs obtained the best results, with a significantly large improvement over the other models, including CNNs. While the size of this data set was small, this study highlights the potential of deep learning in processing activity recognition measures for clinical use. (Zhu et al., 2015) obtained promising results in predicting Energy Expenditure (EE) from triaxial accelerometer and heart rate sensor data during ambulatory activities. EE is considered important in tracking personal activity and preventing chronic diseases such as obesity, diabetes and cardiovascular diseases. They used CNNs and significantly improved performances over regression and a shallow neural network.

THE DEEP NEURAL NETWORK TECHNIQUES IN HEALTHCARE BIOMETRICS

Deep learning is the branch of machine learning that is based on set of algorithms. In terms of data, these deep learning algorithms illustrate high-level abstraction. It construct deep graph with various processing layer, consist of many linear and nonlinear transformation. As the structure of every human brain, the connection of various images and situations to establish a set of instructions those assist us through the daily cognitive process and stems from the neural network. Neurons are connected to each other and transmit electric impulses which are based on all bodily communication. Follow the human brain pattern; thousands of neurons contribute something together as well as individually to lend person's imaging, intellectual and reasoning abilities. This problem solving pattern is achieved through algorithms in case of machine.

Deep Convolutional Network (DCN)

Convolutional Network consists of one or more layers and it is usually designed in a way that takes advantage of the 2-Dimensional structure of a unit. They have learnable weights, thus they are simple to train. They are inspired by the biological arrangement of animals (especially humans) where individual neurons of the animal cortex are organize in such a manner to intercept the visual field which contain maximum potential. The framework for modern convolutional neural networks is diverse and certain designs have also originated distinguish the various features over time. LeNet-5, initiating 7-level convolutional network by Yann LeCun discovers that it used in banks to recognize hand-written numbers on checks in the form of digitalized 32 x 32 pixel images. The architecture of deep Convolutional network with multiple hidden layers are shown in Figure 4. The DCN achieves performance with all tested experts across several tasks and capable of classifying skin cancer with a level of competence comparable to dermatologists (Esteva et al., 2017).

Recurrent Neural Network (RNN)

Recurrent Neural Network (RNN) is type of artificial network where multiple units are linked together and form a directed cycle. However, this network and its various pattern had to go through variation in designs, one of the foremost modifications being the development of the neural history compressor. Here, the neural system for current input level has the potential to predict from the previously existing information. So the higher level RNN also understand a compressed illustration of the RNN provided

Figure 4. Deep Convolutional Neural Network with multiple hidden layers used by healthcare biometrics

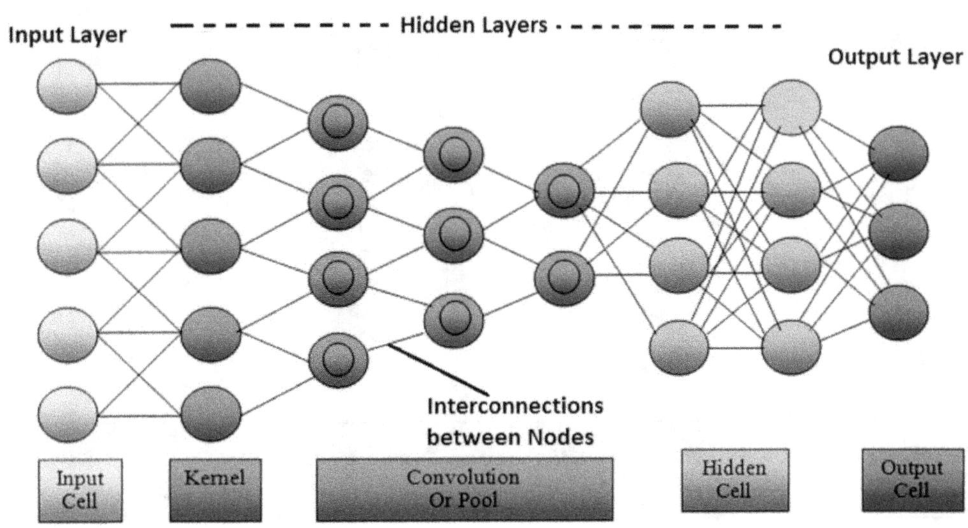

below. With the incoming information sequence having some variety of learnable predictability, it is therefore becomes easier to resolve sophisticated sequence related problem. In 1993, scientists managed the behavior of a deep learning exercise comprising of over a thousand layers, with their results make the half the anticipated time. The Recurrent Neural Network architecture with multiple layers is shown in Figure 5.

Structured predictions over variable length units have got Recurrent Neural Network which is a deep neural network that is produced by the imposition of a group of weights recursively over single structure. Things like scalar predictions are possible by traversing a given structure in a distinguished topological order. Therefore these networks can perceive deep structured data. It was utilized in predicting next word during a conversation as a part of natural language processing. A recurrent neural network is also used for the classification of uterine electrohysterography signals for the prediction of term and preterm delivery for pregnant women (Askar et al., 2016).

Deep Belief Network (DBN)

DBN is the multi-layer neural network that consists of multiple layers of latent variables. These layers have acquired connections each other however the nodes in similar layer do not have connections to each other. They are programmed in a way that they will work as feature detectors on inputs. DBNs can be created to perform classifications, with bigger progression and updation. This deep learning model has been used in iris recognition based biometrics implemented by using contour detection method (Bagar et al.,2016). The Figure 6 shows the architecture for Deep Belief Network. The DBN works effectively for prediction of cardiovascular risk disease (Kim et al., 2017). DBN can be used in facial recognition (biometrics) that is turning into an enormous form in the security industry, wherever CCTV footage is

Figure 5. Architecture of Recurrent Neural Network used by healthcare biometrics

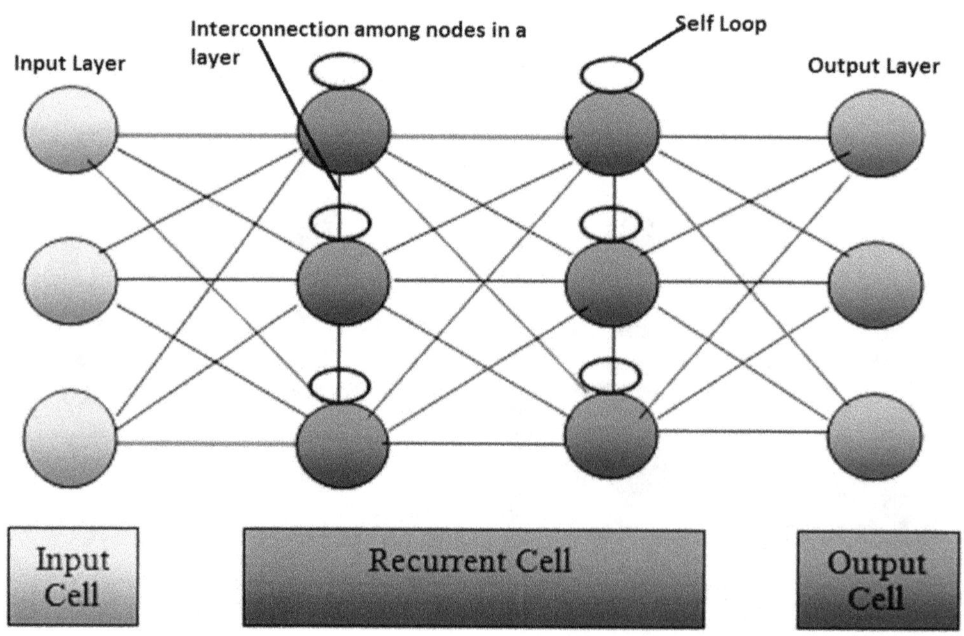

being utilized to spot persons of interest in real-time. This technology is being extensively utilized by security services around the world to observe terror modules. NASA is using DBN to classify terabytes of high-resolution, extremely diverse due to collection, pre-processing, filtering, etc. satellite imaging.

Growth and Technological Innovation of Healthcare

In 2014, the Indian government declared that it planned to utilize its national Aadhaar biometric database to deploy its newly proposed universal healthcare program (King, 2014). Aadhaar is the 12-digit unique identification number issued by the Indian government to each individual resident of India. The Aadhaar project aims to produce a unique identity that captures all the demographic and biometric details of each Indian resident. Recently, the government has been issued Aadhaar numbers more than 89 percent of the total population and over one billion people have currently been listed within the scheme.

According to the national government's agenda, Our Prime Minister has assured that radical reforms in healthcare with the introduction of the "National Health Assurance Mission" (NHAM) theme. The goal of this new program is to produce accessible and reasonable healthcare to each Indian citizen. As to accomplish this goal, the Indian government intends to use Aadhaar as a way of identification for healthcare insurance beneficiaries. The introduction of universal health care to India's citizens will arguably be the most challenging use of the biometric information. India only expends 1.04 percent of gross domestic product on publicly funded health that is one of the lowest amounts within the world. Higher amounts of public health finance are crucial to produce a wider range of essential basic health services, along with access to life-saving medication and expanded health care facilities, like hospitals and health centers.

In the U.S., in contrast, healthcare is one of the most important expenditure categories within the national economy. According to the World Health Organization (WHO), total healthcare spending in the U.S. was 17.1 percent of its gross domestic product (GDP) in 2014, the highest in the world. Despite

Figure 6. Architecture of deep belief neural network used by healthcare biometrics

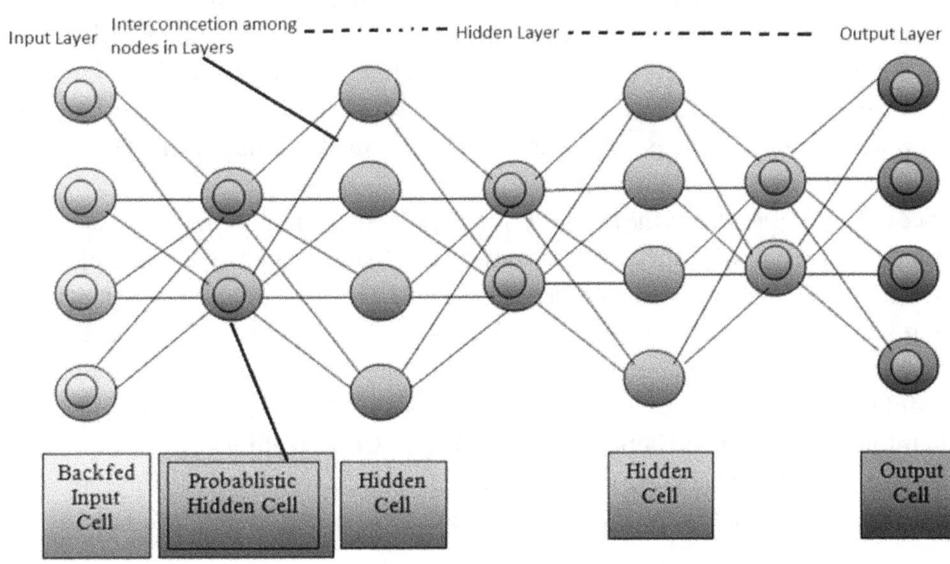

high levels of spending on healthcare, comparatively to other industrialized countries, most analysts rank the U.S. last within the quality of care provided, grounded on economic investment against real health results. Reasons are innumerous for poor performance compared to high levels of investment. The attention system within the U.S. has historically been divided between those who have insurance and those who do not. Dissimilar to other industrial nations, the U.S. has not followed a unique, national healthcare system provided by public insurance, funded through taxation. Alternatively, healthcare provision is split between an ineffective mixture of public and personal insurance.

The Affordable Care Act was planned to raise insurance quality and affordability, lower the uninsurable rate by increasing sum of money and scale back the prices of healthcare.-Republicans have opposed the Affordable Care Act due to the increased tax burden it has created on higher income earners, along with higher premiums and deductibles levied through exchanges, along with the perceived lack of choice due to imposed mandates upon individuals. With the alignment of the Trump administration and the Republican-controlled Congress, there will be pressure to introduce effective healthcare legislation that will address these criticisms. While ideological division will ensure that universal health coverage is not adopted in the U.S., cost containment will continue to be an important consideration, and a primary driver for the use of biometrics, regardless of what reforms are implemented. Due to this fact, the U.S. will dominate the marketplace for biometric health care offerings, and can get advantage from the development of biometric platforms that support multi-modal biometric identity, in conjunction with the growing handiness of biometric solutions to combat fraud and improve patient privacy and safety. Biometrics also will be more and more used for medical observation and mobile health care.

Medical Biometrics

In contrast to health biometrics, personal medical data, which includes digital images and biorhythm recordings, are referred to as "medical biometrics." Such data is produced in ever-increasing quantities and is used for diagnostic and therapy purposes. Medical biometric research aims to use personal medical data sets, such as images and biologically-measurable signals, for solving medical problems and to provide high-performance services in the medical field.

Medical biometric systems integrate multiple technologies from the fields of biology, medicine, consumer electronics, statistics and ubiquitous computing to create systems of computer-aided diagnosis and therapy. Previously such systems were expensive and contained to medical facilities, but increasingly, such systems are becoming miniaturized and integrated into wearable technologies, such as bracelets, headbands and watches. These devices will be able to provide medical diagnostic data through Internet-based cloud applications in the near-term.

For assurance of right patient takes the right service some steps are required. When patient visit hospital or a local clinic for treatment, then they give certain details like patient Name, Date of Birth etc. to the hospital. Then a membership card with a unique membership id is issued to the patient and every time information is read from the barcode or chip which is present on membership card. Whenever patient visit the hospital, fingerprint information which is stored on the card is compared with live fingerprint. In case of successful member verification, the system generates a Claim Verification Code which is based on Biographic Information (Patient Name, Member ID etc.), Context Information (Healthcare Provider ID number, claim date, service date and time etc.) and Biometric Information (fingerprint information based on BioHash technology). Figure 7 shows the various steps for judging a right patient

Figure 7. Medical biometric system using fingerprint

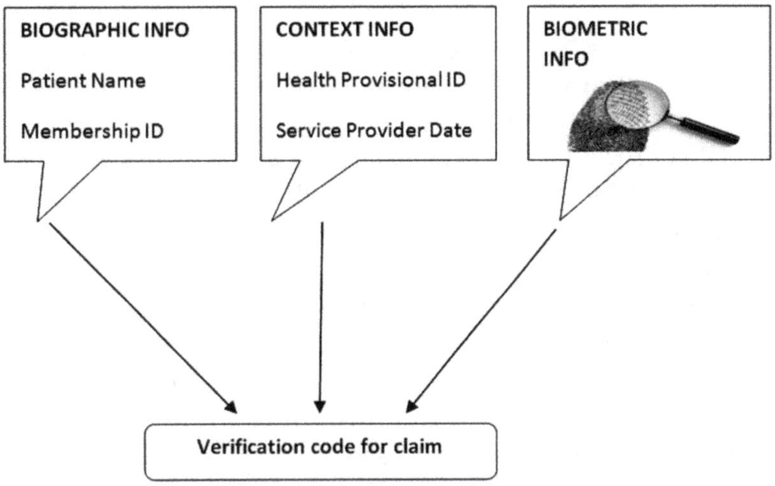

However, Next generation of consumer electronics will focus on measuring biorhythms. Biorhythms are defined simply as the rhythms of life, and include vital body functions, including heart rate and blood pressure. Medical chronobiologists have found that biologic rhythms can affect the severity of disease symptoms, diagnostic test results, and even the body's response to drug therapy.

RECENT RESEARCH IN HEALTHCARE BIOMETRICS

BIO-Key

BIO-key provides fingerprint sensors and biometric software to healthcare and other industries The Company's core software engine is WEB key, which delivers multi-modal authentication via an SDK. It also offers Vector Segment Technology (VST), a highly accurate and scalable fingerprint system for Microsoft Windows, and ID Director, which delivers biometric authentication in the native environment, and is available for healthcare EHR platforms. BIO-key provides healthcare organizations with its NIST tested interoperable software and FIPS-compliant fingerprint readers for online security, workflow, and end-user convenience benefits. The company's technologies also meet DEA and State Board of Pharmacy regulatory standards. Its solutions protect patient record and medications access, and provide efficient and compliant two-factor authentication for EPCS. Healthcare industry customers of BIO-key include Cleveland Clinic, Nationwide Children's Hospital, and Genesis Health Systems. BIO-key has formed strategic partnerships to create a global ecosystem for mobile authentication with companies including mobile technology developer InterDigital, NEXT Biometrics, and IDEX, as well as security infrastructure providers such as IBM, CA, and Oracle. BIO-key traces its roots back to 1993, when it introduced its first fingerprint algorithm. The company went public on OTC Markets in 2003. Based in Wall, New Jersey, BIO-key also has U.S. offices in Minnesota and Massachusetts, and international offices in Hong Kong and China.

GenKey

GenKey provides large-scale biometric identity management solutions for uses including elections and digital healthcare. The company provides a full identification solution, including fingerprint enrolment software, database management, and a verification solution. The company's BioFinger SDK has received NIST MINEX and PFTII certification for performance and interoperability, and its fingerprint image exchange algorithm meets the FBI's WSQ certification and accuracy requirements. It also provides an Automated Biometric Identification System (ABIS), which it released a new version of in early 2017, introducing a label independent matching (LIM) function to detect duplicate fingerprints, among other upgrades. GenKey's healthcare customers include the National Hospital Insurance Fund (NHIF) of Kenya, and Ghana's National Health Insurance Scheme. The company's biometrics-based healthcare solution was shortlisted for the "Solutions of the Year Award" in the technology and solutions providers operating in or exporting to Africa category of the 2016 Africa Healthcare Awards. GenKey holds 39 patents. It was founded in 2001, and is based in Eindhoven, the Netherlands.

ImageWare

ImageWare Systems is a developer of identity management products, specializing in multi-modal biometrics for mobile, healthcare, financial, and enterprise markets. GoVerifyID, the company's flagship mobile and cloud multi-modal biometric authentication service, is among ImageWare solutions providing credentialing to healthcare companies. Integrated IT solutions firm CDW partnered with ImageWare to offer its customers GoVerifyID and the CloudID product suite in late 2015.

Intelligent Fingerprinting

Intelligent Fingerprinting provides a mobile, non-invasive drug screening system based on analysis of fingerprint sweat, a patented diagnostic technique. The system consists of the Intelligent Fingerprint Reader 1000 and single-use collection cartridges. The reader uses proprietary immunoassay reagents to analyze sweat, and does not require specialist collection or handling services, unlike other screening methods. Sample collection takes less than five seconds, and testing for multiple drug groups takes less than 10 minutes. A version of the reader which confirms the identity of the person providing the sample using their fingerprint is in development. The system allows samples to be tested immediately or retained in tamper-proof cartridges for later testing, and also allows drug detection cutoff levels to be customized. The test identifies drugs of abuse in four groups: amphetamines, cannabis, cocaine, and opiates.

IriTech

IriTech provides hardware, software iris recognition products, as well as cloud and mobile solutions. The company's hardware products are the IriShield Series of iris scanners sold in cases or as modules, and the Gemini Camera, which captures face and dual iris at the same time with a single CMOS sensor. It also provides the IriCore iris recognition SDK for Windows and Linux, along with the IrisCoreLite API library, IriMaster middleware server software, and other software products. IriTech provides the healthcare industry with a patient-identification solution which includes patient registration and tracking for treatments, different departments, and checkup arrangement. The solution supports national or private health insurance cards, ambulant treatment documents, wrist bands, and iris biometric identification.

Biometrics and Healthcare Cost Containment

One of the most significant challenges facing healthcare today is the need to reduce costs while improving the quality of care. At the center of this challenge is the ability to efficiently ensure the proper identification of both healthcare professionals and their patients. Not only are there significant risks associated with the improper identification of staff or patients but all of the healthcare constituencies (providers, payers and patients) suffer greatly when inefficient and inadequate identity and access control mechanisms are employed.

Moreover, the failure to correctly identify patients invites fraud and creates another serious problem: patient ID theft. Unlike banking fraud where the risks are mostly financial, patient ID theft can not only lead to costly medical errors, but also limit access to critical services by those who need them most, dramatically increase costs and even result in serious injury or loss of life.

Whether considering developing markets where government-sponsored healthcare is a critical service being offered to citizens, or developed markets where both government and commercial stakeholders struggle to balance costs and care, the central problem is how to ensure that the right people are getting the right services at the lowest possible cost. These challenges place a significant burden on currently deployed identity management technologies. Unfortunately, many of these systems are difficult to deploy, hard to use and inefficient. What is needed to solve this identity management problem is a solution that is both secure and convenient for all constituents.

Rather than placing barriers in the path of patients and healthcare workers in the name of security, providers can choose secure identity solutions that are robust, reliable, efficient and convenient. This is where biometrics can make a significant difference. For several years now biometrics have been deployed in a variety of healthcare use cases ranging from medical dispensing and the electronic prescription of controlled substances (EPCS) to real-time patient identification for medical services.

Patient Registration at Check-In

A fingerprint authentication solution widely used today for patient registration is provided by a world-renowned healthcare provider in U.S, i.e. HID global (Mogli, 2011). A leader in innovation, this large integrated health system constantly assesses new tools and methods to deliver accurate, cost-effective care. As the obvious solution to the problem of patient misidentification, biometric authentication was implemented at patient registration locations to help reduce medical errors, increase productivity and improve patient satisfaction.

In traditional patient registration, the patient provides proof of identity and insurance with a variety of cards and documents. This time-consuming system was replaced at hundreds of points of registration by an advanced biometric authentication system that streamlines the patient check-in process. To register, a patient simply provides their name and date of birth and places their pre-enrolled finger on the fingerprint sensor to verify their identity. In this way, the biometric authentication solution helps to ensure that the patient is linked to their own medical record so accurate insurance information can be confirmed, and the appropriate care can be provided. In some instances, co-pays can be paid and applied using the same system.

This authentication solution, provided by Certify, uses HID Global Lumidigm fingerprint sensors because of the superior performance characteristics of multispectral imaging technology. A successful authentication solution must incorporate a sensor that provides intuitive, single-try capture for all patients, regardless of their age or health, while simultaneously performing liveliness detection to ensure that the fingerprint being presented is not a fake. The sensor technology must also perform highly accurate matching. The Lumidigm technology was selected by this respected healthcare provider because it excels at these performance requirements. The biometric data is managed by the Certify cloud-based service and is kept separate from protected patient information, further protecting patient privacy.

Biometric patient registration at this large provider decreased the number of staff required to check patients in and out, reduced medical errors resulting from improper patient identification, and decreased the number of insurance claims denied as a result of human error or incomplete registration information. Certify notes that this biometric authentication solution reduced patent check-in time by an average of 12 minutes, increasing staffing efficiency and enhancing the patient experience. With biometric authentication, patients spend less time at check in, are not required to produce ID and insurance documentation, and have fewer errors in their records or billing. They can be confident that the provider is delivering appropriate care based upon who they are.

Government Health Benefits From Healthcare Biometrics

The benefits of healthcare biometrics were realized by a Latin America country's large public health system when they used Lumidigm biometrics implemented by HID global (Mayhew, 2018). In addition to the challenges faced by the large U.S. provider discussed above, this government healthcare system was serving patients throughout the country, including at mobile rural clinics. Reducing fraudulent access to healthcare was the main driver of biometric adoption, and reliable fingerprint image capture was critical to the success of the authentication program. The Lumidigm technology successfully operated in real world situations such as for patients with rough and dry skin and in diverse outdoor climates — usage conditions beyond the capabilities of conventional fingerprint sensors. More than 28 million users have benefited from biometric authentication in this program, allowing vulnerable populations in the country to access basic health service. The provider reports that the biometric authentication solution facilitates faster access to medical services by authorized patients, while reducing fraud and government cost. A recent report from the Ponemon Institute states that the primary root cause of patient misidentification is incorrect identification of patients at registration, and that the use of biometrics can reduce patient misidentification (Ponemon Institute Report, 2016). The two examples above bear this claim out: where biometrics has been deployed, there have been significant improvements to costs and patient safety, along with enhanced patient convenience. Ponemon goes on to say that biometric registration could reduce denied insurance claims by 25 percent. Significant improvements have also been realized in control and access to medical dispensing systems, virtually eliminating theft of controlled substances. Mandates by the US Drug Enforcement Agency (DEA) for electronic prescriptions of controlled substances (EPCS) impose compliance requirements which can be met with significant cost and time savings by utilizing these systems

There is a great deal of focus on pooling data from various mobile devices in order to aggregate and make sense of more live health data. *Apple's Research Kit* is aiming to do this in the treatment of Parkinson's disease and Asperger's syndrome by allowing users to access interactive apps (one of which applies Deep Learning for facial recognition) that assess their conditions over time; their use of the app feeds ongoing progress data into an anonymous pool for future study.

Automatic Treatment or Recommendation

IBM Watson Health in partnership with Medtronic developed an automatic system using deep learning methods to help people suffering from diabetes and be able to make more informed decisions (IBM News Room, 2018). Medtronic wants to have their insulin checking pumps work autonomously, monitoring blood-glucose levels and injecting insulin as needed, without disturbing the user's daily life. This, of course, is a microcosm of a much larger picture of autonomous treatment. Imagine a machine that could adjust a patient's dose of pain killers or antibiotics by tracking data about their blood, diet, sleep, and stress. Instead of counting on distractible human beings to remember how many pills to take, a small kitchen table Deep Learning "agent" (think Amazon's Alexa) might dole out the pills, monitor how many you take, and call a doctor if your condition seems dire or you haven't followed its directions.

Table 2. Healthcare programs run by US Government benefitted by healthcare biometrics

Health Care Programs	Description
MEDICARE	Medicare is a national social insurance program, administered by the U.S. federal government since 1966, currently using about 30 private insurance companies across the United States. Medicare provides health insurance for Americans aged 65 and older who have worked and paid into the system. It also provides health insurance to younger people with disabilities, end stage renal disease and amyotrophic lateral sclerosis.
MEDICAID	Medicaid is a social health care program in U.S. for families and individuals with low income and limited resources. Medicaid has been described as a government insurance program for persons of all ages whose income and resources are insufficient to pay for health care.
TRICARE	Tricare is a health care program of the U.S. Department of Defense Military Health system. Tricare provides civilian health benefits for military personnel, military retirees and their dependents, including some members of the Reserve Component of United States Armed Forces. Tricare was formerly known as the Civilian Health And Medical Program of the Uniformed Services (CHAMPUS).
CHILDREN'S HEALTH INSURANCE PROGRAM	Children's Health Insurance Program (CHIP) is a program administered by the United States Department of Health and Human Services that provides matching funds to states for health insurance to families with children. The program is designed to cover uninsured children in families with incomes that are modest but too high to qualify for Medicaid.
VETERANS HEALTH ADMINISTRATION	Veterans Health Administration (VHA) is the component of the United States Department of Veterans Affairs (VA) led by the Under Secretary of Veterans Affairs for Health that implements the medical assistance program of the VA through the administration and operation of numerous VA Medical Centers (VAMC), Outpatient clinics (OPC), Community Based Outpatient Clinics (CBOC) and VA Community Living Centers (VA Nursing Home) Programs.

The legal constraints of putting so much power in the "hands" of an algorithm are not trivial, and like any other innovation in healthcare, autonomous treatments of any kind will likely undergo long trails to prove their viability, safety, and superiority to other treatment methods. HealthCare: voice assistants in support of sick or disabled people are now quite commonly used and their effectiveness will mature even more when these systems start using biometric sensors and the analysis of unstructured data. Artificial intelligence is also already in use in operating rooms with systems that perform as real assistants to surgeons and doctors. More revealing is the use of machine and deep learning systems for the prevention of rare diseases and cancers since AI solutions can analyze data, even unstructured, such as documents, texts, and Scientific publications, with a greater speed not even comparable to that of man. Artificial intelligence in the short term for healthcare will bring about transparency and empowerment of decision-making. The ability to gather and decipher vast amounts of data and then provide a statistically driven answer will completely change the landscape for patients, physicians and hospitals.

Deep Learning can work to recognize trends in data sets, such as patient examination files, which may help a physician recognize the early signs of cancer or a potential risk based on other patient data, family history, habits, etc. Awareness and early diagnosis can help save millions of lives. In the purchasing space, AI will bring transparency of cost and spending trends to help create a tailored supply chain experience, like what Amazon has done for the general consumer. Imagine a completely automated purchasing department that receives the preferred equipment at the best price and at the most efficient time.

Thirty years from now, we could see the majority of patients never leaving their house for treatment. We could see robotic or software programs that provide remote patient monitoring and care. We could see everyone equipped with wearable technology that measures vital health parameters and delivers an instant feedback report. This device could research a database filled with millions of predictive health data points and spit back a life plan to follow: what to eat, how to dress, when to exercise. Furthermore, these devices could connect to the nervous system and transmit neurosignals to modify blood flow, reduce stress (or) even slow the aging process. Fundamentally, there is no limit to the capabilities of a technology that is constantly learning from itself and making improvements."

Mining Medical Data for Better and Quicker Treatment

Medical records such as doctors' reports, test results and medical images are a gold mine of health information. Using GPU-accelerated deep learning to process and study a patient's condition over time and to compare one patient against a larger population could help doctors provide better treatments.

Projects in the forefront include:

- Massachusetts General Hospital's new *clinical data science center* will compare a patient's tests and history with data from a vast population of other patients to improve detection, diagnosis, treatment, and management of disease. NVIDIA is the founding technology partner of the center, which will use our *NVIDIA DGX-1* deep learning supercomputer to deliver insights.
- By analyzing electronic health records, researchers from Sutter Health and the Georgia Institute of Technology showed they could *predict heart failure* as much as nine months before doctors using traditional means.
- At *Children's Hospital Los Angeles,* scientists are gaining insights from 13,000 patient snapshots and 10 years of electronic health records to provide more effective drug treatments in the hospital's pediatric intensive care unit.

Genomics for Personalized Medicine

Genomics data is accumulating in unprecedented quantities; giving scientists the ability to study how genetic factors such as mutations lead to disease. Deep learning could one day lead to what's known as personalized or "precision" medicine, with treatments tailored to a patient's genomic makeup (Offit, 2011).

Although much of the research is still in its early stages, two promising projects are:

- A University of Toronto team is advancing computational cancer research by developing a GPU-powered "genetic interpretation engine" that would more quickly identify cancer-causing mutations for individual patients.
- *Deep Genomics*, a Toronto startup, is applying GPU-based deep learning to understand how genetic variations lead to disease, transforming personalized medicine and therapies.

Deep Learning to Aid Blind People

The large number of people worldwide struggle to manage such tasks as crossing the road, reading a product label, or identifying a face because they're blind or visually impaired (Tiquet, 2016). Deep learning is beginning to change that. *Horus Technology*, winner of NVIDIA's first social innovation award at the 2016 *Emerging Companies Summit*, is developing a wearable device that uses deep learning, computer vision, and GPUs to understand the world and describe it to users. One of the early testers wept after trying the headset-like device, recalled Saverio Murgia, Horus CEO and co-founder. "When you see people get emotional about your product, you realize it's going to change people's lives."

APPLICATION OF DEEP LEARNING IN HEALTHCARE BIOMETRICS AND FOLLOWED BY DISEASE DIAGNOSIS

Now a days, most of diagnostic systems are being developed using deep learning approaches and they are able to produce high performance in diagnosis and prognosis and they are largely supported by healthcare biometric systems.

Cancer Classification With Deep Neural Networks

A research paper published in *Nature* by Stanford University researchers has shown that their convolutional neural network (CNN) achieves performance on par with all tested experts when classifying skin cancer. Google's CNN system has demonstrated the ability to identify deadline skin cancers at an accuracy rate on par with practitioners, potentially extending diagnosis reach outside the clinic and into service-based apps that have been popping up as mobile access expands worldwide. Yet another instance is a program developed at Case Western Reserve University that has been able to outperform physicians at brain cancer diagnoses.

Tumor Segmentation

Researchers from China have used deep learning for segmenting brain tumors in MR images, where it provided more stable results as compared to manually segmenting the brain tumors by physicians, which is prone to motion and vision errors. A team led by (Zhang et al., 2016) of Shanghai University found that deep learning can accurately differentiate between benign and malignant breast tumors on ultrasound shear-wave elastography (SWE), yielding more than 93% accuracy on the elastogram images of more than 200 patients.

Histopathologic Cancer Diagnosis

With the advent of personalized medicine, diagnostic protocols need to focus equally on efficiency and accuracy, thus increasing the workload and complexity of histopathologic (microscopic examination of tissue in order to study the manifestations of disease) in cancer diagnosis. This has led researchers in the Netherlands to use deep learning to improve the efficiency of histopathologic slide analysis, where the workload for pathologists is reduced and the objectivity of diagnoses is increased. The researchers concluded that deep learning could improve the efficacy of prostate cancer diagnosis and breast cancer staging. In another case, Philips and LabPON are planning to create the world's largest pathology database of annotated tissue images for deep learning. One of the things the database will provide is data for research and discovery to develop new insights in disease assessment, including cancer. In an effort to accelerate cancer research, Oak Ridge National Laboratory (ORNL) researchers are applying deep learning toward automating information collected from cancer pathology reports that are documented across a nationwide network of cancer registry programs.

Tracking Tumor Development

Deep learning can be used to measure the size of tumors undergoing treatment and detect new metastases that might be overlooked. This is exactly what researchers from the Fraunhofer Institute for Medical Image Computing in Germany are doing. The more patient CT and MRI scans the deep learning algorithm reads, the more accurate it becomes, which is the core of deep learning technology.

Google's Deep Learning Tumor Prediction Heat Maps

Google Research is also hard at work developing deep learning tools that can "naturally complement pathologists' workflow." They used images to train their deep learning algorithm Inception (aka GoogLeNet) to identify breast cancer tumors that have spread to adjacent lymph nodes. The algorithm reached a localization score of 89%, exceeding the 73% accuracy rate for pathologists.

Prepare for Deep Learning With the Right Hardware

A big part of the increase in computational power since the late 2000s is due to chips designed by NVIDIA to increase video games' visual realism (NVIDIA news, 2015). Using graphics processing units (GPUs) for deep learning turned out to be 20 to 50 times more efficient than using central processing units (CPUs). In August 2016, NVIDIA announced that its quarterly revenue had more than doubled year over year for its data center segment, reaching $151 million. CFO Colette Kress told investors that "The vast majority of the growth comes from deep learning by far". In 2016 Intel purchased deep-learning startups Nervana Systems and Movidius while Google unveiled tensor processing units (TPUs) specifically designed to facilitate deep learning. Today, NVIDIA is taking aim at the medical industry. The company's business development manager Kimberley Powell wants to see NVIDIA's processors used to meet the need for deep learning analysis of medical imaging. Futurism reports that "NVIDIA's hardware has established its silent but prominent role in deep learning's marriage with medicine." The company is building powerful computers designed to quickly and efficiently run the computations that can make diagnosis faster and more accurate. Powell is hoping to get NVIDIA's DGX-1 installed in hospitals and medical research centers across the world. Some hospitals—such as Massachusetts General Hospital's new clinical data science center are already using this new hardware for population health, comparing patients' test results and medical histories history to identify correlations in the data

CHALLENGES IN DEEP LEARNING BIOMETRICS

Deep learning (DL) is essentially a subset of Machine Learning(ML) that extends ML capabilities across multilayered neural networks to go beyond just categorizing data. DL can actually learn—self-train, essentially—from massive amounts of data. With DL, it's possible to combine the unique ability of computers to process massive amounts of information quickly, with the human-like ability to take in, categorize, learn, and adapt. Together, these skills allow modern DL programs to perform advanced tasks, such as identifying cancerous tissue in MRI scans. DL also makes it possible to develop driverless cars and designer medications that are tailored to an individual's genome.

However, in order to work with machines, it is important for them to understand and recognize things just the way the human brain does. For example, we may recognize an apple through its shape and colour. For a robot to go through the same cognitive process, it must be fed with programming structures to recognize the same. This is termed as machine learning and deep learning is basically a highly revered and practiced branch of machine learning. Deep learning can be called a one-stop solution to several advanced scientific creations. A representation or simply an object can be understood in many ways. To simplify the learning task by a machine, the objects can be analyzed and set into an abstract setup comprising of shapes and edges and pixels. This way of analytical breaking down of an object into its most easy-to-understand state, rules out the complete need for handcrafted features or elaborate programming techniques. Thus, your computer or robot can learn the things you want it to, without your supervision and intensive involvement.

Many problems using deep networks require enough examples to fit the parameters of a complex model, which is difficult. Training on insufficient data would also result in overfitting, due to the high degree of expressive power of deep networks. Training a shallow network with 1 hidden layer using supervised learning usually resulted in the parameters converging to reasonable values, but when training a deep network, this turns out with bad local optima. When using back propagation to compute the derivatives, the gradients that are propagated backwards from the output layer to the earlier layers of the network rapidly diminish in magnitude as the depth of the network increases. The weights of the earlier layers change slowly when using gradient descent and the earlier layers fail to learn much leading to "diffusion of gradients." Another challenge associated is data incompleteness and noisy labels, as majority of data may not be labeled, or if labeled, there exist noisy labels. Advanced deep learning methods are required to deal with noisy data and to tolerate some messiness.

HANDLING CHALLENGES FACED BY WORLD'S HEALTH SECTOR USING BIOMETRICS

In recent survey, it is found that around 1.1 billion people in the world, do not have a formal identity, 40% of them are children whose age is under 18 and 78% are residing in underdeveloped regions of Sub-Saharan Africa and South Asia (World Bank, 2017). A lack of personal identity means that these people are consistently overlooked or forgotten by the respective governments and health agencies and ultimately not permitted from accessing essential services i.e. healthcare, banking, education, and consequently trapped in a cycle of poor health, illiteracy and poverty. Now a days, NGOs and governments are increasingly trying to use biometrics as a method of solving largely occurring identification challenges. This technology is being used to authenticate patients and health professionals for healthcare systems and also for other services that can help in increasing the living standards of individuals (Alan & Decker, 2012). It was observed that existing biometric technologies using traditional methods failed in current populations with large number and high levels of manual labour, as biometric feature fingerprints usually become thick and gradually burn over time due to cooking, or get damaged where the farmers usually do harvesting crops by hand. Those biometric systems which recorded less than 1% error rates in the Western countries often had error rates greater than 10% when used in rural areas of Zambia and urban slums located in Bangladesh, and the poorest people were at the higher risk of segregation (Storisteanu et al., 2015).

Biometrics based company, Simprints has designed their biometric systems to overcome from these technical challenges, and also added more functionalities to enhance the usability for operators (individuals) with varying levels of technical education and exposure to the current technology. As current age research at the University of Cambridge have developed biometric systems that are now able to reliably and consistently identify individuals in the diverse and demanding contexts in the availability highly poor resources. The biometric feature i.e. tap of a individual's finger can now be utilized to accurately link a patient to their own medical history so that it may provide the continuity of care for better treatment. Currently Simprints deployed these systems across various different countries which provide high-impact projects in the field of maternal health, education, immunisations, and cash transfers. It also explores the potential of Neonate biometrics which would enable more efficient delivery of vaccine and nutrition programs, aid distribution, and greater continuity of care. Health clinics in Bangladesh revealed a statement during field testing of Simprints biometric system that 1.68% equal error rate (EER) i.e. an algorithm for measurement for accuracy of biometric based security systems that functionally decreases to zero when multiple fingers were used in a system. Digital technology and artificial intelligence-based methods are increasingly used to structure and formulate interventions in global healthcare market. When mutual consent, personal security, and other ethical and privacy-based issues to follow the strict standards, biometric systems can help ensure mothers are able to get essential health services such as antenatal care visits and subsequent treatment from community health workers and professionals. It also ensures that life-saving vaccines are administered globally all the time and try to maintain continuity of proper care for everyone. More than one billion people in developing countries, are now registered in various biometric programmes especially in healthcare sector, biometric based identification process has and will continue to empower interventions in world-wide global health market and beyond.

CONCLUSION

Although machines and robots may not be able to replace doctors and nurses completely in the healthcare sector, machine or deep learning technology and artificial intelligence are increasingly transforming the healthcare industry. This technology not only comprises huge potential, but also offers a complete platform to implement variety of applications in the healthcare sector. Over the last couple of decades, a dramatic change has been observed the way in which healthcare system operates, and this change appears to be continued in the coming years. Emerging from the branch of Artificial Intelligence such as computational machine learning and pattern recognition based intelligent theory, deep learning methods helps in building biometric based automated systems for healthcare that tries to explore and vitally analyze the available medical or biometric based data sets (physiological and behavioral) in order to identify and extract hidden patterns and make useful predictions by using recognition algorithms including those patterns. Now a days, Deep Learning (DL) is increasingly being used in implementation of automated systems in the field of healthcare, especially in medical image registration, medical image analysis and interpretation, image database retrieval, image-guided therapy, image annotation, computer-aided diagnosis, computer aided medical biometrics etc. In this chapter, some of the key healthcare applications of deep learning have been discussed and a direction has been shown how DL techniques have benefited this industry in building biometrics based models. Gradually, computer and AI based technological developments are taking over the problems of healthcare domain at large scale, which were considered or even resolved by using techniques completely based on human operated. The recent technological developments in the field of artificial intelligence including machine and deep learning are able to reduce the burden of requirement of expert doctors and physicians in large number or of highly expertise and aided them in combination in making better diagnoses which indirectly encourage to provide the best possible treatments to their patients.

Healthcare biometric system in turn refers to applications of biometric feature-based identification and interpretation of an individual or their associated problems in medical offices, doctors' chambers, medical hospitals, or for using as a tool in monitoring patients. This may include access control of patients and medical staff, identification and interpretation of patients and their type of diseases, workforce management required for medical hospitals and healthcare centers or storage of patients' record's for future use. Biometric technology is emerged to create revolution and remarkable breakthrough in healthcare industry. Generally, biometric devices take unique information (physical and physiological) about an individual by extracting distinguished features from different parts of the body such as eye (by capturing retinal images), palm geometry, or thumb print, face etc. and use it to identify individuals. Thereafter, this information can be used to ensure identity that you are who you say you are, and then only permission can be granted work with the healthcare information system which you want to access. Biometric technology represents the future for positive healthcare identification and making easy the diagnosis and prognosis process and will enhance the security features of using, storing, and exchanging of personal health information among the health professional systems. The lack of patient identity safeguards usually posses many issues and difficulties for patients and providers, patients may become victims of medical identity theft as their biometric technology based digital records may contain critical health related data and claims that are not theirs, jeopardizing treatments in the future and also may create issue of health care financial limits. Medical records of the patients may have some falsified records which may support fraudulent claims, when they have not received any such care or treatment by the doctors. Fraudsters can misuse providers and patients' identities to falsify healthcare claims, all of these issues may lead to less safe and inefficient patient care which may cause serious problems at patient's life. Biometrics also eliminates the issues and difficulties in identifying individuals with the same name or multiple surnames. It also helps to ensure that each time a patient comes for a healthcare facility, their medical history is properly documented irrespective of any number of hospital visits, no matter what variation of their name is provided during the same.

Accurately identifying patients, and thereafter linking them to their own electronic medical records which may store past medical conditions, drug allergy based information, medications and vaccination related history is very sensitive and critical to health care industry. David Bell, a well known healthcare technologist said in a press release. "While adult biometrics are widely used in financial services, these have not been translated to the health sector where the need to identify a person from birth to old age presents additional challenges; namely, that infants have delicate, rapidly changing features that are difficult to capture." Therefore, now a days, in the healthcare sector, biometrics are frequently used in combination with passwords or smart identification cards in order to secure access to sensitive patient healthcare related records and to assist with patient registration requirements for computerized storage. Biometrics Research Group, Inc. is expecting the use of biometrics to accelerate through the entire global healthcare sector due to the adoption of biometrics in large extent to enhance cost saving methods to fight with possible fraud in both government-insured sectors and private-sector based insurance marketplaces. Growth in the use of biometric technology in the healthcare sector will primarily be driven by the continuance of healthcare reform through legislative repeal and replacement in throughout the world, which will be largely driven by cost containment imperatives. Medical biometrics, led by renowned organizations such as mHealth and wearables, will continue to make in-roads in the consumer technology marketplace in healthcare sector.

REFERENCES

Abdel-Hamid, O., Mohamed, A-R., & Jiang, H. (2014). Convolutional neural networks for speech recognition. *IEEE/ACM Trans Audio Speech Lang Process*, *22*, 1533–1545.

Alan, G., & Decker, C. (2012). Cash at your fingertips: Biometric technology for transfers in developing countries. *The Review of Policy Research*, *29*(1), 91–117. doi:10.1111/j.1541-1338.2011.00539.x

Alipanahi, B., Delong, A., Weirauch, M. T., & Frey, B. J. (2015). Predicting the sequence specificities of DNA-and RNA-binding proteins by deep learning. *Nature Biotechnology*, *33*(8), 831–838. doi:10.1038/nbt.3300 PMID:26213851

Angermueller, C., Parnamaa, T., Parts, L., & Stegle, O. (2016). Deep learning for computational biology. *Molecular Systems Biology*, *12*(7), 878. doi:10.15252/msb.20156651 PMID:27474269

Aronson, S., & Rehm, H. (2015). Building the foundation for genomics in precision medicine. *Nature*, *526*(7573), 336–342. doi:10.1038/nature15816 PMID:26469044

Askar, H. A., Radi, N., & MacDermott, A. (2016). Recurrent Neural Networks in Medical Data Analysis and Classifications. Emerging Topics in Computer Science and Applied Computing, 147-165.

Bagar, M., Ghani, A., Aftab, A., Arbab, S., & Yasin, S. (2016). Deep belief networks for iris recognition based on contour detection. *International Conference on Open Source Systems & Technologies (ICOSST)*. 10.1109/ICOSST.2016.7838580

Bartlett, R. (2017). *Enlitic uses deep learning to make doctors faster and more accurate*. Retrieved from http://www.enlitic.com/index.html

Bellazzi, R., & Zupan, B. (2008). Predictive data mining in clinical medicine: Current issues and guidelines. *International Journal of Medical Informatics*, *77*(2), 81–97. doi:10.1016/j.ijmedinf.2006.11.006 PMID:17188928

Bengio, Y. (2009). Learning deep architectures for AI. *Foundations and Trends® in Machine Learning*, *2*, 1–127.

Bengio, Y. (2012). Practical recommendations for gradient-based training of deep architectures. *Neural Networks*, *2*, 437–478.

Bengio, Y., Courville, A., & Vincent, P. (2013). Representation learning: A review and new perspectives. *IEEE Transactions on Pattern Analysis and Machine Intelligence*, *35*(8), 1798–1828. doi:10.1109/TPAMI.2013.50 PMID:23787338

Bengio, Y., Lamblin, P., & Popovici, D. (2007). Greedy layer-wise training of deep networks. *Advances in Neural Information Processing Systems*, *19*, 153–160.

Bishop, C. M. (2006). Pattern Recognition and Machine Learning (Information Science and Statistics) (2nd ed.). Springer.

Brosch, T., & Tam, R. (2013). *Manifold learning of brain MRIs by deep learning* (Vol. 16). Medical Image Computing and Computer- Assisted Intervention.

Chen, C. K., Lin, C. L., & Chiu, Y. M. (2011). Individual identification based on chaotic electrocardiogram signals. *6th IEEE Conference on Industrial Electronics and Applications (ICIEA)*, 1771-1776.

Chen, Y., Li, L., Zhang, G. Q., & Xu, R. (2015). Phenome-driven disease genetics prediction toward drug discovery. *Bioinformatics (Oxford, England)*, *31*(12), i276–i283. doi:10.1093/bioinformatics/btv245 PMID:26072493

Cheng, J.-Z., Ni, D., Chou, Y.-H., Qin, J., Tiu, C. M., Chang, Y. C., ... Chen, C. M. (2016). Computer-aided diagnosis with deep learning architecture: Applications to breast lesions in US images and pulmonary nodules in CT scans. *Scientific Reports*, *6*(1), 24454. doi:10.1038rep24454 PMID:27079888

Cheng, Y., Wang, F., Zhang, P., & Hu, J. (2015). Risk prediction with electronic health records: a deep learning approach. *ACM International Conference on Knowledge Discovery and Data Mining*, 705–14.

Cho, K., Courville, A., & Bengio, Y. (2015). *Describing multimedia content using attention-based encoder-decoder networks*. arXiv.:1507.01053v1.

Cho, K., van Merrienboer, B., Bahdanau, D., & Bengio, Y. (2014). *On the properties of neural machine translation: encoder-decoder approaches.* arXiv preprint arXiv:1409.1259.

Choi, E., Bahadori, M. T., Schuetz, A., Stewart, W. F., & Sun, J. (2015). *Doctor AI: predicting clinical events via recurrent neural networks.* arXiv: 1511.05942

Choi, Y., Ms, C. Y.-I. C., & Sontag, D. (2016). Learning low-dimensional representations of medical concepts. *Proceedings of the AMIA Summit on Clinical Research Informatics.*

Clark, J. (2016). *Google's Deep Mind forms health unit to build medical software.* Retrieved from https://www.bloomberg.com/news/articles/2016-02-24/google-s-deepmind-forms-health-unit-to-buildmedical-software

Collins, F. S., & Varmus, H. (2015). A new initiative on precision medicine. *The New England Journal of Medicine, 372*, 793–795.

Deep Learning. (2016). *Advances in Medicine | NVIDIA.* Retrieved from http://www.nvidia.com/object/deep-learning-in-medicine.html

Deng, L., & Li, X. (2013). Machine learning paradigms for speech recognition: An overview. *IEEE Transactions on Audio, Speech, and Language Processing, 21*(5), 1060–1089. doi:10.1109/TASL.2013.2244083

Dernoncourt, F., Lee, J. Y., Uzuner, O., & Szolovits, P. (2016). De-identification of patient notes with recurrent neural networks. *Journal of the American Medical Informatics Association, 24*(3), 596–606. PMID:28040687

Esteva, A., Kuprel, B., Novoa, R. A., Ko, J., Swetter, S. M., Blau, H. M., & Thrun, S. (2017). Dermatologist-level classification of skin cancer with deep neural networks. *Nature, 542*(7639), 115–118. doi:10.1038/nature21056 PMID:28117445

Fakoor, R., Ladhak, F., & Nazi, A. (2013). Using deep learning to enhance cancer diagnosis and classification. *International Conference on Machine Learning.*

Farhan, W., Wang, Z., Huang, Y., Wang, S., Wang, F., & Jiang, X. (2016). A predictive model for medical events based on contextual embedding of temporal sequences. *Journal of Medical Internet Research, 4*, e39. PMID:27888170

Fierrez-Aguilar, J., Ortega-Garcia, J., Gonzalez-Rodriguez, J., & Bigun, J. (2004). Kernel-based multimodal biometric verification using quality signals. *Proceedings of the Society for Photo-Instrumentation Engineers, 5404*, 544–554. doi:10.1117/12.542800

Gerstung, M., Papaemmanuil, E., Martincorena, I., Bullinger, L., Gaidzik, V. I., Paschka, P., ... Campbell, P. J. (2017). Precision oncology for acute myeloid leukemia using a knowledge bank approach. *Nature Genetics, 49*(3), 332–340. doi:10.1038/ng.3756 PMID:28092685

Gottlieb, A., Stein, G. Y., Ruppin, E., Altman, R. B., & Sharan, R. (2013). A method for inferring medical diagnoses from patient similarities. *BMC Medicine, 11*(1), 194. doi:10.1186/1741-7015-11-194 PMID:24004670

Gulshan, V., Peng, L., Coram, M., Stumpe, M. C., Wu, D., Narayanaswamy, A., ... Webster, D. R. (2016). Development and validation of a deep learning algorithm for detection of diabetic retinopathy in retinal fundus photographs. *Journal of the American Medical Association, 316*(22), 2402–2410. doi:10.1001/jama.2016.17216 PMID:27898976

Hammerla, N. Y., Halloran, S., & Ploetz, T. (2016). *Deep, convolutional, and recurrent models for human activity recognition using wearables.* arXiv preprint arXiv:1604.08880.

Hannun, A., Case, C., Casper, J., Catanzaro, B., Diamos, G., Elsen, E., . . . Ng, A. Y. (2014). *Deep speech: scaling up end-to-end speech recognition.* arXiv:1412.5567.

Healthcare Biometrics Market by Technology (Single-factor (Fingerprint, Face, Iris, Palm, Behavioral (Signature, Voice)), Multifactor, Multimodal), Application (Workforce Management), End User (Hospitals, Healthcare Institutions) - Global Forecast to 2021. (2016). Retrieved from marketsandmarkets.com

Hinton, G. E., Osindero, S., & Teh, Y.-W. (2006). A fast learning algorithm for deep belief nets. *Neural Computation, 18*(7), 1527–1554. doi:10.1162/neco.2006.18.7.1527 PMID:16764513

Hinton, G. E., & Salakhutdinov, R. R. (2006). Reducing the dimensionality of data with neural networks. *Science, 313*(5786), 504–507. doi:10.1126cience.1127647 PMID:16873662

Hochreiter, S., & Schmidhuber, J. (1997). Long short-term memory. *Neural Computation, 9*(8), 1735–1780. doi:10.1162/neco.1997.9.8.1735 PMID:9377276

Hripcsak, G., & Albers, D. J. (2013). Next-generation phenotyping of electronic health records. *Journal of the American Medical Informatics Association, 20*(1), 117–121. doi:10.1136/amiajnl-2012-001145 PMID:22955496

Hubel, D. H., & Wiesel, T. N. (1968). Receptive fields and functional architecture of monkey striate cortex. *The Journal of Physiology, 195*(1), 215–243. doi:10.1113/jphysiol.1968.sp008455 PMID:4966457

IBM News Room. (2018). *IBM Watson Health Showcases Health Technology Advancements at American Diabetes Association's 78th Scientific Sessions, June 26, 2018. ICD-9 Code.* Retrieved from https://www.cms.gov/Medicare/Coding/ICD9ProviderDiagnosticCodes/index.html(02 June 2014, date last accessed)

Jensen, P. B., Jensen, L. J., & Brunak, S. (2012). Mining electronic health records: Towards better research applications and clinical care. *Nature Reviews. Genetics, 13*(6), 395–405. doi:10.1038/nrg3208 PMID:22549152

Jindal, V., Birjandtalab, J., Pouyan, M. B., & Nourani, M. (2016). An adaptive deep learning approach for PPG-based identification. *38th Annual International Conference of the IEEE Engineering in Medicine and Biology Society (EMBC)*, 6401–6404.

Jordan, M. I., & Mitchell, T. M. (2015). Machine learning: Trends, perspectives, and prospects. *Science, 349*(6245), 255–260. doi:10.1126cience.aaa8415 PMID:26185243

Kelley, D. R., Snoek, J., & Rinn, J. L. (2016). Basset: Learning the regulatory code of the accessible genome with deep convolutional neural networks. *Genome Research, 26*(7), 990–999. doi:10.1101/gr.200535.115 PMID:27197224

Kim, J. M. S., Kang, U., & Lee, Y. (2017). Statistics and Deep Belief Network-Based Cardiovascular Risk Prediction. *Healthcare Informatics Research, 23*(3), 169–175. doi:10.4258/hir.2017.23.3.169 PMID:28875051

King, R. (2014). *India to use Aadhaar to provide universal healthcare.* Retrieved from http://www.biometricupdate.com/201410/india-to-use-aadhaar-to-provide-universal-healthcare

Koh, P. W., Pierson, E., & Kundaje, A. (2016). Denoising genome-wide histone ChIP-seq with convolutional neural networks. *bioRxiv*. doi:10.1101/052118

Kryszczuk, K., Richiardi, J., Prodanov, P., & Drygajlo, A. (2005). Error handling in multimodal biometric systems using reliability measures. *Proc. 13th Eur. Signal Process. Conf.*, 1-4.

Lane, N. D., Bhattacharya, S., Georgiev, P., Forlivesi, C., Jiao, L., & Kawsar, F. (2016). DeepX: a software accelerator for low-power deep learning inference on mobile devices. *ACM/IEEE International Conference on Information Processing in Sensor Networks*, 1–12.

Lane, N. D., & Georgiev, P. (2015). Can deep learning revolutionize mobile sensing? *International Workshop on Mobile Computing Systems and Applications*, 117–22.

Lasko, T. A., Denny, J. C., & Levy, M. A. (2013). Computational phenotype discovery using unsupervised feature learning over noisy, sparse, and irregular clinical data. *PLoS One, 8*(6), e66341. doi:10.1371/journal.pone.0066341 PMID:23826094

LeCun, Y., Bengio, Y., & Hinton, G. (2015). Deep learning. *Nature, 521*(7553), 436–444. doi:10.1038/nature14539 PMID:26017442

Lecun, Y., Bottou, L., Bengio, Y., & Haffner, P. (1998). Gradient-based learning applied to document recognition. *Proceedings of the IEEE, 86*(11), 2278–2324. doi:10.1109/5.726791

Lee, W.-H., & Lee, R. B. (2017). Implicit Smartphone User Authentication with Sensors and Contextual Machine Learning. *47th Annual IEEE/IFIP International Conference on Dependable Systems and Networks (DSN).*

Leung, M. K. K., Delong, A., Alipanahi, B., & Frey, J. (2016). Machine learning in genomic medicine: A review of computational problems and data sets. *Proceedings of the IEEE, 104*(1), 176–197. doi:10.1109/JPROC.2015.2494198

Li, L., Cheng, W.-Y., Glicksberg, B. S., Gottesman, O., Tamler, R., Chen, R., ... Dudley, J. T. (2015). Identification of type 2 diabetes subgroups through topological analysis of patient similarity. *Science Translational Medicine, 7*(311), 311ra174. doi:10.1126citranslmed.aaa9364 PMID:26511511

Liang, Z., Zhang, G., Huang, J. X., & Hu, Q. V. (2014). Deep learning for healthcare decision making with EMRs. *IEEE International Conference on Bioinformatics and Biomedicine*, 556–9.

Libbrecht, M. W., & Noble, W. S. (2015). Machine learning applications in genetics and genomics. *Nature Reviews. Genetics, 16*(6), 321–332. doi:10.1038/nrg3920 PMID:25948244

Lipton, Z. C., Kale, D. C., & Elkan, C. (2015). Learning to diagnose with LSTM recurrent neural networks. *International Conference on Learning Representations*, 1–18.

Liu, S., Cai, W., Pujol, S., Kikinis, R., & Feng, D. D. (2014). Early diagnosis of Alzheimer's disease with deep learning. *International Symposium on Biomedical Imaging*, 1015–18.

Luo, J., Wu, M., Gopukumar, D., & Zhao, Y. (2016). Big data application in biomedical research and healthcare: A literature review. *Biomedical Informatics Insights*, *8*, 1–10. doi:10.4137/BII.S31559 PMID:26843812

Lyman, G. H., & Moses, H. L. (2016). Biomarker tests for molecularly targeted therapies - the key to unlocking precision medicine. *The New England Journal of Medicine*, *375*, 4–6.

Lyons, J., Dehzangi, A., Heffernan, R., Sharma, A., Paliwal, K., Sattar, A., ... Yang, Y. (2014). Predicting backbone Ca angles and dihedrals from protein sequences by stacked sparse auto-encoder deep neural network. *Journal of Computational Chemistry*, *35*(28), 2040–2046. doi:10.1002/jcc.23718 PMID:25212657

Ma, J., Sheridan, R. P., Liaw, A., Dahl, G. E., & Svetnik, V. (2015). Deep neural nets as a method for quantitative structure – activity relationships. *Journal of Chemical Information and Modeling*, *55*(2), 263–274. doi:10.1021/ci500747n PMID:25635324

Mamoshina, P., Vieira, A., Putin, E., & Zhavoronkov, A. (2016). Applications of deep learning in biomedicine. *Molecular Pharmaceutics*, *13*(5), 1445–1454. doi:10.1021/acs.molpharmaceut.5b00982 PMID:27007977

Manning, C. D., Raghavan, P., & Schutze, H. (2008). *Introduction to information retrieval* (Vol. 3). Cambridge, UK: Cambridge University Press. doi:10.1017/CBO9780511809071

Mayhew, S. (2018). *HID Global reports strong biometrics adoption in Latin America*. Retrieved from https://www.biometricupdate.com/201804/hid-global-reports-strong-biometrics-adoption-in-latin-america

Miotto, R., Li, L., & Dudley, J. T. (2016). Deep learning to predict patient future diseases from the electronic health records. *European Conference in Information Retrieval*, 768–774.

Miotto, R., Li, L., Kidd, B. A., & Dudley, J. T. (2016). Deep patient: An unsupervised representation to predict the future of patients from the electronic health records. *Scientific Reports*, *6*(1), 26094. doi:10.1038rep26094 PMID:27185194

Miotto, R., & Weng, C. (2015). Case-based reasoning using electronic health records efficiently identifies eligible patients for clinical trials. *Journal of the American Medical Informatics Association*, *22*(e1), e141–e150. doi:10.1093/jamia/ocu050 PMID:25769682

Mogli, G. D. (2011). Role of Biometrics in healthcare privacy and security management system. *Sri Lanka Journal of Bio-Medical Informatics*, *2*(4), 156–165. doi:10.4038ljbmi.v2i4.2245

Mohan, A., Blough, D. M., Kurc, T., Post, A., & Saltz, J. (2011). Detection of conflicts and inconsistencies in taxonomy-based authorization policies. *IEEE International Conference on Bioinformatics and Biomedicine*, 590–594.

Murphy, K. P. (2012). *Machine learning: A probabilistic perspective*. MIT Press.

Nguyen, P., Tran, T., Wickramasinghe, N., & Venkatesh, S. (2017). Deepr: A Convolutional Net for Medical Records. *IEEE Journal of Biomedical and Health Informatics*, *21*(1), 22–30. doi:10.1109/JBHI.2016.2633963 PMID:27913366

Nurse, E., Mashford, B. S., Yepes, A. J., Kornek, I. K., Harrer, S., & Freestone, D. R. (2016). Decoding EEG and LFP signals using deep learning: heading TrueNorth. *ACM International Conference on Computing Frontiers*, 259–266. 10.1145/2903150.2903159

NVIDIA news. (2015). *NVIDIA Announces Financial Results for Second Quarter Fiscal 2016*. Retrieved from https://nvidianews.nvidia.com/news/nvidia-announces-financial-results-for-second-quarter-fiscal-2016

Offit, K. (2011). Personalized medicine: New genomics, old lessons. *Human Genetics*, *130*(1), 3–14. doi:10.100700439-011-1028-3 PMID:21706342

Park, Y., & Kellis, M. (2015). Deep learning for regulatory genomics. *Nature Biotechnology*, *33*(8), 825–826. doi:10.1038/nbt.3313 PMID:26252139

Pham, T., Tran, T., Phung, D., & Venkatesh, S. (2016). *DeepCare: a deep dynamic memory model for predictive medicine*. arXiv:1602.00357.

Piwek, L., Ellis, D. A., Andrews, S., & Jainson, A. (2016). The rise of consumer health wearables: Promises and barriers. *PLoS Medicine*, *13*(2), e1001953. doi:10.1371/journal.pmed.1001953 PMID:26836780

Ponemon Institute Report. (2016). *National Patient Misidentification Report*. Ponemon Institute LLC.

Prasoon, A., Petersen, K., Igel, C., Lauze, F., Dam, E., & Nielsen, M. (2013). *Deep feature learning for knee cartilage segmentation using a triplanar convolutional neural network* (Vol. 16). Medical Image Computing and Computer- Assisted Intervention.

Ravi, D., Wong, C., Lo, B., & Yang, G. Z. (2017). A deep learning approach to on-node sensor data analytics for mobile or wearable devices. *IEEE Journal of Biomedical and Health Informatics*, *21*(1), 56–64. doi:10.1109/JBHI.2016.2633287 PMID:28026792

Razavian, N., Marcus, J., & Sontag, D. (2016). Multi-task prediction of disease onsets from longitudinal laboratory tests. *Proceedings of the 1st Machine Learning for Healthcare Conference*, 73–100.

Rumelhart, D. E., Hinton, G. E., & Williams, R. J. (1986). Learning representations by back-propagating errors. *Nature*, *323*(6088), 533–536. doi:10.1038/323533a0

Safie, S. I., Soraghan, J. J., & Petropoulakis, L. (2011). Electrocardiogram (ECG) Biometric Authentication Using Pulse Active Ratio (PAR). *IEEE Transactions on Information Forensics and Security*, *6*(4), 1315–1322. doi:10.1109/TIFS.2011.2162408

Salah, A. A. (2010). Machine Learning for Biometrics. In Handbook of Research on Machine Learning Applications and Trends: Algorithms, Methods, and Techniques. Centre for Mathematics and Computer Science (CWI). Doi:10.4018/978-1-60566-766-9.ch026

Salakhutdinov, R., Mnih, A., & Hinton, G. (2007). Restricted Boltzmann machines for collaborative filtering. *Proceedings of the 24th International Conference on Machine Learning*, 791–798.

Sathyanarayana, A., Joty, S., Fernandez-Luque, L., Ofli, F., Srivastava, J., Elmagarmid, A., ... Taheri, S. (2016). Correction of: Sleep quality prediction from wearable data using deep learning. *JMIR Mhealth Uhealth, 4*(4), e130. doi:10.2196/mhealth.6953 PMID:27885989

Schmidhuber, J. (2015). Deep learning in neural networks: An overview. *Neural Networks, 61*, 85–117. doi:10.1016/j.neunet.2014.09.003 PMID:25462637

Shakil, K. A., Zareen, F. J., Alam, M., & Jabin, S. (2017). BAMHealthCloud: A Biometric Authentication and Data Management System for Healthcare Data in Cloud. *Journal of King Saud University - Computer and Information Sciences*, 1-9.

Shameer, K., Badgeley, M. A., Miotto, R., Glicksberg, B. S., Morgan, J. W., & Dudley, J. T. (2017). Translational bioinformatics in the era of real-time biomedical, healthcare and wellness data streams. *Briefings in Bioinformatics, 18*(1), 1105–1124. doi:10.1093/bib/bbv118 PMID:26876889

Smolensky, P. (1986). *Information processing in dynamical systems: Foundations of harmony theory (No. CU-CS-321-86).* Colorado Univ at Boulder Dept of Computer Science.

SNOMED CT. (2016). *U.S. National Library of Medicine*. Retrieved from https://www.nlm.nih.gov/healthit/snomedct/

Storisteanu, D. M., Norman, T. L., Grigore, A., & Norman, T. L. (2015). Biometric fingerprint system to enable rapid and accurate identification of beneficiaries. *Global Health, Science and Practice, 3*(1), 135–137. doi:10.9745/GHSP-D-15-00010 PMID:25745127

Tatonetti, N. P., Ye, P. P., Daneshjou, R., & Altman, R. B. (2012). Data-driven prediction of drug effects and interactions. *Science Translational Medicine, 4*(125), 125ra31. doi:10.1126citranslmed.3003377 PMID:22422992

Tiquet, A. (2016). *Horus Among Seven Prize-Winning Startups at Emerging Companies Summit*. Retrieved from https://blogs.nvidia.com/blog/2016/04/07/horus-help-blind-people-see/

Toh, K.-A., Yau, W.-Y., Lim, E., Chen, L., & Ng, C.-H. (2004). *Fusion of auxiliary information for multi-modal biometrics authentication in Bio-metric Authentication*. Berlin, Germany: Springer. doi:10.1007/978-3-540-25948-0_92

UMLS (Unified Medical Language System). (n.d.). Retrieved from https://www.nlm. nih.gov/research/umls/

Vincent, P., Larochelle, H., Lajoie, I., Lajoie, I., Bengio, Y., & Mazagol, P. A. (2010). Stacked denoising autoencoders: Learning useful representations in a deep network with a local denoising criterion. *Journal of Machine Learning Research, 11*, 3371–3408.

Wang, B., Mezlini, A. M., Demir, F., Fiume, M., Tu, Z., Brudno, M., ... Goldenberg, A. (2014). Similarity network fusion for aggregating data types on a genomic scale. *Nature Methods, 11*(3), 333–337. doi:10.1038/nmeth.2810 PMID:24464287

Wang, F., Zhang, P., Wang, X., & Hu, J. (2014). Clinical risk prediction by exploring high-order feature correlations. *AMIA Annual Symposium*, 1170–1179.

Williams, R. J., & Zipser, D. (1989). A learning algorithm for continually running fully recurrent neural networks. *Neural Computation, 1*(2), 270–280. doi:10.1162/neco.1989.1.2.270

World Bank Group. (2017). *ID4D Identification for Development 2017 Annual Report.* World Bank.

Wu, G., Wang, J., Zhang, Y., & Jiang, S. (2018). A Continuous Identity Authentication Scheme Based on Physiological and Behavioral Characteristics. *Sensors (Basel), 18*(1). doi:10.339018010179 PMID:29320463

Xu, R., Li, L., & Wang, Q. (2014). dRiskKB: A large-scale disease-disease risk relationship knowledge base constructed from biomedical text. *BMC Bioinformatics, 15*(1), 105. doi:10.1186/1471-2105-15-105 PMID:24725842

Yoo, Y., Brosch, T., Traboulsee, A., Li, D. K. B., & Tam, R. (2014). Deep learning of image features from unlabeled data for multiple sclerosis lesion segmentation. *International Workshop on Machine Learning in Medical Imaging*, 117–124.

Zhang, Q., Xiao, Y., Dai, W., Suo, J., Wang, C., Shi, J., & Zheng, H. (2016). Deep learning based classification of breast tumors with shear-wave elastography. *Ultrasonics, 72*, 150–157. doi:10.1016/j.ultras.2016.08.004 PMID:27529139

Zhou, J., & Troyanskaya, O. G. (2015). Predicting effects of noncoding variants with deep learning-based sequence model. *Nature Methods, 12*(10), 931–934. doi:10.1038/nmeth.3547 PMID:26301843

Zhu, J., Pande, A., Mohapatra, P., & Han, J. J. (2015). Using deep learning for energy expenditure estimation with wearable sensors. *17th International Conference on E-health Networking, Application Services (HealthCom),* 501–6.

Chapter 5
Voice Liveness Detection for Medical Devices

Bin Hao
University of Louisiana at Lafayette, USA

Xiali Hei
University of Louisiana at Lafayette, USA

ABSTRACT

Many healthcare providers integrate biometric recognition/verification schemes into patient identification or other information security systems. While overcoming the disadvantages of using passwords, PINs, and tokens which may be forgotten, or stolen, biometric systems are susceptible to spoofing attacks, or presentation attacks. Liveness detection is an effective mechanism used to defeat a presentation attack. This chapter focuses on voice liveness detection in automatic speaker verification (ASV) systems. The authors explain the spoofing attacks to ASV systems comprising impersonation, voice conversion, speech synthesis, and replay and then present four types of liveness detection (anti-spoofing) methods used to mitigate ASV spoofing attacks: challenge-response-based methods, acoustic feature-based methods, hardware-based methods, and multi-modal biometric-based methods. This chapter analyzes the advantages and disadvantages of each kind of liveness detection method and proposes the possible application of voiceprint-based liveness detection schemes in the insulin pump system.

INTRODUCTION

According to the analysis report from Crystal Market Research (CMR), the global healthcare biometrics market will reach around $12 billion by 2025. This growth is due to the fact that more and more healthcare providers integrate biometric recognition/verification schemes to patient identification or other information security systems. Biometric identification systems establish the identity of the users based on their extracted physiological features including face, fingerprint, iris, retina, vein (finger or palm), palm geometry, etc. or behavioral features including voice, signature, keystroke dynamics, etc.

DOI: 10.4018/978-1-5225-7525-2.ch005

While overcoming the disadvantages of using passwords, PINs, and tokens which may be forgotten, or stolen, biometric systems are susceptible to spoofing attacks, or Presentation Attacks (PA) which outwit a biometric sensor by presenting a counterfeit biometric evidence of a legitimate user using methods such as artifact, mutilations, replay, etc. to achieve impersonation or concealment.

Liveness detection is an effective mechanism used to detect a presentation attack. Recently, liveness detection has become a hot research topic in fingerprint recognition, iris recognition, and automatic speaker verification (ASV) communities. Research results from LivDet Iris 2017, LivDet Fingerprint 2017, and ASVspoof 2017 show that liveness detection or presentation attacks detection (PAD) systems still need to make more advancements, especially when under unknown attacks.

This chapter mainly focuses on voice liveness detection in ASV systems. An AVS system extracts the vocal characteristics of an individual to establish the identity either by imposing the fixed vocabulary constraints (text-dependent) or in a dynamic way (text-independent) i.e. without imposition of vocabulary constraints on the individuals. Currently, ASV is mature technique ready for commercial application in user authentication. But it is confirmed that ASV is vulnerable to spoofing attacks which undermine users' confidence on it. ASV spoofing attacks comprise (Wu et al., 2015a): impersonation whereby an attacker attempts to mimic a target legitimate user's voice; voice conversion whereby an attacker resembles the speech of a target legitimate user using another user's speech; speech synthesis whereby an attacker using text-to-speech (TTS) technique generates intelligible, natural-sounding artificial speech with inputs of arbitrary text; and replay whereby an attacker tries to pass the authentication of ASV by providing a pre-recorded speech sample collected from genuine target legitimate user.

According to state-of-the-art research results, there are mainly four kinds of voice liveness detection methods as anti-spoofing Countermeasures (CM) to mitigate ASV spoofing attacks.

- **Challenge-Response-Based Methods:** Aley-Raz et al. (2013) proposed a speaker recognition system with liveness detection scheme. This mechanism requires users' explicit cooperation (provide a sentence selected randomly from a closed set of sentences for liveness detection) when the voice authentication is processing.
- **Acoustic-Feature-Based Methods:** The basic idea of acoustic-feature-based methods comes from the observation that design of spoofing countermeasures should focus on the search for discriminative features rather than the design of complex classifiers. Many features such as Mel Frequency Cepstral Coefficients (MFCCs) and Linear Prediction Cepstral Coefficients (LPCCs), have been used for training classifiers (Font et al., 2017). For backend classifiers, Gaussian Mixture Model (GMM), GMM with Universal Background Model (GMM-UBM), Support Vector Machine (SVM), Deep Neural Networks (DNN)-based classifiers are the most often choices.
- **Hardware-Based Methods:** Chen et al. (2017) proposed a robust smartphone based voice impersonation defense system that captures the magnetic field emitting from the loudspeaker and extracts special physical characteristics of the magnetic field to discriminate between a live human and a loudspeaker (impersonation attacker). Zhang et al. (2016) proposed a liveness detection scheme (VoiceLive) that uses phoneme sound localization, i.e., the time-difference-of-arrival (TDoA) changes to the two microphones equipped in a smartphone, to determine whether a passphrase is spoken by a live person or replayed by loudspeaker. Zhang et al. (2017) proposed VoiceGesture that leverages the response of human speech production system to external stimuli to implement liveness detection.

- **Multi-Modal Biometric Based Methods:** No recognition system based on feature extraction of single biometric modality can guarantee 100% accuracy due to several influential factors, such as noise, spoofing attacks, etc. Voice liveness detection system is not an exception. Experimental results show that Multi-modal biometric based systems can offer higher recognition accuracy and lower equal error rate (EER) although it increases the cost of the system implementation.
  - Source of features
    - **Multiple Different Feature Extractions:** For speaker recognition/verification, different features are often complementary and can be combined to promote accuracy. Two or more features extracted from fragments of same voice acquired by single sensor (e.g. microphone) are then processed as the inputs of same or different types of classifiers the scores of which are fused to achieve the final decision.
    - **Multi-Factor Authentication (MFA):** While having high cost and high system design complexity, MFA is used to protect the security and privacy of the most sensitive data, e.g., important medical information. MFA system can integrate several factors, e.g., passwords/PIN, OTP, biometrics (voice, finger, iris ...) etc.
  - Operation modes
    - **Serial/Cascaded Mode:** Multiple features are processed one by one. The output of first feature verification is the input of the second processing. For voice authentication, the first process is ASV which verifies that the voice acquired by a microphone comes from a legitimate speaker (genuine or replayed), and the second process is Countermeasure (CM) which achieves liveness detection distinguishing a live person from a replayed loudspeaker.
    - **Parallel Mode:** Multiple features are processed simultaneously and the outputs of the multiple processed are determined in same one decision maker to obtain the final result: reject or accept.
  - Fusion methodologies
    - For each feature, an optimal set of parameters is achieved by training each corresponding classifier. The scores of different classifiers are then fused using some strategy in the final classifier. The score fusion can mitigate the overfitting problem when training the classifiers using different features, and results in a robust spoofing detection system with low EER.

The objectives of this chapter comprise:

- Analyzing the spoofing attacks to ASV systems.
- Summarizing the state-of-the-art liveness detection (anti-spoofing) methods, analyze their advantages and disadvantages.
- Proposing possible acoustic channel based liveness detection and access control solutions for one kind of medical system, i.e., insulin pump system used to treat diabetes.
- Providing suggestions about future research opportunities within voice liveness detection.

BACKGROUND

ASV System

Automatic Speaker Verification (ASV) system aims to accept or reject a claimed speaker (identity) and comprises two types: test-dependent system using fixed or prompted phrases same for both enrollment and verification, and test-independent system operating on arbitrary utterances (Sizov et al., 2015). Speaker Verification is one type of Speaker Recognition the other two types of which are Speaker Identification which determines who is speaking given a set of enrolled speakers and Speaker Diarization which categorizes input voice stream into homogeneous segments according to the speaker's identity. This chapter focuses on ASV system which may be used for access control and authentication in medical devices. A typical ASV system, shown in Figure 1 (Ratha et al., 2001), comprises two subsystems: front-end and back-end.

Front-End

The front-end subsystem acquires voice from the speaker, implements feature extraction, and generates a feature matrix from the voice. Each column (feature vector) of the feature matrix corresponds to one frame of the voice.

There are many features used in ASV and anti-spoofing (countermeasure) system. Threefold information including voice timbre, prosody, and language content are in a speech signal, which are corresponding to three types of features comprising short-term spectral, prosodic, and high-level idiolectal features (Wu et al., 2015a). Short-term spectral features are extracted from frames with a length of tens of milliseconds. Mel-frequency cepstral coefficients (MFCCs) and linear predictive cepstral coefficients (LPCCs) are popular spectral features. Prosodic features used to characterize speaking style and intonation, are extracted from longer segments such as syllables and word-like units. While having the advantage of being less sensitive to channel effects, these features such as pitch, energy, and duration are sparse and their extraction requires relatively large amounts of training data. Extracted from a lexicon or other discrete tokens, high-level features characterize behavior or lexical cues of the speaker.

To enhance the robustness of the feature extractor, *feature normalization* is used to normalize the features before feeding them onto the classifiers. Some frequently used feature normalization methods are as follows (Kinnunen & Li, 2010):

- *Cepstral Mean Subtraction (CMS) or Cepstral Mean Normalization (CMN)* subtracts the mean vector of the features over the whole voice to make features from different channels become zero-mean and therefore reduces the channel effects. Then the variances are divided by the standard deviation of all features to achieve equalization.
- *RelAtive SpecTrAl (RASTA) filtering* adopts a bandpass filter along the temporal trajectory of each feature in the log-spectral or cepstral domain, which suppresses modulation frequencies outside of typical speech signals.
- *Feature Warping* aims at adjusting the short-term feature distribution to match a reference distribution such as Gaussian by warping the cumulative distribution function of the features.

- *Feature Mapping* transforms the features obtained from different channel conditions into a channel-independent feature space to reduce channel variability using a set of channel-dependent GMMs adapted from a channel-independent root model.

Back-End

The back-end classifies the features using the trained classifiers based on speaker models, then outputs a verification result (accept or reject) using decision logic. The classifier training mainly consists of two processes: *speaker modeling* and *model normalization.*

Speaker modeling intending to train a speaker model is of the following types:

- *Gaussian Mixture Model (GMM)* (Reynolds & Rose, 1995) used as a two-class classifier (genuine or spoofed) in speaker verification systems (and countermeasures) is a weighted combination of multivariate Gaussian distributions.
- *GMM with Universal Background Model (GMM-UBM)* (Reynolds et al., 2000) adopts GMM for likelihood functions, a universal background model (UBM) to represent alternative speakers, and a kind of Bayesian adaptation such as maximum a posteriori (MAP) to generate speaker-specific models from the UBM.
- *Support Vector Machine (SVM)* is a binary classifier which searches a separating hyperplane to model the decision boundary between two classes, one of which contains the target speaker training vectors (labeled as +1) and the other of which contains the training vectors from a spoofing attacker (labeled as -1). The combination of GMM and support vector machine (SVM) classifiers generates models with high generalization performance (Campbell et al., 2006).
- *Deep Neural Networks (DNN)*-based classifiers including Convolutional Neural Network (CNN) and Recurrent Neural Network (RNN) have been used in speaker recognition/verification (Lei et al., 2014). The DNNs can combine the feature extraction and speaker modeling and make joint optimization as a whole. Besides, DNNs are also used to fuse subsystems adopting other speaker models.

Model normalization aiming to model the inter-speaker/intra-speaker variability and to compensate for the inter-session variability mainly consists of:

- *Nuisance Attribute Projection (NAP)* (Burget et al., 2007) intends to remove nuisance dimensions from the SVM expansion space by seeking an appropriate projection matrix carrying out an orthogonal projection in the channel's complementary space, i.e., the speaker space.
- *Joint Factor Analysis (JFA)* (Kenny et al., 2008) seeks to model speaker-and channel-dependent (or session dependent) variability by Gaussian distribution on supervectors, each of which can be decomposed into a sum of a speaker supervector and a channel supervector.
- *I-vector* (Dehak et al., 2011) defines a low-dimensional speaker-and channel-dependent space, the total variability space, which models both speaker and channel variabilities using a simple factor analysis. In total variability space, a given speech utterance is represented by total factors or i-vectors and the channel compensation is carried on in low-dimensional space.

- *Within Class Covariance Normalization (WCCN)* (Hatch et al., 2006) is applied in SVM modeling to minimize the expected error rate of false acceptances and false rejections during the SVM training step by minimizing a set of upper bounds defined on the classification error metric.
- *Linear Discriminant Analysis (LDA)* (Vogt & Kajarekar, 2008) is used for dimensionality reduction by seeking new orthogonal axes to better distinguish one class from the others.
- *Probabilistic Linear Discriminant Analysis (PLDA)* (Li et al., 2012) is a probabilistic extension of linear discriminant analysis (LDA) and usually based on i-vectors when used in speaker recognition context.

ASV Security

Direct Attacks

Direct attacks (attack point 1 in Figure 1, Ratha et al., 2001), called spoofing attacks, can be directly applied to microphone sensor. The types of direct attacks consist of impersonation and replay attacks.

Indirect Attacks

Indirect attacks (attacking points 2-9 in Figure 1, Ratha et al., 2001) explore the vulnerabilities of the ASV system. The attacks need system level privilege and manipulate the ASV components including feature extraction (attacking point 3), speaker models database (attacking point 5), classifier (attacking point 7), and decision logic (attacking point 9). Besides, the indirect attacks also can explore the vulnerabilities in communication or transmission channels (attacking points 2, 4, 6, 8). Note that the voice conversion and speech synthesis attacks are generally performed in communication channel (attacking point 2). But they can be considered directed attacks when a converted voice or a synthesized speech is replayed in front of a microphone.

Figure 1. Components of ASV system with 9 attack points

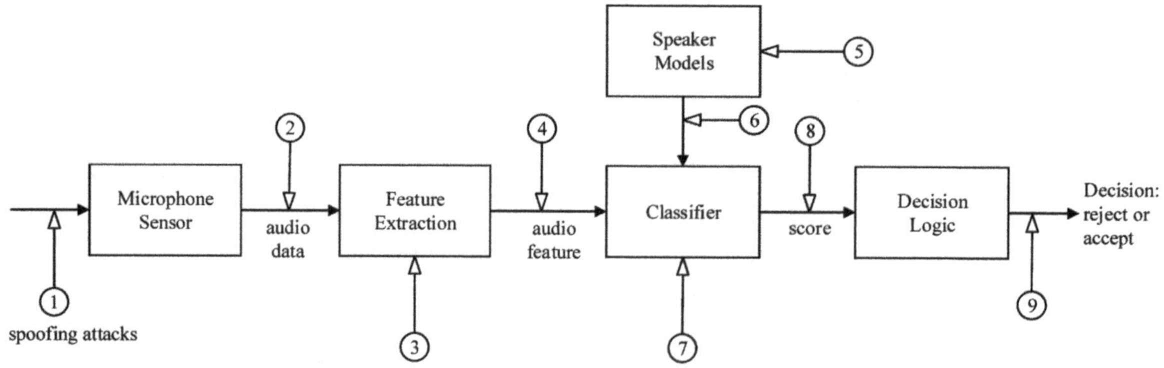

Presentation/Spoofing Attack

Presentation Attack (PA) Definition (ISO Standards): Presentation of an artefact or human characteristic to the biometric capture subsystem in a fashion that could interfere with the intended policy of the biometric system.

Spoofing Attack presents a counterfeit biometric evidence of a target user to outwit the biometric system. In the context of this chapter, Spoofing Attack implies an attacker presents a recorded, converted, or synthesized speech sample at microphone level (attack point 1 in Figure 1) or the communication/transmission level immediately after microphone (attacking point 2 in Figure 1) in order to outwit the speaker verification system to accept the attacker/impostor as a claimed identity. The attacks and defenses at attacking points 3-9 are beyond the scope of this chapter. In this chapter, Presentation Attack and Spoofing Attack are interchangeable when referred to.

Countermeasures/Anti-Spoofing

Countermeasure, often separated from stand-alone ASV system, is anti-spoofing mechanism to defend speaker verification system from spoofing attacks. When evaluating the performance of a countermeasure method, there are three types of speakers/attackers to be considered: genuine speaker, zero-effort impostor, and spoofed attacker. The first two types including genuine speaker and zero-effort impostor are considered when evaluating a stand-alone ASV system. The spoofed attacker can use recorded, converted, or synthesized speech sample to outwit the ASV system and the countermeasure added to the ASV.

Metrics, Protocols, and Datasets

Metrics

In the context of the ASV, there are two types of input: genuine and zero-effort impostor speech while for CM, there is another type: spoofed speech. Both the ASV and CM should accept the genuine and reject the (zero-effort or spoof) impostor, resulting in one of four outputs illustrated in Table 1. The metrics often used to evaluate the performance of ASV and/or countermeasure system are as follows (This chapter presents results only in terms of FAR, FRR, and EER).

- **False Acceptance Rate (FAR):** Percentage of impostor trials (incorrectly) accepted by the system.
- **False Rejection Rate (FRR):** Percentage of genuine trials (incorrectly) rejected by the system.
- **False Positive Rate (FPR):** equivalent to FAR.
- **False Negative Rate (FNR):** equivalent to FRR.
- **True Positive Rate (TPR):** Percentage of genuine trials (correctly) accepted by the system.
- **True Negative Rate (TNR):** Percentage of impostor trials (correctly) rejected by the system.
- **Equal Error Rate (EER):** The EER corresponds to the threshold θ_{EER} at which FAR and FRR are equal or approximately equal. The threshold θ_{EER} is determined in development phase using a reference database $\mathcal{D}_{\text{base}}$ by

$$\theta_{EER} = \arg\min_{\theta} \left| \mathrm{FRR}\left(\theta, \mathcal{D}_{\mathrm{base}}\right) - \mathrm{FAR}\left(\theta, \mathcal{D}_{\mathrm{base}}\right) \right| \tag{1}$$

- **Half Total Error Rate (HTER)** (Chingovska et al., 2014): Measuring both the verification performance and the spoof ability of the system. The HTER is computed using the evaluation database $\mathcal{D}_{eval}$ by

$$\mathrm{HTER}\left(\theta_{EER}, \mathcal{D}_{eval}\right) = \frac{\mathrm{FAR}\left(\theta_{EER}, \mathcal{D}_{eval}\right) + \mathrm{FRR}\left(\theta_{EER}, \mathcal{D}_{eval}\right)}{2} \tag{2}$$

- **Spoof False Acceptance Rate (SFAR):** Percentage of spoof trials (incorrectly) accepted by the system. SFAR is used to measure the vulnerability of the (CM) system.
- **Detection Error Trade-Off (DET) Curve:** The DET curve is a plot of FRR vs. FAR, and the trade-off between high security (high FRR and low FAR) and high convenience (high FAR and low FRR).

Datasets

Datasets for ASV

The TIMIT corpus (Garofolo et al., 1993) is designed for the development and evaluation of automatic speech recognition systems. There are 630 speakers in TIMIT, each reading ten phonetically rich sentences. Besides time-aligned orthographic, phonetic and word transcriptions, TIMIT contains a 16-bit, 16 kHz speech waveform file for each utterance.

The RSR2015 (Larcher et al., 2014) is designed for the development, training, and testing of automatic text-dependent speaker verification systems. It includes utterances of different lengths, short pass-phrases as well as random digit sequences in order to support different types of protocols. There are 300 speakers in RSS2015 and a total of 657 utterances in 9 sessions per speaker.

Datasets for Countermeasure

SAS (Wu et al., 2015b) used in ASVspoof 2015, a common database of both genuine speech and spoofed speech, is collected from 106 speakers (45 male and 61 female) and generated using ten mainstream speech-synthesis and voice-conversion spoofing-attack algorithms. The statistics of the corpus is shown in Table 2.

Table 1. Trial decisions of ASV and CM

	Decision	
	Acceptance	**Rejection**
Genuine trial	Correct acceptance	False rejection
Imposter (zero-effort) trial	False acceptance (ASV)	Correct rejection (ASV)
Spoof (voice conversion, speech synthesis and replay) trial	False acceptance (CM)	Correct rejection (CM)

Table 2. Statistics of the ASVspoof 2015 corpus

Subset	Number of Speakers		Number of Utterances	
	Male	*Female*	*Genuine*	*Spoof*
Training	10	15	3750	12625
Development	15	20	3497	49875
Evaluation	20	26	9404	184000
Total	45	61	16651	246500

Source: (Wu et al., 2016)

The datasets of ASVspoof 2017 challenge, the primary technical goal of which is to evaluate spoofing (replay) attack detection accuracy of the countermeasures (Kinnunen et al., 2017a). The datasets consisting of genuine (Lee et al., 2015) and replay/spoof recordings (Kinnunen et al., 2017b), with the former coming from recent text-dependent RedDots corpus and the latter from the replay version of the former. The datasets are separated into three subsets comprising Training, Development and Evaluation set, details of which are listed in Table 3. In ASVspoof 2017 challenge, the Training and Development subsets are provided for the design of replay countermeasures and the Evaluation subset is used to test the accuracy and generalization capacity of submitted replay detectors.

Protocols

The protocols present the separation of the corpus and the usage and intention of each partitioned dataset.

For an unbiased protocol, all the speakers will be divided into three non-overlapping subsets named training, development, evaluation. The training set can be used to train ASV (background models such as UBM) or countermeasure. The development set can be used to determine the decision threshold and possible calibration parameters. The evaluation set is used to validate the performance.

SPOOFING ATTACKS

This section presents 4 types of spoofing attacks: impersonation whereby an attacker attempts to mimic a target legitimate user's voice; voice conversion whereby an attacker resembles the speech of a target

Table 3. Statistics of the ASVspoof 2017 corpus

Subset	Number of Speakers	Number of Utterances	
		Genuine	*Spoof*
Training	10	1507	1507
Development	8	760	950
Evaluation	24	1294	11988
Total	42	3561	14445

Source: (Kinnunen et al., 2017a)

legitimate user using another user's speech; speech synthesis whereby an attacker using text-to-speech (TTS) technique generates intelligible, natural-sounding artificial speech with inputs of arbitrary text; and replay whereby an attacker tries to pass the authentication of ASV by providing a pre-recorded speech sample collected from genuine target legitimate user (Wu et al., 2015a).

Impersonation

During impersonation attack, an impostor adapts his/her voice to mimic a target speaker's voice timbre and prosody to spoof an ASV system. Previous researches were often constructed in small datasets consisting of a small number of speakers and the results about the threats of impersonation attack are somewhat inconsistent. Some researches show the FAR sharply increases after adding impersonation attack to the standard ASV while others show no obvious increment. Al thought having no convincing evidence about the effectiveness of impersonation, there are some findings about the characteristics of impersonation. Some impostors have high similarities in voice timbre and prosody with other speakers and some speakers are prone to be impersonated by impostors. The impersonation attack can be launched by elaborately selecting an impostor who has voice similar with that of the target speaker.

Voice Conversion

Voice conversion manipulates the speech of a given speaker in order that it resembles that of another target speaker. A typical voice conversion two processes spectral mapping which relates to voice timbre, and prosody conversion which relates to prosodic features, such as fundamental frequency (F0) and duration.

There are three types of methods including unit-selection, statistical parametric, and frequency warping to implement spectral mapping (Wu et al., 2015a). The unit-selection voice conversion method generates converted speech by using selected speech frames of the target speaker. The statistical parametric approaches such as GMM-based and neural network-based, map the spectral features of an input speech towards features representative of the target speaker using linear or nonlinear conversion functions. The frequency warping methods warp the frequency distribution of a source spectrum to match that of another target. Figure 2 shows the vulnerability of an ASV system using GMM-UBM speaker model and ASVspoof 2015 (partial) training set in case of voice conversion attacks, which displays that the score distributions for genuine and converted speech have considerable overlap, and means that the FAR will definitely increase.

Speech Synthesis

Speech synthesis, also referred to as text-to-speech (TTS), can generate human-like speech from any arbitrary text. A typical speech synthesis system mainly consists of two phases: text analysis process which converts text into linguistic specification such as phonemes, and speech waveform generation process which generates speech waveform using linguistic specification (Wu et al., 2015a).

Currently, there are two major speech waveform generation methods: unit selection-based speech synthesis and statistical parametric speech synthesis. The unit selection-based speech synthesis selects speech diphones from a large database of speech according to their prosodic/phonetic match with the target description of the target and concatenates the best sequence of diphones to produce the target utterance (Hunt & Black, 1996). The statistical parametric speech synthesis such as Hidden Markov

Figure 2. Score distribution of genuine and converted speech for a GMM-UBM system

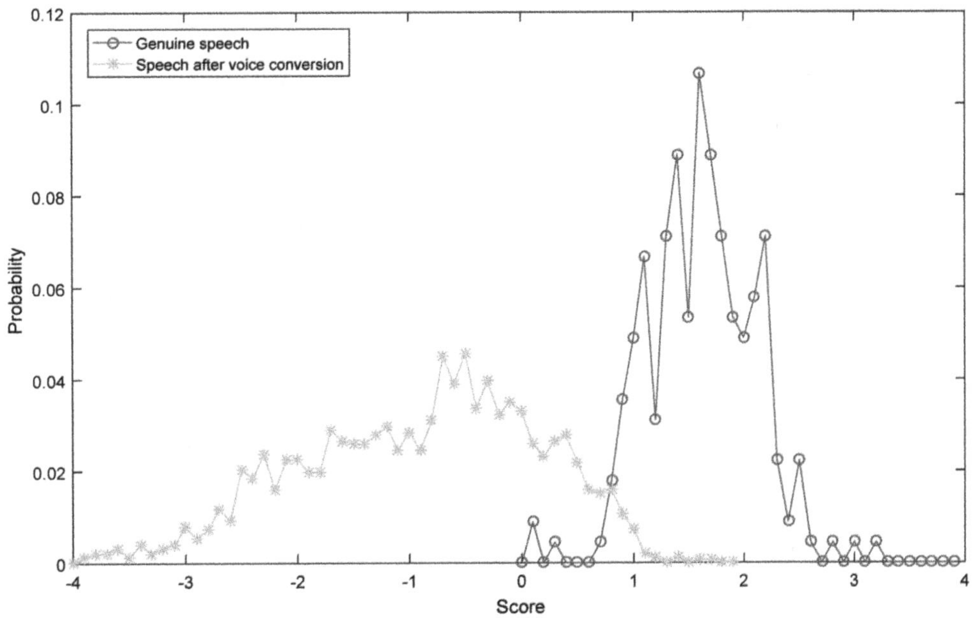

Model (HMM)-based can learn speech models from a small amount of speaker-specific data using speaker adaptation techniques such as maximum-likelihood linear regression (MLLR) (Zen et al., 2009).

Figure 3 shows the vulnerability of an ASV system using GMM-UBM speaker model and ASVspoof 2015 (partial) training set in case of speech synthesis attacks, which displays that the score distributions for genuine and synthesized speech have considerable overlap, and means that the FAR will definitely increase.

Replay

In the context of replay attacks, an adversary tries to spoof an ASV system using a pre-recorded speech sample collected from a genuine target speaker. Replay attack can be considered as the most accessible spoofing method due to the high penetration rate of high quality and low-cost recording devices, such as smartphones. Figure 4 shows a comparison of a genuine speech and its corresponding replayed version. It is obvious that the spectrogram of the replay speech is almost indistinguishable from the original one. There is a high probability that the spectral features extracted from a replayed spectrogram will match the feature distribution of the target speaker.

Figure 3. Score distribution of genuine and synthesized speech for a GMM-UBM system

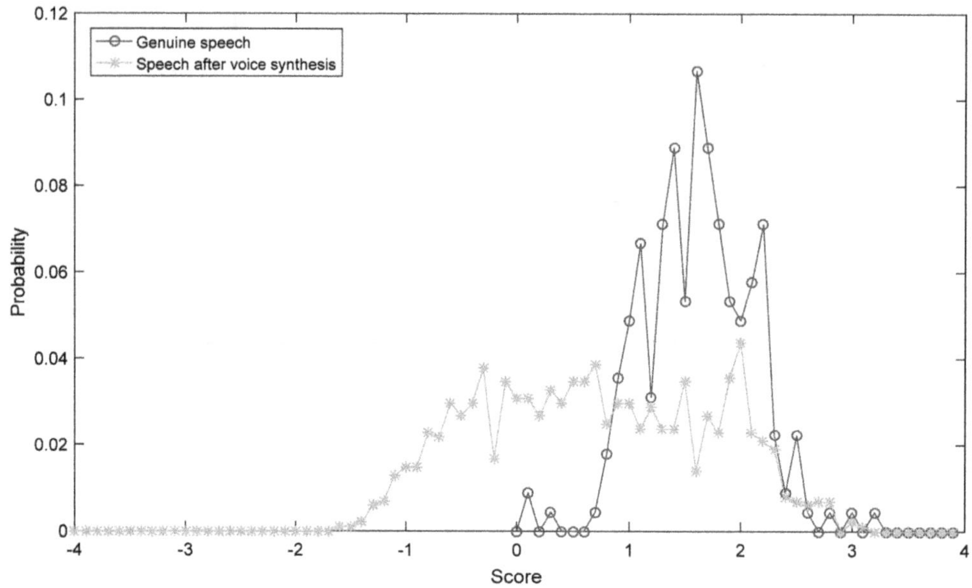

Figure 4. Comparison of a genuine speech and its corresponding replayed speech

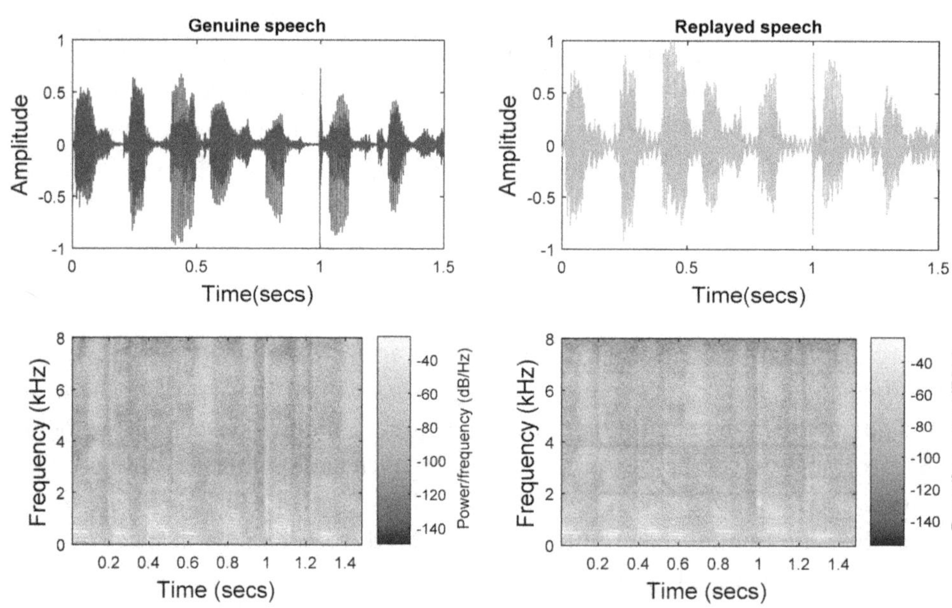

LIVENESS DETECTION SOLUTIONS

Challenge-Response Based Methods

Aley-Raz et al. (2013) proposed a speaker recognition system with liveness detection scheme. The system comprises three phases: text-dependent enrollment, text-dependent verification, and text-independent liveness detection.

In the enrollment phase, the user provides a specific passphrase (text-dependent audio) as the prompt of the system. The voiceprint generator generates a text-dependent voiceprint stored in a voiceprint database.

In the verification phase, the speaker repeats the passphrase (verification audio) registered in the enrollment phase as the prompt. The text-dependent (TD) recognizer generates a TD verification score by comparing the verification audio representative and a voiceprint from the voiceprint database.

In the text-independent liveness detection phase, the speaker is prompted to provide a sentence selected randomly from a closed set of sentences. The utterance validation module checks whether or not the speaker indeed provided the requested sentence, and generate an utterance validation score. The Text-independent (TI) voiceprint generator generates TI voiceprint for the TD verification audio provided in verification phase. The TI recognizer compares the TI voiceprint of the verification audio with that of the random sentence and generates a TI verification score (Score 1) which denotes whether the two audios compared come from the same speaker. For high accuracy, the Text-independent (TI) voiceprint generator also generates TI voiceprint for the random sentence provided in the initial liveness detection phase. The TI recognizer compares the TI voiceprint of the random sentence with that of the verification audio and generates a TI verification score (Score 2).

Taking as inputs the utterance validation score (Score 1 and Score 2), the liveness detection score and decision generator generates a liveness detection score. Finally, the session decision generator combines the liveness detection score and the TD verification score and generates a decision, which denotes the authenticity and liveness of the speaker.

Acoustic Feature Based Methods

The acoustic feature based methods focus on searching for discriminative features rather than the design of complex classifiers.

Sahidullah et al. (2015) evaluated the performance of different features for synthetic speech detection task using GMM and SVM based classifiers on the ASVspoof 2015 corpus. The tested features are categorized into three classes: short-term power spectrum features, short-term phase features, and spectral features containing long-term characteristics. The selected feature vectors may contain static coefficients and/or dynamic coefficients, i.e. deltas (Δ) and double-deltas ($\Delta\Delta$). For known attacks, the best performance (lowest EER 0.07%) is achieved by countermeasure with SCFC feature and GMM-ML classifier. For unknown attacks, the countermeasure with LFCC and GMM-ML classifier gets the best result.

Font et al. (2017) performed evaluations on a set of features using GMM-ML classifier for replay attack detection task. The experiments are performed on ASVspoof 2017 corpus. The classifier is trained on the training set and evaluated on the development and evaluation set, respectively. The selected optimal parameters set includes the number of static coefficients and corresponding delta (Δ) and double delta ($\Delta\Delta$) coefficients. The best performance on development set is achieved by Inverted Mel Frequency Cepstral Coefficients (IMFCCs) while the best performance on evaluation set is achieved by Subband Spectral Centroid Magnitude Coefficients (SCMCs).

Hardware-Based Methods

Magnetic Force Based Liveness Detection

Chen et al. (2017) proposed a robust smartphone based voice impersonation defense system that captures the magnetic field emitting from the loudspeaker and extracts special physical characteristics of the magnetic field to discriminate between a live human and a loudspeaker (impersonation attacker). The voice impersonation attacks considered by the authors include machine-based voice impersonation (an attacker replays a voice recording, a synthesized speech or a converted voice using a loudspeaker) and human-based voice impersonation (an attacker given the acquired sample mimics the target speaker's voice without the help of hardware devices). The experimental results show that the proposed system achieves very high accuracy with 0% EER in defending against voice impersonation.

While capable of defeating the vast majority of voice impersonation attacks, this method, however, requires the users to rotate the smartphone around their heads with predefined trajectory while speaking the given passphrase.

Phoneme Localization Based Liveness Detection

Zhang et al. (2016) proposed a liveness detection scheme (*VoiceLive*) that uses phoneme sound localization, i.e., the time-difference-of-arrival (TDoA) changes to the two microphones equipped in a smartphone, to determine whether a passphrase is spoken by a live person or replayed by loudspeaker.

The core idea of the liveness detection system relies on performing TDoA ranging for a sequence of phoneme sounds at the two microphones of the phone. A user speaks an utterance to the phone closely placed to the user's mouth; the phoneme sounds of the utterance are sequentially received on the two microphones of the phone, respectively. The phone deduces TDoA of each phoneme received by the two microphones. A sequence of TDoA differences (TDoA dynamic) to the two microphones will be produced and then compared with the one extracted during enrollment. If the similarity score is greater than the pre-defined threshold, a live user is declared. Otherwise, a playback attack is detected. Experiments with 12 participants and three different types of phones show that the system achieves over 99% detection accuracy at around 1% EER.

While achieving high accuracy, the system needs the support of advanced smartphone audio hardware, i.e. equipped with two microphones for stereo recording (one on the top and the other at the bottom) and capable of recording at standard 48 kHz and 192 kHz sampling rates, which may be not supported in (embedded) medical devices.

Articulatory Gesture Based Liveness Detection

Zhang et al. (2017) proposed *VoiceGesture* that leverages the response of human speech production system to external stimuli to implement liveness detection. The scheme is smartphone based and achieves high accuracy in detecting live users while not requiring users' additional cumbersome operations. The proposal leverages the (discriminable) characteristics of articulatory motions when speaking/replaying a passphrase to distinguish a human speaker pronouncing phonemes relying on multiple articulators' coordinated multidimensional movements from a loudspeaker producing phonemes relying solely on one dimension movements of the diaphragm (i.e., forward and backward).

During the voice authentication, the system uses the audio subsystem in a smartphone as a Doppler radar where a high-frequency acoustic tone at 20 kHz (inaudible to the human ear) is transmitted from the built-in speaker and the reflections are received by the microphone. The Doppler frequency shifts at around 20 kHz generated by the movements of a user's articulators when speaking a passphrase/utterance and the user's voice sample are simultaneously recorded. Due to at different frequency bands, the user voice and the Doppler frequency shifts are easily separated. The user's voice is used for voice authentication while the Doppler frequency shifts for liveness detection.

Evaluated by extensive experiments with 21 participants and three different types of phones, the scheme achieves around 1% EER with the users holding the phones in their habitual ways, i.e. to user's ear or in front of the mouth.

Multi-Modal Biometric Based Methods

Multi-Factor Authentication (MFA)

These systems use two (2FA) or more uncorrelated biometric traits captured through different acquiring sensors. Czyzewski et al. (2017) introduced a new multimodal corpus of English speech recordings. The corpus comprising 31 hours of recordings was created to support the development of audio-visual speech recognition systems (AVSR). The authors also developed an audio-visual automatic speech recognition (ASR) engine and performed experiments to employ the multimodal corpus. The ASR system used MFCC features in the acoustic layer and Active Appearance Models (AAM) in visual layer (lip detection). The experiments showed that by addition of visual features the accuracy of the proposed system improves by 5% in Word Error Rate (WER). Although created for speech recognition, the corpus was labeled for each speaker at word speaker, it can be used for speaker verification.

Multiple Features Fusion Methods

Score Fusion

For each feature, an optimal set of parameters is achieved by training each corresponding classifier. The scores of different classifiers are then fused using some strategy in the final classifier. The score fusion can mitigate the overfitting problem when training the classifiers using different features, and results in a robust spoofing detection system with low EER.

Serial/Cascaded Fusion

Multiple features are processed one by one. The output of first feature verification is the input of the second processing. For voice authentication, the first process is ASV which verifies the voice acquired by a sensor (such as a microphone) comes from a legitimate speaker (the voice may be genuine or replayed), and the second process is Countermeasure (CM) which achieves liveness detection distinguishing a live person from a replayed loudspeaker. Experiments show that multiple features cascaded fusion can greatly improve spoofing detection performance while keeping ASV performance unchanged.

De Leon et al. (2012) and Alegre et al. (2013) proposed countermeasures to protect ASV against speech synthesis attacks and speech conversion attacks, respectively. The cascaded fusion approach is illustrated in Figure 5. If an utterance passes the verification of ASV ($Score \geq \theta_{ASV}$), it would continue to be verified by CM. If passing the verification of CM ($Score \geq \theta_{CM}$), a live target (legitimate) speaker is declared.

Parallel Fusion

In parallel fusion approach, multiple features are processed simultaneously and the outputs of the multiple processed are determined in same one decision maker to obtain the final result: reject or accept. The traditional parallel fusion is achieved by merging the outputs of ASV and CM in score level, which is illustrated in Figure 6. The fusion of scores can be performed by computing a weighted sum. The weights can be determined using logistic regression, which is a powerful score fusion technique used to combine heterogeneous speaker classifiers.

Figure 5. Illustration of cascaded fusion of ASV and CM

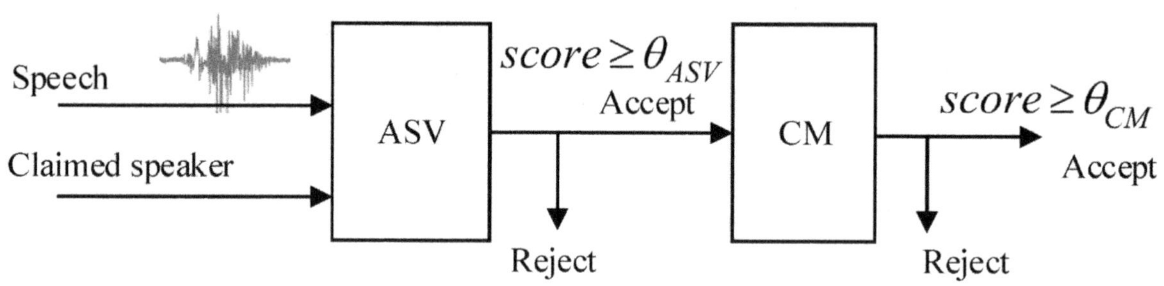

Figure 6. Illustration of parallel fusion of ASV and CM

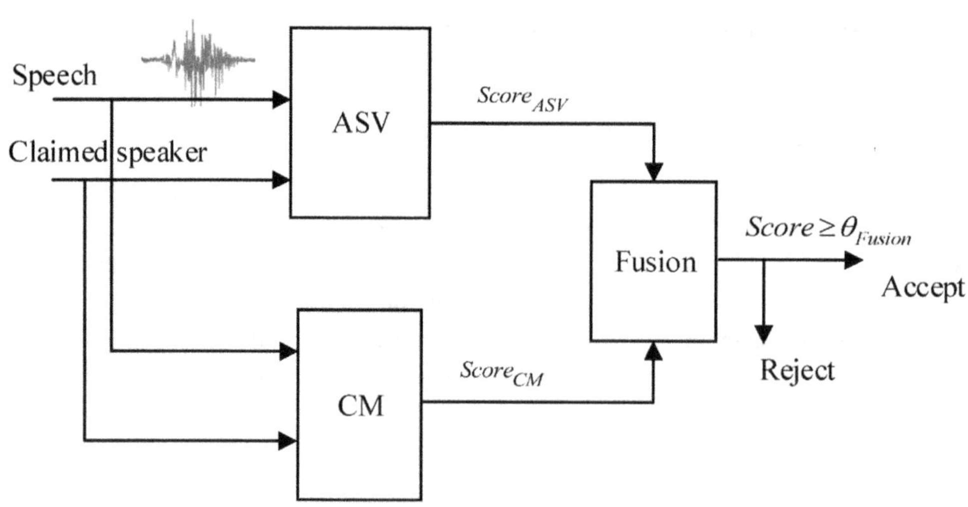

Fusion-Based Countermeasures of Speech-Synthesis and Voice-Conversion Spoofing Attacks

While the fusion can be performed between ASV and CM, it also can be performed among CMs. This section analyses the design and performance of CMs submitted to ASVspoof 2015, which focus exclusively on speech synthesis and voice conversion attacks detection (Wu et al., 2016).

Spoofing Attacks

The ASVspoof 2015 corpus was generated by 10 state-of-the-art spoofing-attack algorithms (S1-S10) from the speech-synthesis and voice-conversion communities (Wu et al., 2016). Five types of voice conversion algorithms are adopted to generate conversion corpus: (1) *Frame-selection-based* schemes generate converted speech by selecting target speech frames; (2) *Slope-shifting-based* schemes shift the slope of the source spectrum towards that of the target by adjusts the first Mel-cepstral coefficient; (3) *GMM-based* schemes adopt joint density Gaussian mixture models (GMMs) and maximum likelihood parameter generation to convert voice; (4) *Tensor-based* schemes generate converted speech based on tensor representation of speaker space; (5) *KPLS-based* schemes implement a non-linear transformation by adopting kernel-based partial least square (KPLS) without using dynamic information. Two types of speech synthesis algorithms are chosen to generate synthesis corpus: (1) *HMM-based* schemes learn speech models from a small amount of speaker-specific data using speaker adaptation techniques such as maximum-likelihood linear regression (MLLR); (2) *Unit selection-based* schemes select speech diphones according to their prosodic/phonetic match with the target and concatenates the best sequence of diphones to produce the target utterance.

Countermeasures Profiles

Most of the CMs adopted multiple types of features such as combination of MFCCs, Mel-frequency principal coefficients (MFPCs), Log magnitude spectrum (LMS), Phase-based features including group delay (GD) and baseband phase difference (BPD), Relative phase information (RPI), Relative phase shift (RPS), etc. Most of the countermeasures used score fusion, that is, the scores of multiple classifiers are fused to obtain the final score while some used *i*-vector concatenation in which multiple *i*-vectors were concatenated into one vector. In summary, all the countermeasures focused on developing new features rather than on new classifiers.

Datasets, Protocols, and Metrics

The datasets comprise three distinct subsets: training set, development set, and evaluation set. The training set includes genuine and spoofed utterances generated from S1-S5 called known attacks. The development set includes both genuine utterances from another subset of speakers and spoofed utterances also generated from known attacks. The evaluation set also includes genuine utterances and spoofed utterances from both *known* (S1-S5) and *unknown* attacks (S6-S10).

The EER is used to evaluate the relative performance of the countermeasures. The evaluation considered two types of errors: *false alarm* when a spoofed sample is incorrectly classified as a genuine one, and *miss* when a genuine sample is incorrectly classified as a spoofed one. Then two error rates including false alarm and miss rates are computed, respectively. The EER can be computed when false alarm rate equals or approximately equals to miss rate. The primary metric is the "threshold-free" EER computed using false alarm and miss rates as follows.

$$P_{\text{fa}}\left(\theta\right) = \frac{\#\{Score_{\text{spoof}} > \theta\}}{\#\left\{Score_{\text{spoof}}\right\}} \tag{3}$$

$$P_{\text{miss}}\left(\theta\right) = \frac{\#\left\{Score_{\text{genuine}} \leq \theta\right\}}{\#\left\{Score_{\text{genuine}}\right\}} \tag{4}$$

$$\theta_{EER} = \arg\min_{\theta}\left|P_{\text{fa}}\left(\theta\right) - P_{\text{miss}}\left(\theta\right)\right| \tag{5}$$

The EER computed by $\text{EER} = P_{\text{fa}}\left(\theta_{EER}\right) = P_{\text{miss}}\left(\theta_{EER}\right)$ on evaluation set will be used to rank the countermeasures.

Countermeasures Performance

The performance results of the CMs show that the EERs for unknown attacks are significantly higher than known attacks (Wu et al., 2016). Some system obtained the best performance for unknown attacks but the EER is five times the result for known attacks. The results also show the over-fitting problem of the countermeasures to known attacks. For example, some system obtained lowest EER for known attacks but for unknown attacks it achieved EER more than 2 times of the best performance achieved by other system. The best EER for unknown attacks (S10) was as high as 8.490%, which confirms the importance of developing generalized countermeasures.

Fusion-Based Countermeasures of Replay Attacks

This section analyses the design and performance of countermeasures submitted to ASVspoof 2017, which focused exclusively on replay attacks detection (Kinnunen et al. 2017).

Countermeasures Profiles

Similar with ASVspoof 2015, most of the countermeasures adopted multiple types of features (Kinnunen et al. 2017) and used score fusion. They also focused on developing new features rather than on new classifiers.

Datasets, Protocols, and Metrics

The datasets also contains training set, development set, and evaluation set. The training set is used to train CMs. The development set is used to design and optimize the CMs. The evaluation set, including new speakers, environments, replay-recording conditions substantially different from those in training set and development set, is used to evaluate both the performance and generalization ability of the designed countermeasures. The evaluation concentrates on stand-alone CM without an ASV. For each trial, the CM should compute a score, such as log-likelihood ratio, the high value of which indicates genuine speech whereas the low value indicates a replayed impostor sample. The primary metric is the "threshold-free" EER computed using false alarm and miss rates by the equations (3)-(5).

Countermeasures Performance

The best performance system achieves the EER of 6.73%. Comparing the results with that of ASVspoof 2015 the best performance of which is 1.211 (all average), shows that the detection of replay attacks has more difficulties than the detection of speech synthesis and voice conversion spoofing attacks. Finding a countermeasure with good generalization ability also remains an open problem.

Comparison of Serial and Parallel Fusion

Sahidullah, et al. 2016 presented a study of fused countermeasures evaluated on ASVspoof 2015 corpus and also evaluated different ASV and countermeasure (CM) integration strategies.

The objectives of ASV and CM systems (and their integration) are shown in Table 4 where +1 denotes genuine and -1 denotes attacks for the corresponding system. There is a different decision on zero-effort impostor for ASV and CM. While a zero-effort impostor should be rejected as a speaker impersonation by the ASV system, it should be accepted as a genuine trial by the CM.

There are two fusion modes: cascaded fusion (Figure 5) and parallel fusion (Figure 6). Sahidullah, et al. 2016 evaluated the results of fused ASVs, fused CMs, fused-fused ASVs and CMs (integration of fused ASVs and fused CMs). This chapter only present the integration of stand-alone ASV and CM (ASV1+CM5) to demonstrate the performance comparison of the cascaded and parallel fusion. More details about fused-fused integration are found in Sahidullah, et al. 2016. ASV1 uses the MFCC features and a Gaussian mixture model – universal background model (GMM-UBM) model. The speaker models are trained using TIMIT and RSR2015 corpus. CM5 uses the Constant Q cepstral coefficients (CQCCs) and a two class GMM classifier and is evaluated using ASVspoof 2015 dataset.

The ASV performance is evaluated in FAR and FRR while the performance of CM is evaluated in EER where FAR (Z) denotes FAR for zero-effort impostors, FAR (K) for known spoofing impostors, FAR(U) for unknown spoofing impostors, EER (K) for (CM) known spoofing impostors, and EER (U) for (CM) unknown spoofing impostors. For the standalone CMs, the performance for unknown spoofing attacks is poorer than for known attacks. For the ASV, both the FARs of known and unknown attacks are significantly higher than the FAR of zero-effort attacks, which shows that the standalone ASV is vulnerable to speech synthesis and voice conversion attacks. The authors show that the performance of

Table 4. Definition of positive and negative trials for ASV and CM tasks

Task	Genuine	Zero-Effort Impostor	Spoofing Impostor
Stand-alone ASV	+1	-1	-1
Stand-alone CM	+1	+1	-1
ASV with CM	+1	-1	-1

Source: (Sahidullah, et al. 2016)

the two fusions (cascaded and parallel) of ASVs and CMs is significantly better than that of standalone ASV in case of spoofing attacks. For the cascaded fusion system, the FRR and FAR (Z) are almost the same as for the standalone ASV1 while FAR (K) and FAR (U) are sharply reduced. The FRR and FAR (Z) for the cascaded fusion system is considerably lower than that of the parallel fusion system while FAR (K) and FAR (U) is higher for cascaded fusion system.

VOICEPRINT-BASED LIVENESS DETECTION FOR INSULIN PUMP SYSTEM

This section presents the possible application of voiceprint-based liveness detection schemes in one kind of medical system, i.e., insulin pump system which is used for treatment of diabetes.

System Overview of a Real Time Insulin Pump System

Figure 7 demonstrates the components a Medtronic Paradigm real-time insulin pump system (Hei et al. 2013). The remote control operated by the user sends instructions, e.g., suspend or resume basal rate, to the insulin pump via wireless link 1. The OneTouch meter acquires BG readings from an implanted sensor via wireless 3 and transmits the BG information to the insulin pump via wireless link 2. The transmitter connected with sensor transmits historical BG information to the Carelink USB via wireless link 4 and sends current BG levels to the pump via link 6. The Carelink USB gathers reports on BG trends and patterns from the insulin pump via wireless 5 and uploads data (including data from link 4) to a web-based management system utilizing a laptop or PC.

Figure 7. A real time insulin pump system
Source: Hei et al. (2013)

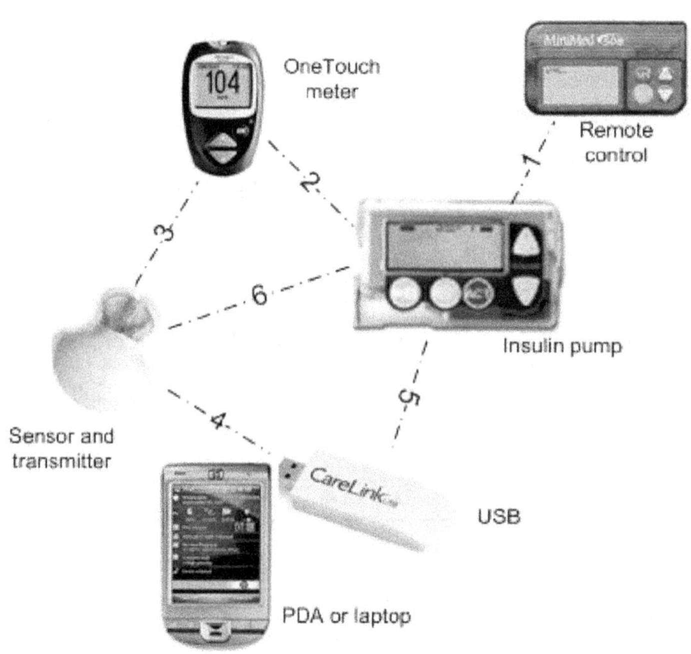

Attack and Authentication for Insulin Pump System

The wireless communication in the system is not secured by encryption. Li et al. (2011) implemented different types of attacks on an insulin pump using reverse-engineering technology, and then proposed a cryptographic solution (rolling code) and body-coupled communication to protect the wireless links. Marin et al. (2016) extended their attacks by fully reverse-engineering the wireless communication protocol in the insulin pump system, and proposed an AES-MAC based cryptographic solution based on sharing of two symmetric keys. Inchingolo et al. (2001) proposed authentication schemes using additional external devices, which may be forgotten, lost or stolen, and could potentially disclose a patient's status. Hei, X. & Du, X. (2011) and Hei et al. (2010) proposed various access control schemes for wireless medical devices. Based on general well-studied radio signal channels, these works can be attacked by remote attackers having sound knowledge of the radio propagation patterns. The chapter presents solutions that utilize an acoustic channel as a source of proximity declaration to establish secure communication between unacquainted devices in proximity.

Anti-Replay Speaker Verification

This chapter investigates acoustic-channel-based access control approach to defend against passive and active attacks through link 3 (between the insulin pump and Carelink USB), which can be easily launched by remote adversaries who can passively eavesdrop the data from an insulin pump and remotely change settings (such as dosage level) on the pump, which brings potential threats to patient's privacy and safety. The proposed scheme does not need permanent key sharing or external devices except needing two audio sensors to be embedded in the two devices, respectively. When the Carelink USB wants to request data or modify the dosage setting of the insulin pump, the pump needs a legitimate user to grant her or his permission to the Carelink USB by speaking random passphrases. After successfully passing the speaker verification, the two devices can start to construct a secure channel, which will be presented in next section.

In embedded systems like insulin pump system, several issues and challenges should be solved when implementing the above verification: First, the insulin pump has limited computing power and memory; second, the verification requires the pump to store speaker models, which need enough storage capacity; third, the accuracy of the verification must be high enough to guarantee the safety. For the first issue, the scheme should choose the feature whose extraction needs lower computation. To lower memory occupation, the number of filter bands (feature dimension) should be reduced while keeping appropriately high verification performance (e.g., low EER and high verification accuracy). For the second issue, the scheme needs not to build a large-scale system to verify many speakers. The speaker model is lightweight because each system only has one patient/speaker. The accuracy of the scheme is easier to achieve a higher value than in large-scale systems since concentrating on only one speaker and optimizing the verification for this particular scenario.

In the insulin pump and Carelink USB, an automatic speaker verification (ASV) system like the one shown in Figure 1 is deployed. The standalone ASV system is vulnerable to replay attacks presented in section "SPOOFING ATTACKS". Some dedicated countermeasures (CMs) aiming to detect replay attacks have been proposed, which is described in section "LIVENESS DETECTION SOLUTIONS". Since genuine utterances should be accepted by both ASV and CM and either ASV or anti-replay CM should reject the utterances from impostors, a cascaded combination of ASV and CM can be a solution for the insulin pump to resist the replay attack. An example cascaded fusion framework is shown in Figure 5. If an utterance passes the verification of ASV, it would continue to be verified by CM. If passing CM verification, a live target (legitimate) speaker is declared.

Voiceprint and Proximity-Based Device Paring

After the target speaker passes verification, an attacker still has chance to modify the dosage before delivery. To avoid this threat, device paring or key agreement protocol should be implemented between the two devices. Many device paring schemes have been proposed (Roeschlin et al., 2018 & Fomichev et al., 2018). The scheme deployed in embedded devices should have low computation and communication complexity. This chapter proposes voiceprint and proximity-based device paring scheme ensuring that access to the insulin pump can be granted to the Carelink USB only when the latter is in close proximity to both the insulin pump and the target speaker.

Firstly, each device needs to extract a binary characteristic sequence, i.e., voiceprint, from the sampled audio. An energy-difference-based scheme can be used to extract voiceprint (Haitsma, J. & Kalker, T., 2002 and Schürmann, D. & Sigg, S., 2013). Because the pump and CareLink USB may embed different types of audio sensors, and even the same type of sensors have different hardware characteristics, the voiceprints computed in these two devices are similar but hardly identical, which means they cannot be directly used as a common key. To ensure proximity, the voiceprints extracted in each of the two devices must have high similarity. A fuzzy commitment scheme (Schürmann, D. & Sigg, S., 2013) can be used to verify the similarity between the voiceprints and generate a common key between the two devices. Besides, after checking the voiceprints similarity, Diffie-Hellman key exchange protocol can be used to generate the common key by considering the voiceprint as random numbers (a,b in the common key g^{ab}). An example of the voiceprint and proximity-based device paring scheme proceeds as follows in Figure 8.

FUTURE RESEARCH DIRECTIONS

- **Countermeasures With Good Generalization Ability:** The results of previous countermeasures show that the performance for unknown attacks is significantly lower than known attacks. The results also show the over-fitting problem of the countermeasures to known attacks. This confirms the importance of developing more generalized countermeasures.
- **Countermeasures for Embedded Devices:** Most of the current systems are designed for PC or server and may be not suitable for medical devices which are often embedded and have limited computing power and memory. It is necessary to design acoustic channel verification scheme the feature extraction and classifier model of which are optimized for embedded devices while having high accuracy to guarantee the safety of the users.
- **Replay Attacks Countermeasures:** The evaluations of countermeasures using acoustic feature based methods show that the best performance achieved by the detection of replay attacks is remarkably lower than that of the detection of speech synthesis and voice conversion spoofing attacks. While hardware-based methods achieve higher performance, they need additional hardware, which may bring more complexities of system design and high cost for medical devices.

Figure 8. An illustration of the voiceprint and proximity-based device paring scheme

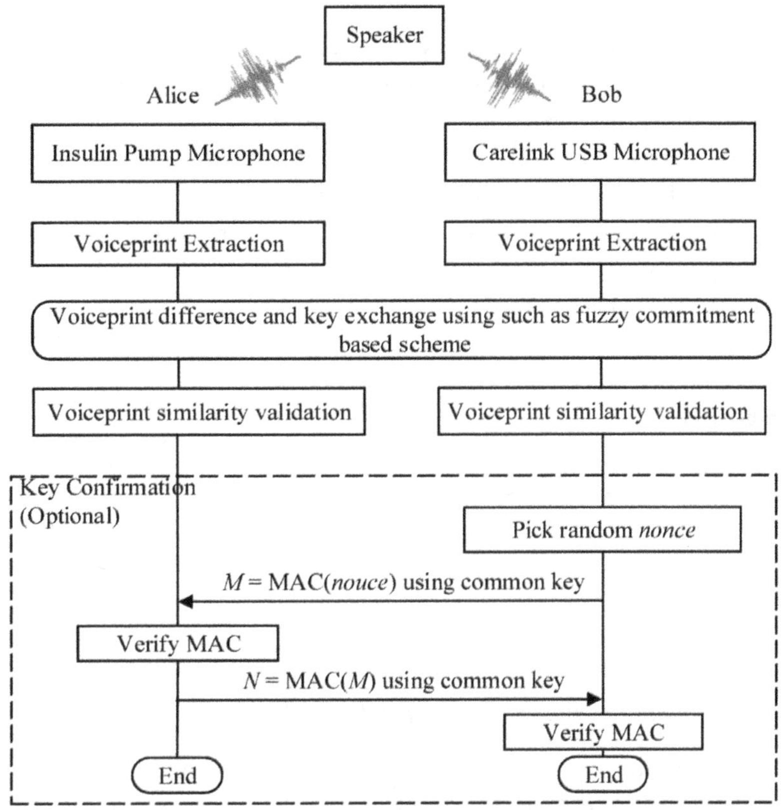

Finding acoustic feature based countermeasures with higher performance for the detection of replay attacks is a challenging work.

- **Multi-Factor Authentication (MFA):** No ASV system with liveness detection countermeasure based on single biometric modality can guarantee 100% accuracy due to several influential factors, such as noise, spoofing attacks, etc. While previous experiments presented in this chapter show that the fusion of face and speaker verification systems achieves a significant performance improvement when comparing to the best uni-modal system, the countermeasures based on combinations of voice and other biometric traits (finger, iris, etc.) are also anticipated.
- **Text-Dependent Countermeasures:** Most of the previous proposals focus on text-independent ASVs which verify the identity of the target speaker without constraints on passphrases, which can be used for secure transactions (identifying the person making the transaction). Text-dependent ASV requires the speaker saying exactly the enrolled passphrase(s) and can achieve higher recognition accuracy with shorter utterances, which is suitable for access control and authentication. To broaden the application areas of speaker verification, more work should focus on text-dependent countermeasures in the future.
- **Evaluation of Integrated ASV and Countermeasure:** Current work mainly focuses on the evaluation of a standalone ASV or countermeasure. The future work should pay more attention to the development of an integrated ASV and countermeasure system and evaluate its total performance for spoofing attacks presented in this chapter.

CONCLUSION

This chapter mainly focuses on voice liveness detection in ASV systems. The authors explain the spoofing attacks to ASV systems comprising impersonation, voice conversion, speech synthesis, and replay and then presented four types of liveness detection (anti-spoofing) methods used to mitigate ASV spoofing attacks: Challenge-response based methods, Acoustic feature based methods, Hardware-based methods, and Multi-modal biometric based methods. The authors analyze advantages and disadvantages of each kind of liveness detection method, and propose possible application of voiceprint-based liveness detection schemes in insulin pump system. At Last, the authors provide suggestions about future research directions within voice liveness detection.

ACKNOWLEDGMENT

This research was supported in part by US NSF under grant CNS-1812553.

REFERENCES

Alegre, F., Amehraye, A., & Evans, N. (2013). Spoofing countermeasures to protect automatic speaker verification from voice conversion. *Proceedings of IEEE ICASSP*, *13*, 3068–3072.

Aley-Raz, A., Krause, M., & Salmon, I. (2013). *Device, system, and method of liveness detection using voice biometrics.* United States Patent, No. USOO8442824B2.

Burget, L., Matějka, P., Schwarz, P., Glembek, O., & Cernocky, J. H. (2007). Analysis of feature extraction and channel compensation in a GMM speaker recognition system. *IEEE Transactions on Audio, Speech, and Language Processing*, *15*(7), 1979–1986. doi:10.1109/TASL.2007.902499

Campbell, W. M., Sturim, D. E., & Reynolds, D. A. (2006). Support vector machines using GMM supervectors for speaker verification. *IEEE Signal Processing Letters*, *13*(5), 308–311. doi:10.1109/LSP.2006.870086

Chen, S., Ren, K., & Piao, S. (2017). You can hear but you cannot steal: Defending against voice impersonation attacks on smartphones. In *Proceedings of IEEE 37th International Conference on Distributed Computing Systems* (pp. 183-195). Atlanta, GA: IEEE.

Chingovska, I., Anjos, A., & Marcel, S. (2014). Biometrics evaluation under spoofing attacks. *IEEE Transactions on Information Forensics and Security*, *9*(12), 2264–2276. doi:10.1109/TIFS.2014.2349158

Czyzewski, A., Kostek, B., Bratoszewski, P., Kotus, J., & Szykulski, M. (2017). An audio-visual corpus for multimodal automatic speech recognition. *Journal of Intelligent Information Systems*, *49*(2), 167–192. doi:10.100710844-016-0438-z

De Leon, P., Pucher, M., Yamagishi, J., Hernaez, I., & Saratxaga, I. (2012). Evaluation of speaker verification security and detection of HMM-Based synthetic speech. *IEEE Transactions on Audio, Speech, and Language Processing*, *20*(8), 2280–2290. doi:10.1109/TASL.2012.2201472

Dehak, N., Kenny, P. J., Dehak, R., Dumouchel, P., & Ouellet, P. (2011). Front-end factor analysis for speaker verification. *IEEE Transactions on Audio, Speech, and Language Processing*, *19*(4), 788–798. doi:10.1109/TASL.2010.2064307

Fomichev, M., Álvarez, F., Steinmetzer, D., Gardner-Stephen, P., & Hollick, M. (2018). Survey and Systematization of Secure Device Pairing. *IEEE Communications Surveys and Tutorials*, *20*(1), 517–550. doi:10.1109/COMST.2017.2748278

Font, R., Espín, J. M., & Cano, M. J. (2017). Experimental analysis of features for replay attack detection-results on the ASVspoof 2017 Challenge. *Proceedings of INTERSPEECH*, *17*, 7–11. doi:10.21437/Interspeech.2017-450

Garofolo, J. S., Lamel, L. F., & Fisher, W. M. (1993). *TIMIT acoustic-phonetic continuous speech corpus*. Philadelphia: Linguistic Data Consortium. doi:10.6028/NIST.IR.4930

Haitsma, J., & Kalker, T. (2002). A highly robust audio fingerprinting system. *Proceedings of the 3rd International Conference on Music Information Retrieval*.

Hatch, A. O., Kajarekar, S., & Stolcke, A. (2006). Within-class covariance normalization for SVM-based speaker recognition. *Proceedings of INTERSPEECH*, *06*, 1471–1474.

Hei, X., & Du, X. (2011). Biometric-based two-level secure access control for implantable medical devices during emergencies. *Proceedings - IEEE INFOCOM*, *11*, 346–350.

Hei, X., Du, X., Lin, S., & Lee, I. (2013). PIPAC: Patient Infusion Pattern based Access Control Scheme for Wireless Insulin Pump System. In Proceedings of IEEE INFOCOM'13 (pp. 3030-3038). Turin, Italy: IEEE.

Hei, X., Du, X., Wu, J., & Hu, F. (2010). Defending Resource Depletion Attacks on Implantable Medical Devices. *Proceedings of IEEE Globecom*, *10*, 1–5.

Hunt, A. J., & Black, A. W. (1996). Unit selection in a concatenative speech synthesis system using a large speech database. *Proceedings of IEEE ICASSP*, *96*, 373–376.

Inchingolo, P., Bergamasco, S., & Bon, M. (2001). Medical data protection with a new generation of hardware authentication tokens. *Proceedings of Mediterranean Conf. on Medical and Biological Engineering and Computing*.

Kenny, P., Ouellet, P., Dehak, N., Gupta, V., & Dumouchel, P. (2008). A study of interspeaker variability in speaker verification. *IEEE Transactions on Audio, Speech, and Language Processing*, *16*(5), 980–988. doi:10.1109/TASL.2008.925147

Kinnunen, T., & Li, H. (2010). An overview of text-independent speaker recognition: From features to supervectors. *Speech Communication*, *52*(1), 12–40. doi:10.1016/j.specom.2009.08.009

Kinnunen, T., Sahidullah, M., Delgado, H., Todisco, M., Evans, N., Yamagishi, J., & Lee, K. A. (2017a). The ASVspoof 2017 Challenge: Assessing the limits of replay spoofing attack detection. *Proceedings of INTERSPEECH*, *17*, 2–6. doi:10.21437/Interspeech.2017-1111

Kinnunen, T., Sahidullah, M., & Falcone, M. (2017b). RedDots replayed: A new replay spoofing attack corpus for text-dependent speaker verification research. *Proceedings of IEEE ICASSP, 17*, 5395–5399.

Larcher, A., Lee, K., & Martinez, P. L. S. (2014). Extended RSR2015 for text-dependent speaker verification over VHF channel. *Proceedings of INTERSPEECH, 14*, 1322–1326.

Lee, K., Larcher, A., & Wang, G. (2015). The reddots data collection for speaker recognition. *Proceedings of INTERSPEECH, 15*, 2996–3000.

Lei, Y., Scheffer, N., & Ferrer, L. (2014). A novel scheme for speaker recognition using a phonetically-aware deep neural network. *Proceedings of IEEE ICASSP, 14*, 1695–1699.

Li, C., Raghunathan, A., & Jha, N. K. (2011). Hijacking an insulin pump: Security attacks and defenses for a diabetes therapy system. In *Proceedings of the 13th IEEE Intl. Conf. on e-Health NAS* (pp. 150-156). IEEE.

Li, P., Fu, Y., & Mohammed, U. (2012). Probabilistic models for Inference about Identity. *IEEE Transactions on Pattern Analysis and Machine Intelligence, 34*(1), 144–157. doi:10.1109/TPAMI.2011.104 PMID:21576751

Marin, E., Singelée, D., Yang, B., Verbauwhede, I., & Preneel, B. (2016). On the Feasibility of Cryptography for a Wireless Insulin Pump System. In *Proceedings of ACM CODASPY* '16 (pp. 113-120). ACM. 10.1145/2857705.2857746

Ratha, N. K., Connell, J. H., & Bolle, R. M. (2001). Enhancing security and privacy in biometrics-based authentication systems. *IBM Systems Journal, 40*(3), 614–634. doi:10.1147j.403.0614

Reynolds, D. A., Quatieri, T. F., & Dunn, R. B. (2000). Speaker verification using adapted Gaussian Mixture Models. *Digital Signal Processing, 10*(1-3), 19–41. doi:10.1006/dspr.1999.0361

Reynolds, D. A., & Rose, R. C. (1995). Robust text-independent speaker identification using Gaussian Mixture Speaker Models. *IEEE Transactions on Speech and Audio Processing, 3*(1), 72–83. doi:10.1109/89.365379

Roeschlin, M., Martinovic, I., & Rasmussen, K. B. (2018). Device pairing at the touch of an electrode. *Proceedings of NDSS'18.*

Sahidullah, M., Delgado, H., Todisco, M., Yu, H., Kinnunen, T., Evans, N., & Tan, Z.-H. (2016). Integrated spoofing countermeasures and automatic speaker verification: An evaluation on ASVspoof 2015. *Proceedings of INTERSPEECH, 16*, 1700–1704. doi:10.21437/Interspeech.2016-1280

Sahidullah, Md., Kinnunen, T., & Hanilci, C. (2015). A comparison of features for synthetic speech detection. *Proceedings of INTERSPEECH, 15*, 2087–2091.

Schürmann, D., & Sigg, S. (2013). Secure communication based on ambient audio. *IEEE Transactions on Mobile Computing, 12*(2), 358–370. doi:10.1109/TMC.2011.271

Sizov, A., Khoury, E., & Kinnunen, T. (2015). Joint speaker verification and anti-spoofing in the *i*-vector space. *IEEE Transactions on Information Forensics and Security, 10*(4), 821–832. doi:10.1109/TIFS.2015.2407362

Vogt, R., & Kajarekar, S. (2008). Paper. *Proceedings of Odyssey*, 21–24.

Wu, Z., Evans, N., Kinnunen, T., Yamagishi, J., Alegre, F., & Li, H. (2015a). Spoofing and countermeasures for speaker verification: A survey. *Speech Communication*, *66*, 130–153. doi:10.1016/j.specom.2014.10.005

Wu, Z., Khodabakhsh, A., & Demiroglu, C. (2015b). Spoofing and Anti-Spoofing (SAS) corpus v1.0 [dataset]. University of Edinburgh, The Centre for Speech Technology Research (CSTR) doi:. 10.7488/ds/252

Wu, Z., Yamagishi, J., Kinnunen, T., Hanilci, C., Sahidullah, M., Sizov, A., ... Delgado, H. (2016). ASVspoof: The automatic speaker verification spoofing and countermeasures challenge. *IEEE Journal of Selected Topics in Signal Processing*, *11*(4), 588–604. doi:10.1109/JSTSP.2017.2671435

Zen, H., Tokuda, K., & Black, A. W. (2009). Statistical parametric speech synthesis. *Speech Communication*, *51*(11), 1039–1064. doi:10.1016/j.specom.2009.04.004

Zhang, L., Tan, S., & Yang, J. (2017). Hearing your voice is not enough: An articulatory gesture based liveness detection for voice authentication. *Proceedings of CCS*, *17*, 55–71.

Zhang, L., Tan, S., Yang, J., & Chen, Y. (2016). VoiceLive: A phoneme localization based liveness detection for voice authentication on smartphones. *Proceedings of CCS*, *16*, 1080–1091. doi:10.1145/2976749.2978296

ADDITIONAL READING

Chakroborty, S., Roy, A., & Saha, G. (2007). Improved closed set text-independent speaker identification by combining MFCC with evidence from flipped filter banks. *International Journal of Signal Processing*, *4*(2), 114–121.

Haitsma, J., & Kalker, T. (2002). A highly robust audio fingerprinting system. In *Proceedings of the 3rd International Conference on Music Information Retrieval*, Paris, France.

Huang, X., Acero, A., & Hon, A. H. (2001). *Spoken language processing: A guide to theory, algorithm and system development* (1st ed.). Prentice Hall.

Jain A.K., Ross A., & Prabhakar S. (2004). An introduction to biometric recognition. *IEEE Transactions on Circuits and Systems for Video Technology, Special Issue on Image- and Video-Based Biometrics, 14*(1), 4-20.

Quatieri, T. F. (2001). *Discrete-time speech signal processing: Principles and practice.* Upper Saddle River, NJ: Prentice Hall PTR.

Rabiner, L. R. (1989). A tutorial on hidden Markov models and selected applications in speech recognition. *Proceedings of the IEEE*, *77*(2), 257–286. doi:10.1109/5.18626

KEY TERMS AND DEFINITIONS

Automatic Speaker Verification (ASV): An AVS system extracts the vocal characteristics of an individual to establish the identity either by imposing the fixed vocabulary constraints (text-dependent) or in a dynamic way (text-independent) i.e. without imposition of vocabulary constraints on the individuals.

Countermeasure (CM): Anti-spoofing mechanism to defend speaker verification system from spoofing attacks.

Impersonation: An impostor adapts his/her voice to mimic a target speaker's voice timbre and prosody to spoof an ASV system.

Present Attack: Presentation of an artefact or human characteristic to the biometric capture subsystem in a fashion that could interfere with the intended policy of the biometric system.

Spoofing Attack: An attacker presents a counterfeit biometric evidence of a target user to outwit the biometric system. In the context of this chapter, spoofing attack implies an attacker presents a recorded, converted, or synthesized speech sample in order to outwit the speaker verification system to accept the attacker as a claimed identity.

Speech Synthesis: An impostor generates human-like speech from any arbitrary text in order that it resembles that of another target speaker.

Voice Conversion: An impostor manipulates the speech of a given speaker in order that it resembles that of another target speaker.

Voice Liveness Detection (VLD): Mechanism distinguishing a live target person from impostors who present replayed, synthesized, or converted speech to the ASV.

Chapter 6
EarLocalizer: A Deep-Learning-Based Ear Localization Model for Side Face Images in the Wild

Aman Kamboj
National Institute of Technology, India

Rajneesh Rani
National Institute of Technology, India

Aditya Nigam
Indian Institute of Technology, India

ABSTRACT

With much concern over security, it has become essential to maintain the identity and track of an individual's activities in the modern healthcare sector. Although there are biometric authentication systems based on different modalities, recognition of a person using the ear has gained much attention as ears are unique. Ear localization is a first step for ear-based biometric authentication systems, and this needs to be accurate, since it plays a crucial role in the overall performance of the system. The localization of ear in the side face images captured in the wild possess great challenges due to varying angles, light, scale, background clutter, blur and occlusion, etc. In this chapter, the authors have proposed EarLocalizer model to localize the ear, which is inspired by Faster-RCNN. The model is evaluated on two wild ear databases, UBEAR-II and USTB-III, and has achieved an accuracy of 95% and 99.08%, respectively, at IOU (Intersection over Union) = 0.5. The results of the proposed model signify that the model is invariant to the environmental conditions.

DOI: 10.4018/978-1-5225-7525-2.ch006

INTRODUCTION

The traditional patient tagging system in hospitals are not error-free as there are chances of wrong patient tagging and the mismatch between hospitals records and actual patient tag. This leads to the improper identification, wrong medicine given to the wrong patient, and wrong blood transfusion, which leads to many incidents where patients have lost their lives. Although, the hospitals are migrating for electronic health records (EHR), but it raises privacy concern to medical data theft as the information is accessible online to the healthcare professional, and a fraudulent can misuse the information and may corrupt the data. However, In the last decade, it has been observed that there is a rapid growth in the development of the biometric for the healthcare industry is mainly to combat fraud and to improve patient privacy and safety. As per the research firm Visiongain (Cidon, 2018) and a report ("Boom in healthcare biometric market," 2018), it has been estimated that this growth will increase over 300% in the next 5 years. The Grand View research has also predicted that growth in the global market will reach up to $14.5bn by 2025 and highlighted the main contributors as BIO-Key, NEC, Fujistu, IRIS ID etc. So, this trillion-dollar industry is always under the threat of malicious attack and from people with wrong intent. Although, the Health Insurance Portability and Accountability Act of 1996 HIPPA act (Gold, 2013) has imposed new rules for patient privacy and safety and it leads to all the healthcare facilities to be as per compliance of new rules and regulations and encouraged healthcare sector to utilize biometric technologies at larger scale. To keep the hospital premises safe and to deliver an error-free treatment to the patient, it becomes very crucial and important to maintain the identity of an individual patient and working staff in the hospital. Although there are smart cards being used to know the identity of an individual, but there are chances that the items under possession like token and smart card may be lost or stolen. Knowledge-based methods like pin and password are also being used to provide secure access, but there are chances that the secret information may be shared, or the person may forget. Replacing smart card and knowledge-based methods with the biometric of an individual brings a lot of advantages as they cannot be lost or stolen or shared. Moreover, it is hard to spoof an individual's biometrics, as they are unique, and the information is always carried by the person.

An Overview of Biometric Traits

Biometrics traits (Li & Jain, 2009) of an individual are broadly classified into two categories based on physiological such as the face, fingerprints, iris, palmprint, knuckle print, ear etc. and behavioral such as voice, gait, and signature as shown in Figure 1.

(R & G, 2011), (Jaswal, Nigam, & Nath, 2017) discussed various issues and challenges of each biometric traits as follows:

- **Face:** It is one of the most common and well known biometric trait. The face is a non-intrusive biometric as images can be captured at distance or extracted from video frames without any user co-operation.
    - **Challenges:** Face changes with age and is affected by the expression, eyeglasses, cosmetics and it also raises privacy concern in the medical field, identical twins have the same face, so it violates the property of uniqueness.

Figure 1. Biometrics traits for person authentication

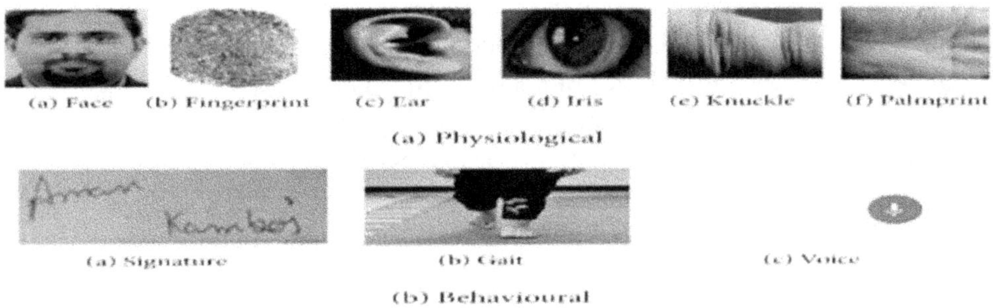

- **Finger-Prints:** Fingerprints are being widely used for person authentication. They are made up of ridges which contains a huge amount of discriminative information. Images are captured using a cheap sensor.
  - **Challenges:** Fingerprints are affected by skin conditions and not acceptable for people working in the industry and for children because of their size changes quickly. Fingerprints of a person can be easily spoofed using an artificial finger.
- **Iris:** This is one of the best biometric trait known for most secure person recognition and resistance to an impostor.
  - **Challenges:** Iris scan needs costly hardware to capture images. Iris scan is affected by health conditions and easily spoofed by contact lenses. It is an intrusive biometric.
- **Palm-Print:** The palm of a human contains discriminative features which are stable and unique. Images can be captured using low-cost sensor non-intrusively.
  - **Challenges:** Difficult to use for some person and pose a great level of challenges under variation of illumination conditions, scale, and translation. An injury can affect the recognition of a person from palm-print.
- **Knuckle-Print:** The outer part of a finger is considered as finger-knuckle. They contain discriminative line-based features. It can be captured easily using the sensor.
  - **Challenges:** Major problems caused by variation of lighting conditions, scale, and translation.
- **Ear:** The human ears are unique and possess several advantages over other biometrics. The ear is a non-intrusive biometric and has a stable structure which is not affected by age. The images can be captured using digital cameras so there is no need to purchase separate hardware.
  - **Challenges:** The performance of the ear is affected by occlusions like hairs, earrings, images of different scales and varying illuminations conditions.

Every trait has its own merits and demerits depending upon the application in which to use. However, (Jain, Bolle, & Pankanti, 2006) identified the following seven properties that a biometric trait should possess to make itself suitable for successful authentication. The comparative study of these properties is for different traits is shown in Table 1.

- **Universality:** The biometric trait should be possessed by everyone and it should not be lost.
- **Uniqueness:** These characteristics of a trait are very primary and indicates that how uniquely the system can identify a user from a group of users.

- **Permanence:** The biometric trait needs to remain constant with time. It should be temporally stable and should not be affected by age or skin disease.
- **Collectability:** The biometric trait should be acquired from anyone at any place.
- **Performance:** These characteristics of biometric trait suggest how well the security system work and what is the speed of recognition and accuracy of a given trait.
- **Acceptability:** How well the biometric trait is accepted by society.
- **Circumvention:** The biometric trait should not be spoofed or forged easily. These characteristics of traits indicate how easily it will lead to failure at the authentication time.

The Architecture of Biometric Authentication System

A biometric-based authentication system is a multistage process as shown in Figure 2. In the initial step, a database of the desired trait of the different person is prepared in an unconstrained environment. In the next step, images are processed for the same dimensions and low-quality images are discarded. In the segmentation step, a region of interest (ROI) is identified in the image and segmented out. In the next step, images are normalized, since the images are captured under varying conditions, so there is a need to align them along the major axis for matching purpose. In the next step, unique features are learned by the model and they are stored in the database. This complete process is known as enrollment. At the time of authentication, the image goes through similar stages and extracted features are matched against those stored in the database and a match score is calculated. If the match score is more than some threshold value, then the user is authenticated as genuine otherwise an imposter.

There are three possible modes in which a biometric system can be operated, which are known as enrolment, identification, and authentication. In the enrolment mode, a sensor is used to acquire an image from the user and its quality is tested. If the acquired sample image has quality above than a certain threshold, then unique features are extracted, and a unique number is assigned and stored in the database. Every user of the system is required to be enrolled first. Based on the required application a biometric system can be an identification or an authentication system. An identification system is a one to many matchings in which a newly captured image is given as an input and unique feature are extracted and they

Table 1. Comparative study of biometric properties

Biometric	Universality	Uniqueness	Permanence	Collectability	Performance	Acceptability	Circumvention
Face	HIGH	LOW	MEDIUM	HIGH	LOW	HIGH	LOW
Fingerprint	MEDIUM	HIGH	HIGH	MEDIUM	HIGH	MEDIUM	HIGH
Hand Geometry	MEDIUM	MEDIUM	MEDIUM	HIGH	MEDIUM	MEDIUM	MEDIUM
Iris	HIGH	HIGH	HIGH	MEDIUM	HIGH	LOW	HIGH
Signature	LOW	LOW	LOW	HIGH	LOW	HIGH	LOW
Voice	MEDIUM	LOW	LOW	MEDIUM	LOW	HIGH	LOW
Gait	MEDIUM	LOW	LOW	HIGH	LOW	HIGH	MEDIUM
Ear	MEDIUM	MEDIUM	HIGH	MEDIUM	MEDIUM	HIGH	HIGH

Source: (Mali & Bhattacharya, 2013)

Figure 2. Biometric authentication system

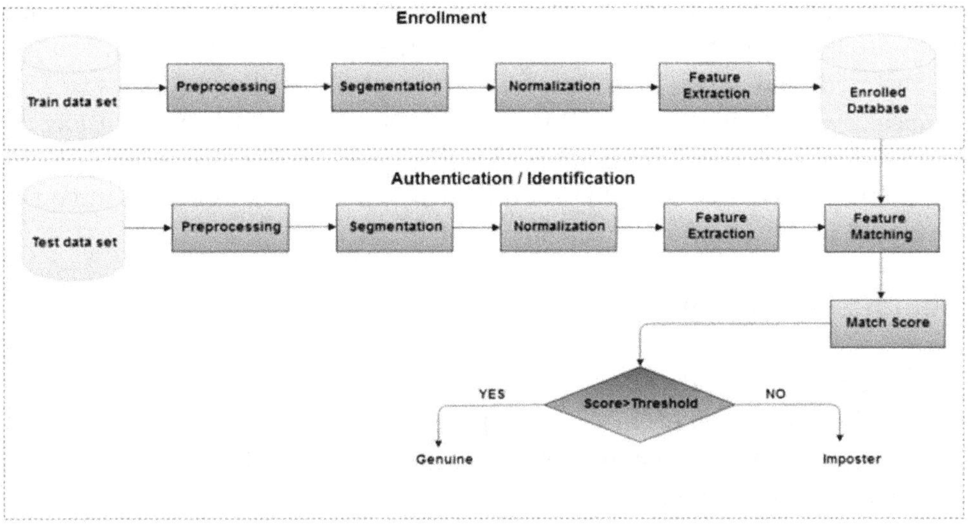

are matched against the stored features. The system returns top best matches. They are broadly classified as a positive and negative identification system. The positive identification system ensures that whether the user information is present in the database. Whereas, the negative identification system ensures that the user information is not present in the database. An authentication system is one to one matching whose main task is to confirm the person who is posing a claim, in which the extracted features from the newly captured image are matched against stored features. If the match score is more than a precomputed threshold score then the user claim is verified and considered as a genuine, otherwise an imposter.

Issues and Challenges in Healthcare Biometrics

Shawl (2013), Segun and Olawale (2017), and Mogli (2012) discussed the issues and challenges of biometrics in the healthcare sector. The existing biometric authentication systems are mainly based on face, iris, fingerprints and are not suitable for the hospitals in the rural areas. As in the majority of the Asian countries like India more than 70% of the population resides in rural areas, works in agriculture sector and use their hands very roughly, due to this their fingerprints get damaged; Moreover, finger ridges are not well developed on early childhood and skin changes in the elderly, even in the hospital most of the time staff wear gloves. Although, psychologically people (generally undocumented immigrants) have a negative perception of the fingerprint as they associate it with law enforcement and decline to participate. Thereby, it becomes very difficult to use their fingerprints for authentication. In iris recognition system a person need to be still during the enrolment and verification, otherwise, the system will not be able to scan properly, and it leads to non-matching. The iris scan system also does not perform well, in case the health condition of the patients is very poor. An intruder can easily spoof the system using contact lenses. Moreover, due to the higher cost of iris scanner, it become very difficult to deploy them at large scale. In the face recognition system, due to privacy concern many patients are reluctant to provide their face images as an individual are afraid of information being leaked. The facial character of a person changes rapidly with age. Face image also affected due to beard and spectacles and cannot distinguish between twins. Religious and cultural barrier also possess a great level of challenge. In emergency cases due to the very poor health condition of a patient, they cannot be able to produce their identity and sometimes it becomes very difficult to recognize an individual from the face. Hence, the identification of an individual is very important, and it need to be error free. Although the existing solutions for biometrics are for the constrained environment which need much user cooperation and does not consider the problems of real-world conditions as in real world there are effects of illumination, background clutter, scale and angle variations. Thereby, there is a need for a new biometric trait that can help to overcome these challenges and work well in the unconstrained environment.

Ear as a Healthcare Biometric

Iannarelli (1989) conducted a study for the uniqueness of the ear. He gathered ear images of 10,000 random persons and found that all of them are unique. The study motivated the researchers to use ear as a biometric for authentication. Moreover, the ear is a very useful trait for person authentication as it possesses several advantages over the other traits. The ear is a non-intrusive biometric, which can be captured easily at a distance, whereas fingerprint, iris, palm-print etc. are intrusive biometrics and need much user cooperation. Thereby, the ear can be useful as a surveillance to suspect the unlawful activities. Unlike the face, the ear has a stable structure and is not affected by age. So, it can helpful to maintain the information of a person over the period. The ear is a useful biometric trait for people who have lost their fingers or works in labor or agriculture sector. Moreover, Ear images can be easily captured using mobile cameras. Hence, there is no need to purchase separate hardware for image acquisition of an individual. In the modern health care system, it can be used in the following perspectives:

- The ear is useful for recognition of newborn babies to avoid child swapping cases happening in metro cities.
- The ear can be helpful for corps identification when facial parts are damaged.
- The ear can be used in hospitals to maintain patients record on the site and to speed up the identification process.
- Useful when an unconscious person is not able to produce identity.
- Ear based biometric system can be integrated with a smartphone device and an individual patient can be monitored remotely by a doctor in rural areas.
- Ear based biometric can be integrated with CCTV cameras installed at hospital premises to track fraudulent activities.
- Ear based biometric system can be installed at the entrance of the hospitals and allows the entry of a legitimate person.
- A multimodal biometric system can be developed using ear and side face image to enhance the security.
- Ear based biometric system can be used to allow only doctors and staff in certain areas and secure medical equipment and facilities.
- Helpful for the healthcare club, wellness center, blood donation center, organ transplant center etc.

In the past, there is much progress in the research to use ear as a biometric. A huge amount of work has been already reported and has achieved significant performance on benchmarked datasets. However, much of the work in the literature is performed on the constrained environment. In which, images are captured under some controlled settings and they are much different from real-world imaging conditions. The unconstrained environment possesses great challenges due to varying conditions such as lighting, scale, pose, background clutter, blur, occlusion (due to piercing and hairs). Due to reasons, there is a need to develop a robust method that can recognize the person in the wild.

The rest of the paper is organized as follows: the section literature review discusses the related work and highlights the existing research gaps; The architecture of the model is discussed in section proposed method; The next section explain the database used and various parameters for model evaluation. The section results and discussion explain the performance of the proposed model and the next section conclude the overall work.

Literature Review

There has been a lot of research and innovation in the healthcare biometrics with the aim to enhance security, to reduce cost and patient safety. The information technology (IT) has played a major role in the development of healthcare biometrics and improved operational efficiency and risk management by ensuring the accurate identification of patients and delivering an error-free treatment. With the help of biometrics in healthcare, the overall fraud is reduced as it has improved the patient care and protected the patient information. Moreover, it has prevented card sharing and identity theft as the person itself need to be present at the time of authentication. Researchers have provided various approaches to use different human traits as biometrics based on physiological characteristics such as face (Nakada, Wang, & Terzopoulos, 2017; Hayat, Khan, & Bennamoun, 2017), fingerprint (Jain, Arora, Cao, Best-Rowden, & Bhatnagar, 2017; Rani & Dahiya, 2017), palmprint (Jaswal, Nath, & Kaul, 2015; Aishwarya, Gowri, & Saranya, 2016), knuckle print (D. Zhang, Lu, & Zhang, 2018; Zhai et al., 2018), ear (Paper et al., 2015) (Halawani, 2016) and behavioral characteristics such as voice (Broman, Marshall, Bellamkonda, Levya, & Hanson, 2015; Safavi, Gan, Mporas, & Sotudeh, 2016), gait (Muramatsu, Shiraishi, Makihara, Uddin, & Yagi, 2015), and signature (Morocho, Morales, Fierrez, & Vera-Rodriguez, 2016). The traits like fingerprint, face, and iris have been already in use and have their demerits discussed earlier in this chapter. So, our goal is to use ear as a biometric trait for healthcare which has been ignored earlier. The human ear has much discriminative power and has been well proved for biometric authentication. The well-known and recent approaches for ear based biometric authentication system are discussed as below:

Burge and Burger (2000) made their first attempt for passive biometric based on ear. The authors introduced a novel graph-based approach for ear recognition. The method was sensitive to light and occlusion. Ansari and Gupta (2007) proposed efficient approach for localization of ear in 2D profile face images. They used the outer helix curve as a unique ear feature for localization and tested the performance on IITK dataset of 700 images and achieved 93% accuracy. (Anwar, Ghany, & Elmahdy, 2015) proposed a method for ear recognition using geometrical features. They used the snake model to detect the ear in profile face and applied a median filter to remove noise. Further, a canny edge detector is used to find the largest boundary and unique features are extracted and classified using nearest neighbor. Their method achieved an accuracy of 98% on the IITD database. (Omara, Li, Zhang, & Zuo, 2016) proposed a novel geometric feature for ear recognition. The proposed method resolved the issue faced by geometric methods using EHL (ear hight line), further they used three ratio-based features which are invariant to scale. They tested their results on USTB-I and IITD database and achieved a recognition rate of 98.33% and 99.60% respectively. (Sarangi, Panda, Mishra, & Dehuri, 2017) proposed ear localization technique based on modified hausdorff distance (MHD). They used this metric to discard the effect of outliers inside face image, to localize the ear efficiently. They created ear template of different shapes and sizes and then MHD is used to localize the ear. The method was tested the on UND-E and CVL database and found very promising results. (Chowdhury, Islam, & Gao, 2017) proposed robust ear recognition approach using a neural network. The method first detects the location of the ear in facial images and then unique edge feature is extracted for matching. They used local features as they are invariant to light illumination. The neural network is used to match the unique features. The model performance was evaluated on three databases viz; IITD, USTB, UND-F. Their proposed method achieved an accuracy closed to 98%, which is more than accuracy achieved by SVM, AdaBoost, BPNN, MNN. (Y. Zhou & Zaferiou, 2017) introduces deformable models for ear alignment and recognition in the wild. They presented first ear database with 55 landmark points on 600 images and a database of 2038 labeled images for ear recognition in the unconstrained environment. They conducted extensive experiments for ear alignment and recognition using both learned and handcrafted features. The experimental results reveal that holistic and AMM are better approaches for ear alignment in the wild. (Y. Zhang & Mu, 2017) proposed multiple scale faster region based convolutional neural network. The three regions of different scales are detected to identify the context information of ear and then they used the ear region filtering approach to eliminates the false regions. The proposed method was tested on UND-J2 and UBEAR databases and achieved ear detection accuracy of 100% and 98.22% respectively and their proposed method is invariant to environmental conditions. (Emersic, Gabriel, Struc, & Peer, 2017) proposed pixel-wise convolutional encoder-decoder for ear segmentation. The network gives pixel-wise detail of the ear. Unlike the other approaches, those return bounding box for ear location the proposed method segments the ear at the pixel level and thus provides much finer detail about the location of the ear. They tested the performance on AWE dataset, which is captured under the unconstrained environment and achieved good detection accuracy.

(Pflug, Busch, & Ross, 2014) proposed 2D ear classification based on unsupervised clustering technique to partition the image into multiple clusters. The input ear image is compared with a large set of gallery images and assigned a class label whose centroids are closest to corresponds clusters. The experiment was conducted on three different heterogeneous datasets UND-J2, AMI, and IITK. They compared classification performance of three texture descriptors: Histograms of Oriented Gradients, uniform Local Binary Patterns and Local Phase Quantization. The experimental results suggested that performance of local phase quantization texture descriptor along with PCA gives better results than other texture descriptors. (Tian & Mu, 2017) proposed ear recognition using CNN features. The neural network consists of 3 convolutional layers, 3 max-pooling layers, and two fully connected layers. The deep neural networks automatically extract features from the input image. The experiment is performed on the USTB-III ear database of 79 different subjects. They used some augmentation scheme to enlarge database size. The method achieved 98.27% accuracy for recognition. (Liu, Zhang, & Zhang, 2015) proposed ear parotic face angel for 3-D ear recognition. The ear-parotic face angle is an angle between the normal vector of the ear plane and the normal vector of the parotic face-plane. This feature is unique and stable for 3-D ears. This feature provides indexing identification and hierarchical verification which enhance both the speed and accuracy of the 3D ear recognition. Ear samples from 250 individuals using 3D ear laser scanner. This is an independent external feature and is based upon spatial orientation information of the ear relative to the face. The experimental results show that the search range is reduced to 9.69% from original and EER has been reduced from 2.8% to 2.3% by combing the angle feature with an iterative closest point(ICP)method. (Islam, Davies, Bennamoun, & Mian, 2011) presented efficient recognition and detection of 3D ears. They used cascaded Ada-boost classifier with appropriate haar-features for accurate and very fast detection of ears form 2D profile face. They used 3D local feature matching via an iterative closest point algorithm to obtain a robust and fully automatic system. The proposed system provides a detection rate of 99.9% and an identification rate of 95.4% and Equal Error Rate (EER) of 4.1%. on 5000 images of UND-F collection. (J. Zhou, Cadavid, & Abdel-Mottaleb, 2012) proposed a 3-D ear recognition system using local and holistic features. The system is comprised of four steps. Ear image segmentation, local feature extraction, holistic features extractions, and fusion of local and holistic features. The ear image segmentation is performed using shape-based features. They used a histogram of the indexed shape of ear localization. For local features extraction, they used surface patch histogram of indexed shapes, for holistic features they used voxelization scheme. The match score is obtained from both local and holistic features to obtain a final match score. The experiment is performed on an UND-G dataset of range images for 415 subjects. The recognition accuracy of the proposed method is 98.3% and an equal error rate is 1.7%.

Researcher have provided different approaches for ear detection and recognition but there are certain research gaps which need to overcome:

- Lack of large ear datasets in the wild.
- Majority of the ear localization approaches are not evaluated based on IOU (intersection over union) parameters, as this indicates how much predicted bounding boxes are fit to ground truth boxes.
- Much of the work is performed on the constrained environment.
- The performance of the existing ear localization approaches is not up to par.

With the advancement of technology, there is a need to adopt new methods to fulfill these research gaps.

Deep Learning for Computer Vision

Deep Learning is the most widely and rapidly emerging field of Artificial Intelligence (AI). In the last decade, there is much advancement in AI due to deep learning. The AI starts in 1950, but it has become more significant in today world as it requires a huge amount of data to learn and higher hardware computation power. Deep learning has become the major area of research and development for computer vision task as it has improved the overall performance in all application areas. In 1995, Yann Lecun, known as the father of Convolutional Neural Networks (CNN), at Bell laboratory observed that object recognition and detection by traditional hand-crafted feature detectors can be improved. The traditional hand-crafted feature detectors were coded for detecting edges, their histograms, a combination of edges and so on. Yann observed that given the complex nature of this world we could never hand design a feature detector that is perfect or even reaches near perfection. So, Yann for the first time gave the idea of a convolution neural network that can learn the domain features automatically and eliminated the need for handcrafted features. This idea of automatic feature learning doesn't only make the system faster but when the system learns by itself, it can extract such combinations of input features that a hand-designed feature detector could never do. However, the computational power and a data requirement for a model that learns using deep learning is higher as compared to a model that learns using machine learning concepts. (LeCun et al., 1995) has proposed a model LeNet for handwriting recognition system. The LeNet performed much better than all the handwriting recognition systems and is still in use. (Krizhevsky, Sutskever, & Hinton, 2012) developed a CNN that he referred to as AlexNet for visual recognition of objects. The network was trained on 1.2 million images of 1000 different categories in the ImageNet Large Scale Visual Recognition Challenge (ILSVRC-2012) and proved to be much better among all the participants. AlexNet was trained based on Graphics Processing Unit (GPU's) and was the only submission based on GPUs. The model has reduced the overall error rate from 28.2% to 16.4%. The year 2012 was marked the best year where a CNN has surpassed all other methods for classification, localization, and detection task. (Guo et al., 2016) discussed the detailed review of state-of-the-art deep learning models like ZF-Net, VGG-16, GoogLeNet and Resnet-50, RCNN etc.

PROPOSED METHOD: DEEP LEARNING BASED EARLOCALIZER

This section describes the architecture details of the model. The model is used to localize the ear in the side face image of a person in an unconstrained environment. The architecture of the model is inspired by Faster RCNN (Ren, He, Girshick, & Sun, 2017), which consist of several components (As shown in Figure 3) viz; Shared layers, A region proposal network, ROI (Region of Interest) pooling layer, classification and regression heads and its layer level specification is given in Table 2. The model faster RCNN has better performance than RCNN (Girshick, Donahue, Darrell, & Malik, 2014). As the RCNN model required the forward pass for every region proposal due to selective search and it has to train three modules (CNN feature extractor, a classifier for prediction of class labels and regressor for bounding box generation) separately. So, these limitations of RCNN make it hard to train. However, in faster RCNN these limitations have been addressed.

Shared Layers

The deep learning neural networks are consisting of a series of layers which majorly consist of feature extraction layers (convolutional layers, polling layer, rectified linear unit), batch normalization layer and dropout. The convolution layer has filters which convolved over the input image and produce an output tensor. Its main objective is to extract features from the given input volume. The polling layer is used to reduce the overall spatial space of the input. The pooling can be performed using functions like min-pooling, max-pooling or average pooling etc. The rectified linear unit is used to interduce non-linearity in the model by using function max (0, x) and it turns all negative values to zeros. The dropout layer is used to avoid the problem of overfitting in which a network work well on training dataset but fails on test data. To overcome this problem the dropout layer randomly set some neurons values to zero. This will help the network to not depends upon specific features and model will search for more general features. The batch normalization layer is used to avoid the problem of vanishing gradients in this situation the values of gradients become very small and model is not able to learn the features, so every batch of the images is normalized with zero mean and unit variance.

These shared layers act as a base for the proposed model and these layers are generally taken from state-of-the-art computer vision models (for image classification) like ResNet-50, VGG-16, Alex-Net etc. As it is known that for the training of any deep learning model from scratch, one needs millions of input data. To overcome this problem, the pretrained-weight (on ImageNet dataset) are used during training. After experimentation over ear datasets, we have found that VGG-16 (Simonyan & Zisserman, 2014) model performs well for ear segmentation task. VGG-16 is also preferred over ResNet-50 as there are less number of feature extraction layers, that help to train the network in less time compared to Resnet-50. Hence, in the proposed model we have used VGG-16 as a base model to extract discriminative features

Figure 3. Architecture of EarLocalizer

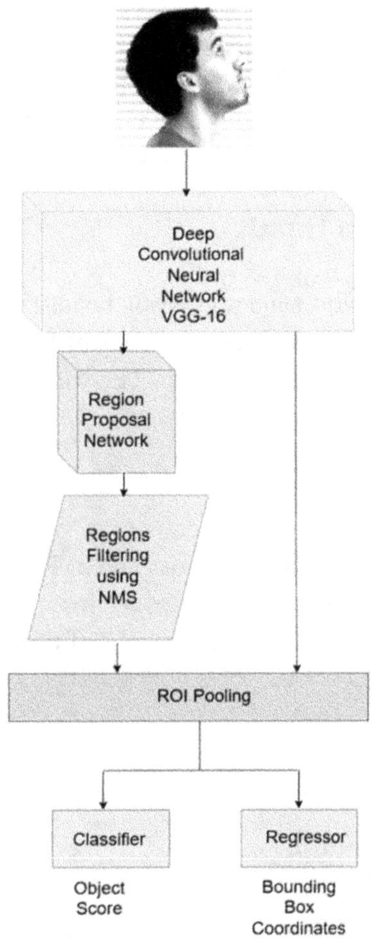

from the input image which are later used by region proposal network for prediction of bounding boxes and classifier and regression heads for the prediction of ear localization and its coordinates respectively.

Region Proposal Network

The Region Proposal network (RPN) as shown in Figure 4, takes the input feature map provided by shared layers and produces the probable regions having an ear and their coordinates. RPN contains a sliding window of size 3*3, which is convolved on feature map and gives K anchor boxes of different scales [128*128, 256*256, 512*512] and ratios [1:1, 2:1, 2:2] to select the best fit boxes for ground truth boxes and refine the coordinates of these anchor boxes by regressing against ground truth coordinates. These anchor boxes are selected based on similarity measure against ground truth boxes. The upshot of the RPN is 2*K (objectness score) for each anchor K and 4*K for the coordinates of K anchor boxes. However, the regions produced by this layer are filters using non-maximum suppression algorithm and it keeps only those anchor boxes having higher objectness score.

ROI Pooling Layer

The output of both shared layer and RPN layer are given to the ROI pooling layer. Since the regions produced by the RPN are variable in size this leads to the different size of the feature map, hence it's not efficient to work on different size feature maps. To classify the proposed region, ROI pooling layer converts it into a fixed vector of size (14*14) and later max pooling is applied in every region.

Classification and Regression Heads

The filtered region is fed into classification and regression heads for the prediction of the class score and bounding box coordinates. The classification layer consists of fully connected layers in which neurons of each layer are connected to neurons in another layer. The softmax activation is used in last fully connected layer for the classification of learned features.

Loss Function of EarLocalizer

The model has two types of loss: 1) Classification Loss and 2) Regression Loss and is calculated as per equation (1). At the training time, the network computes the loss function and tries to minimize this. The loss of EarLocalizer was reached to 0.050 after 100 epochs of training on UBEAR-I model.

Figure 4. RPN (Region Proposal Network)

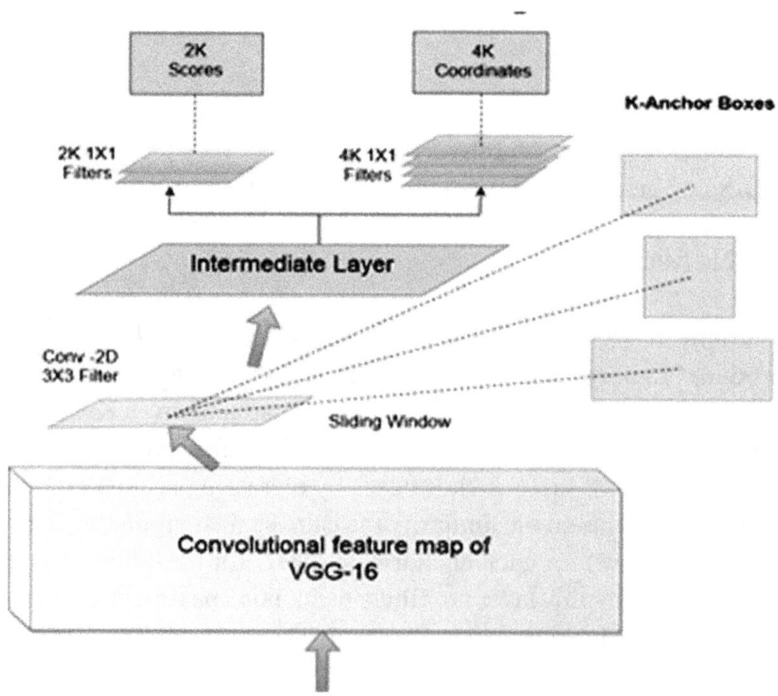

Table 2. Layer-specific details of the model

Major Components of Network		Layer Specification		
	Type, Number of Filters	**Filter Size**	**Batch Normalization**	**Activation Function**
Shared Layers	VGG-16 Layers			
Region Proposal Network	Conv2D, 512	3*3	Yes	Rectified Linear Unit
	Per Anchor 4	1*1*512	For Bounding Box Regression	
	Per Anchor 1	1*1*512	For Classification Score	
ROI Polling	**Resize ever block to 14*14 and Max pooling**			
Classification Head	Conv2D, 512	1*1	Yes	Rectified Linear Unit
	Conv2D, 512	3*3	Yes	Rectified Linear Unit
Regression Head	Conv2D, 2048	1*1	Yes	Rectified Linear Unit
	Conv2D, 2048	1*1	Yes	Rectified Linear Unit

$$L(p_i,b_i)=\frac{1}{N_{cls}}\sum_{i}^{m}L_{cls}(p_i,p_g)+\lambda\frac{1}{N_{reg}}\sum_{i}^{m}L_{reg}(b_i,b_g) \quad (1)$$

$$smooth_{L1}(x)=\begin{Bmatrix} 0.5*x^2 if \mid x \mid < 1 \\ \mid x \mid -0.5 \ otherwise \end{Bmatrix} \quad (2)$$

Here, i: is the index of an anchor in mini-batch; N_{cls}: is number of anchors in mini-batch; L_{cls}: is a loss between the object or not; p_i: is predicted the probability of an anchor i being an object ; p_g: is the ground truth label it is 1 for positive anchors, 0 for negative anchors; λ is a constant weight; N_{reg} is a number of anchors locations mini-batch; b_i is representing 4 coordinates of the predicted bounding box; b_g is representing four coordinates of ground truth bounding box; L_{reg} = $smooth_{L1}(b_i-b_g)$ is a loss function defined in equation (2).

EXPERIMENTAL SETUP AND PERFORAMNCE PARAMETERS

The proposed Earlocalizer model is trained on benchmark ear datasets UBEAR-1 with 4000 images and results are evaluated on USTB-III and UBEAR-II databases. The detail regarding databases and performance evaluation parameters is discussed as below:

Databases

There are several benchmark datasets exist for ear recognition, but most them are in the constrained environment. In this chapter we have evaluated our model on two most popular databases UBEAR and USTB which are captured in the unconstrained environment and discussed as below:

UBEAR Database

This database is captured from the moving subjects in the video. It contains 8000 images from 128 subjects in the unconstrained environment (Raposo, Hoyle, Peixinho, & Proença, 2011). The images are in gray-scale format each with size 1280*960. The database has two subsets: UBEAR-1 and UBEAR-II each contains approximately 4000 images. The subset UBEAR-1 also contains ear segmentation mask and for subset UBEAR-II, the authors have manually prepared segmentation bounding boxes. The Fig.5 show some sample images of the UBEAR database.

USTB Database

This database is prepared from students and teachers at USTB in Nov 2004 (USTB, 2004). The database has three subsets viz; USTB-I, USTB-II, USTB-III. In this chapter, the model is evaluated on USTB-III subset, as the other two subsets have cropped ear images. The database has the right-side profile face images of 79 different subjects. From each subject, 20 images are captured at a various angle of rotation and there is occlusion in the images due to hairs. The distance between the camera and subject is fixed to 1.5 meters. The image has a resolution of 768*576, with 24-bit color. The sample images are shown in Figure 6.

Performance Parameters

The performance of the proposed model is tested using slandered parameters viz; Accuracy, Precision, Recall, and F1-Score, which are discussed as below:

- **Accuracy:** It measures the actual performance of the system, which indicates the number of images in which ear is correctly localized to the total number of images and is calculated as per the equation (1).

$$Accuracy = \frac{TP + TN}{TP + TN + FP + FN} \quad (1)$$

Figure 5. Sample images of UBEAR Database

Figure 6. Sample images of USTB-III Database

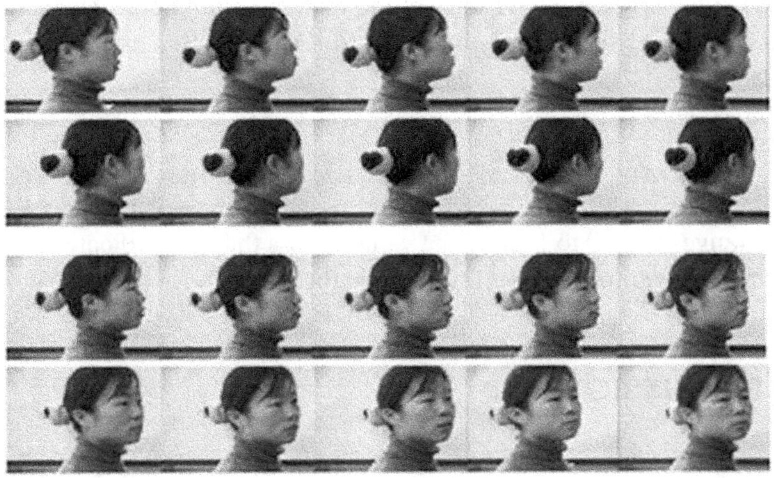

- **Precision:** It measures how accurate is the prediction, it is the ratio of true positive to the sum of true positive and false positive bounding boxes. It measures how much percentage of your positive prediction is correct and is calculated as per equation (2).

$$\text{Precision} = \frac{TP}{TP + FP} \tag{2}$$

- **Recall:** It measures how good the model finds all the positive bounding boxes with respect to all the possible positives and is calculated as per equation (3).

$$\text{Recall} = \frac{TP}{TP + FN} \tag{3}$$

- **F1-Score:** It measures the overall accuracy of the model and is calculated as per equation (4).

$$\text{F1Score} = 2 * \frac{\text{Precision} \ * \ \text{Recall}}{\text{Precision} \ + \ \text{Recall}} \tag{4}$$

Here,

TP (True Positive) = The images in which ear is correctly detected.
FP (False Positive) = The images in which ear is detected mistakenly.
FN (False Negative) = These are the images in which false region is detected as an ear.
TN (True Negative) = 0, as there is only one object i.e. ear in an image.

- **IOU:** An Intersection Over Union (IOU), it measures how much there is an overlap between the bounding box predicted by the model and ground truth bounding box. The values of different performance parameters are calculated with respect to IOU as per equation (5).

$$IOU = \frac{Area \text{ of Overlap}}{Area \text{ of Union}} \tag{5}$$

The value of IOU ranges from 0 to 1. Here the value of less than 0.4 indicates poor overlapping. The value from 0.5 to 0.65 indicates average overlapping and the value above 0.7 indicates good overlapping.

RESULTS AND DISCUSSION

The EarLocalizer model is evaluated on the two ear datasets USTB-III and UBEAR-II, which contains side face images of a person in the wild. The model is evaluated based on performance parameters discussed in the previous section.

Qualitative Evaluation of EarLocalizer

The proposed model is evaluated on some sample images taken from UBEAR database as shown in Figure 7. It has been observed that the model has localized the ear in the images under various challenging conditions such as varying angle, light, and occlusion. The performance of the model indicates that it can work in the real scenario and overcome the challenges of the unconstrained environment.

Quantitative Evaluation of EarLocalizer

As shown in Figure 9a, the EarLocalizer has achieved good ear localization accuracy. At IOU=0.5, the model has achieved an accuracy of 99.08%, 95.06% on USTB-III and UBEAR-II dataset respectively.

Figure 7. EarLocalizer qualitative results on UBEAR database

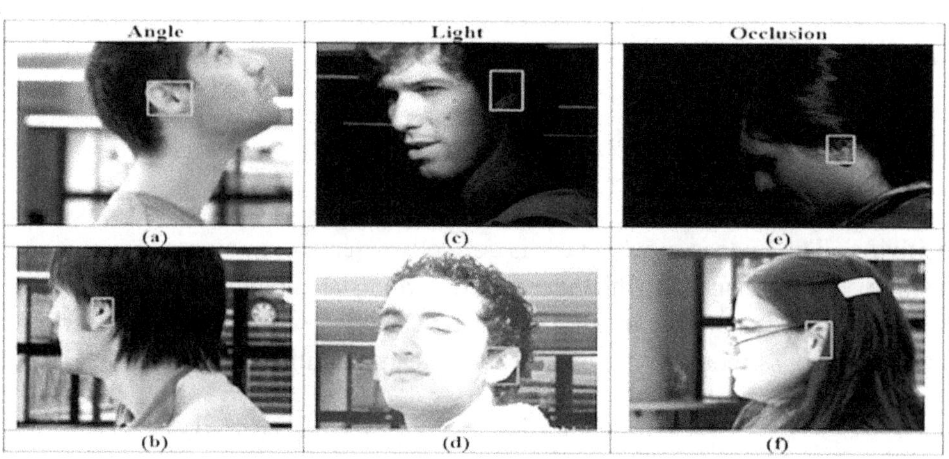

The proposed model has shown promising results on both the datasets. It has been observed that there is a decrease in the performance of the model with the further increase in IOU. The precision and recall values (shown in Figure 9b and Figure 9c) for both the datasets stays above 99% till an IOU=0.5, after this there is a slight decline in the performance. The Figure 9d indicates the overall performance of the system with respect to increase in an IOU. The results are also shown in Table 3 at different values of IOUs. The model has shown little poor performance on UBEAR-II database as the images are captured completely in the wild and in many cases, the ear is not visible properly due to environmental conditions, occlusion, and huge angle variations.

Miss-Classified Images: The Figure 8, shows some miss-classified images by Ear Localizer. As shown in Figure 8a and Figure 8b, there is huge angle variations in the images due to this only the outer helix of the ear is visible. In Figure 8c and Figure 8d the ear is highly occluded by hairs, so only the part of the ear is visible. The images are shown in Figure 8e and Figure 8f suffered from huge light variations, so the structure of the ear is not visible clearly. The Figure 8g and Figure 8h shows that the images are blurred and of poor quality. It has been observed that the EarLocalizer model is not able to predict the exact location of the ear in the cases when the images are having extreme variations and the ear is not visible or highly occluded due to hairs.

CONCLUSION AND FUTURE WORK

Biometrics are becoming popular with the advancement in technology and it became very essential to use these technologies in the hospitals to provide error-free treatment and to improve security and safety. There exist many biometrics traits but the use of an ear as a biometric has been ignored although it possesses several advantages over other biometric traits. The localization of the ear in side face images captured in the wild is a challenging problem, due to environment variations like angle, occlusion, scale, and light etc. So, in this chapter, the authors have proposed EarLocalizer model to localize the ear, which is highly inspired by Faster-RCNN. The proposed model has achieved good ear localization accuracy, of 99.8% and 95.06% on USTB-III and UBEAR-II datasets respectively at an IOU 0.5. In the future, authors like to incorporate this ear localization method into a pipeline of recognition, to build a robust model that can recognize the person from the ear in the wild and work in real scenarios.

Table 3. EarLocalizer Results on USTB-III and UBEAR-II Datasets

Database	**Accuracy**			**Precision**			**Recall**			**F1-Score**		
IOU Threshold	**0.5**	**0.6**	**0.7**	**0.5**	**0.6**	**0.7**	**0.5**	**0.6**	**0.7**	**0.5**	**0.6**	**0.7**
USTB-III	**99.08**	98.77	93.24	**99.85**	99.54	93.96	**99.85**	99.54	93.96	**99.85**	99.54	93.96
UBEAR-II	**95.06**	92.45	83.06	**99.23**	96.51	86.71	**99.23**	96.51	86.71	**99.23**	96.51	86.71

Figure 8. Miss classified images by EarLocalizer; Here, the bounding box in red indicates the ground truth and in blue is predicted by the model

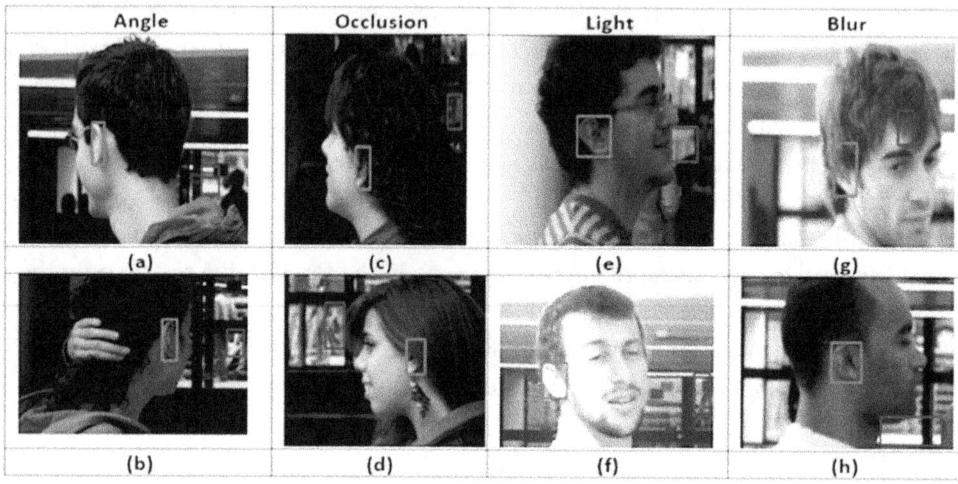

Figure 9. Quantitative evaluation of Ear Localizer

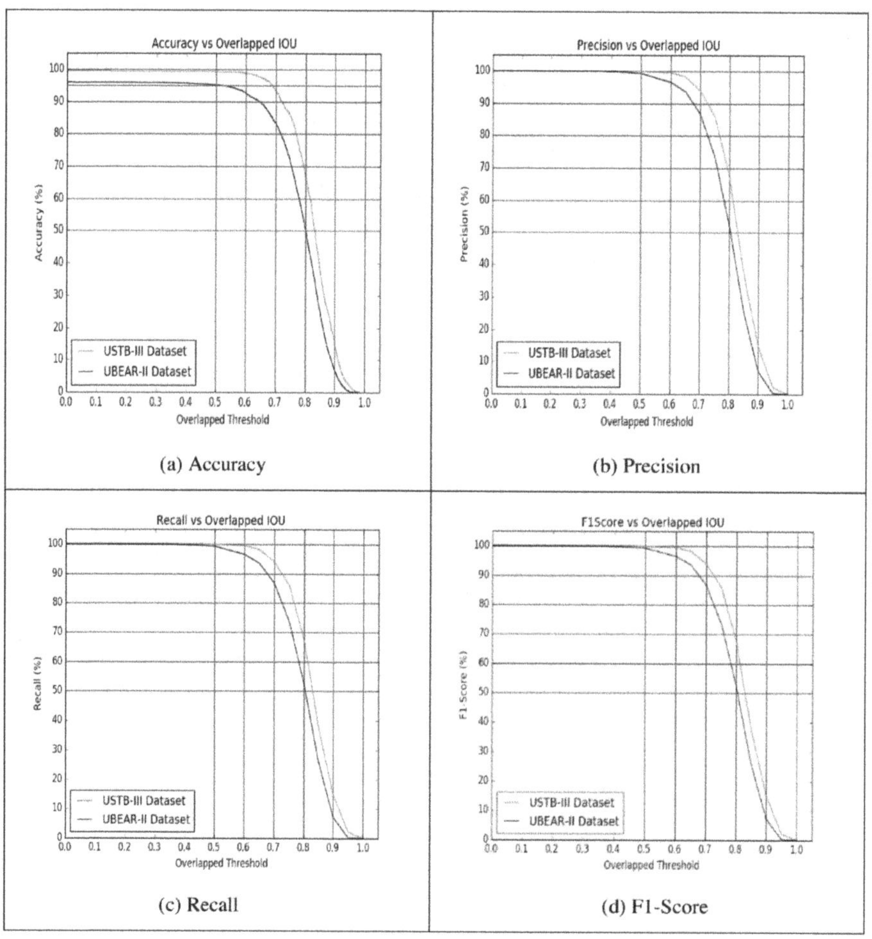

REFERENCES

Aishwarya, D., Gowri, M., & Saranya, R. K. (2016). Palm print recognition using liveness detection technique. In *Science Technology Engineering and Management (ICONSTEM), Second International Conference on* (pp. 109–114). IEEE.

Ansari, S., & Gupta, P. (2007). Localization of Ear using Outher Helix Curve of the Ear. *IEEE Proceedings of the International Conference on Computing: Theory and Applications (ICCTA'07)*, 1–5.

Anwar, A. S., Ghany, K. K. A., & Elmahdy, H. (2015). Human Ear Recognition Using Geometrical Features Extraction. *Procedia Computer Science, 65*, 529–537. doi:10.1016/j.procs.2015.09.126

Boom in healthcare biometric market. (2018). *Biometric Technology Today, 2018*(4), 3.

Broman, V., Marshall, V., Bellamkonda, S., Levya, M., & Hanson, C. (2015, August 18). *Speaker recognition and denial of a transaction based on matching a known voice print*. Google Patents.

Burge, M., & Burger, W. (2000). Ear biometrics in computer vision. *Proceedings 15th International Conference on Pattern Recognition. ICPR-2000*, 2, 822–826. 10.1109/ICPR.2000.906202

Chowdhury, M., Islam, R., & Gao, J. (2017). Robust ear biometric recognition using neural network. In *Industrial Electronics and Applications (ICIEA), 2017 12th IEEE Conference on* (pp. 1855–1859). IEEE.

Cidon, D. (2018). Making IT better: how biometrics can cure healthcare. *Biometric Technology Today, 2018*(7), 5–8.

Emersic, Z., Gabriel, L. L., Struc, V., & Peer, P. (2017). *Pixel-wise Ear Detection with Convolutional Encoder-Decoder Networks*. Retrieved from http://arxiv.org/abs/1702.00307

Girshick, R., Donahue, J., Darrell, T., & Malik, J. (2014). Rich feature hierarchies for accurate object detection and semantic segmentation. In *Proceedings of the IEEE conference on computer vision and pattern recognition* (pp. 580–587). IEEE.

Gold, S. (2013). Healthcare biometrics: solving the staff and patient security governance challenge. *Biometric Technology Today, 2013*(8), 5–9.

Guo, Y., Liu, Y., Oerlemans, A., Lao, S., Wu, S., & Lew, M. S. (2016). Deep learning for visual understanding: A review. *Neurocomputing, 187*, 27–48. doi:10.1016/j.neucom.2015.09.116

Halawani, A. (2016). *Human Ear Localization: A Template-Based Approach*. Academic Press. doi:10.18178/ijsps.4.3.258-262

Hayat, M., Khan, S. H., & Bennamoun, M. (2017). Empowering simple binary classifiers for image set based face recognition. *International Journal of Computer Vision, 123*(3), 479–498. doi:10.100711263-017-1000-3

Iannarelli, A. V. (1989). *Ear Identification*. Paramont Publishing Company.

Islam, S. M. S., Davies, R., Bennamoun, M., & Mian, A. S. (2011). Efficient Detection and Recognition of 3D Ears. *International Journal of Computer Vision, 95*(1), 52–73. doi:10.100711263-011-0436-0

Jain, A. K., Arora, S. S., Cao, K., Best-Rowden, L., & Bhatnagar, A. (2017). Fingerprint recognition of young children. *IEEE Transactions on Information Forensics and Security, 12*(7), 1501–1514. doi:10.1109/TIFS.2016.2639346

Jain, A. K., Bolle, R., & Pankanti, S. (2006). *Biometrics: personal identification in networked society* (Vol. 479). Springer Science & Business Media.

Jaswal, G., Nath, R., & Kaul, A. (2015). Texture based palm Print recognition using 2-D Gabor filter and sub space approaches. In *Signal Processing, Computing and Control (ISPCC), 2015 International Conference on* (pp. 344–349). IEEE.

Jaswal, G., Nigam, A., & Nath, R. (2017). DeepKnuckle: Revealing the human identity. *Multimedia Tools and Applications, 76*(18), 18955–18984. doi:10.100711042-017-4475-6

Krizhevsky, A., Sutskever, I., & Hinton, G. E. (2012). Imagenet classification with deep convolutional neural networks. In Advances in neural information processing systems (pp. 1097–1105). Academic Press.

LeCun, Y., Jackel, L. D., Bottou, L., Cortes, C., Denker, J. S., Drucker, H., ... Simard, P. (1995). Learning algorithms for classification: A comparison on handwritten digit recognition. *Neural Networks: The Statistical Mechanics Perspective, 261*, 276.

Li, S. Z., & Jain, A. K. (2009). *Encyclopedia of Biometrics: I-Z* (Vol. 1). Springer Science & Business Media. doi:10.1007/978-0-387-73003-5

Liu, Y., Zhang, B., & Zhang, D. (2015). Ear-parotic face angle: A unique feature for 3D ear recognition. *Pattern Recognition Letters, 53*, 9–15. doi:10.1016/j.patrec.2014.10.014

Mali, K., & Bhattacharya, S. (2013). Comparative study of different biometric features. *International Journal of Advanced Research in Computer and Communication Engineering, 2*(7), 2776–2784.

Mogli, G. (2012). Role of Biometrics in healthcare privacy and security management system. *Sri Lanka Journal of Bio-Medical Informatics, 2*(4), 156. doi:10.4038ljbmi.v2i4.2245

Morocho, D., Morales, A., Fierrez, J., & Vera-Rodriguez, R. (2016). Towards human-assisted signature recognition: improving biometric systems through attribute-based recognition. In *Identity, Security and Behavior Analysis (ISBA), 2016 IEEE International Conference on* (pp. 1–6). IEEE.

Muramatsu, D., Shiraishi, A., Makihara, Y., Uddin, M. Z., & Yagi, Y. (2015). Gait-based person recognition using arbitrary view transformation model. *IEEE Transactions on Image Processing, 24*(1), 140–154. doi:10.1109/TIP.2014.2371335 PMID:25423652

Nakada, M., Wang, H., & Terzopoulos, D. (2017). AcFR: active face recognition using convolutional neural networks. In *Computer Vision and Pattern Recognition Workshops (CVPRW), 2017 IEEE Conference on* (pp. 35–40). IEEE.

Omara, I., Li, F., Zhang, H., & Zuo, W. (2016). A novel geometric feature extraction method for ear recognition. *Expert Systems with Applications, 65*, 127–135. doi:10.1016/j.eswa.2016.08.035

Paper, C., Tiwari, S., Shankaracharya, S., Campus, T., Kumar, S., Naya, I., … Cad, R. (2015). *Ear Recognition for Newborns*. Academic Press.

Pflug, A., Busch, C., & Ross, A. (2014). 2D ear classification based on unsupervised clustering. In *Biometrics (IJCB), 2014 IEEE International Joint Conference on* (pp. 1–8). IEEE.

R, K., & G, R. (2011). *Biometrics-A Vision for the Future*. Retrieved from www.hcl.com

Rani, K., & Dahiya, S. (2017). Comparative analysis of fingerprint recognition using neural network and genetic algorithm. *Indian Journal of Scientific Research*, 84–90.

Raposo, R., Hoyle, E., Peixinho, A., & Proença, H. (2011). UBEAR: A dataset of ear images captured on-the-move in uncontrolled conditions. In *Computational Intelligence in Biometrics and Identity Management (CIBIM), 2011 IEEE Workshop on* (pp. 84–90). IEEE.

Ren, S., He, K., Girshick, R., & Sun, J. (2017). Faster R-CNN: Towards Real-Time Object Detection with Region Proposal Networks. *IEEE Transactions on Pattern Analysis and Machine Intelligence*, *39*(6), 1137–1149. doi:10.1109/TPAMI.2016.2577031 PMID:27295650

Safavi, S., Gan, H., Mporas, I., & Sotudeh, R. (2016). Fraud detection in voice-based identity authentication applications and services. In *Data Mining Workshops (ICDMW), 2016 IEEE 16th International Conference on* (pp. 1074–1081). IEEE.

Sarangi, P. P., Panda, M., Mishra, B. S. P., & Dehuri, S. (2017). An automated ear localization technique based on modified Hausdorff distance. In *Proceedings of International Conference on Computer Vision and Image Processing* (pp. 229–240). Springer. 10.1007/978-981-10-2107-7_21

Segun, O. F., & Olawale, F. B. (2017). *Healthcare data breaches: Biometric technology to the rescue*. Academic Press.

Shawl, D. (2013). *Biometrics–Implementing into the Healthcare Industry Increases the Security For The Doctors, Nurses, and Patients* (Master's thesis). Davenport University.

Simonyan, K., & Zisserman, A. (2014). *Very deep convolutional networks for large-scale image recognition.* ArXiv Preprint ArXiv:1409.1556.

Tian, L., & Mu, Z. (2017). Ear recognition based on deep convolutional network. In *Proceedings - 2016 9th International Congress on Image and Signal Processing, BioMedical Engineering and Informatics, CISP-BMEI 2016* (pp. 437–441). 10.1109/CISP-BMEI.2016.7852751

USTB. (2004). *University of Science & Technology Beijing (USTB).* Retrieved from http://www1.ustb.edu.cn/resb/

Zhai, Y., Cao, H., Cao, L., Ma, H., Gan, J., & Zeng, J., … Zhi, Y. (2018). A Novel Finger-Knuckle-Print Recognition Based on Batch-Normalized CNN. In *Chinese Conference on Biometric Recognition* (pp. 11–21). Springer. 10.1007/978-3-319-97909-0_2

Zhang, D., Lu, G., & Zhang, L. (2018). Local Features for Finger-Knuckle-Print Recognition. In *Advanced Biometrics* (pp. 111–130). Springer. doi:10.1007/978-3-319-61545-5_6

Zhang, Y., & Mu, Z. (2017). Ear Detection under Uncontrolled Conditions with Multiple Scale Faster Region - Based Convolutional. *Neural Networks*. doi:10.3390ym9040053

Zhou, J., Cadavid, S., & Abdel-Mottaleb, M. (2012). An efficient 3-D ear recognition system employing local and holistic features. *IEEE Transactions on Information Forensics and Security*, *7*(3), 978–991. doi:10.1109/TIFS.2012.2189005

Zhou, Y., & Zaferiou, S. (2017). Deformable Models of Ears in-the-Wild for Alignment and Recognition. *Proceedings - 12th IEEE International Conference on Automatic Face and Gesture Recognition, FG 2017 - 1st International Workshop on Adaptive Shot Learning for Gesture Understanding and Production, ASL4GUP 2017, Biometrics in the Wild, Bwild 2017, Heteroge*, 626–633. 10.1109/FG.2017.79

ADDITIONAL READING

Abaza, A., Hebert, C., & Harrison, M. (2010). Fast learning ear detection for real-time surveillance. IEEE 4th International Conference on Biometrics: Theory, Applications and Systems, BTAS 2010.

Arbab-Zavar, B., & Nixon, M. (2011). On guided model-based analysis for ear biometrics. *Computer Vision and Image Understanding*, *115*(4), 487–502. doi:10.1016/j.cviu.2010.11.014

Bustard, J., & Nixon, M. (2010). Toward unconstrained ear recognition from two-dimensional images. *IEEE Transactions on Systems, Man, and Cybernetics. Part A, Systems and Humans*, *40*(3), 486–494. doi:10.1109/TSMCA.2010.2041652

Chan, T., & Kumar, A. (2012). Reliable ear identification using 2-D quadrature filters. *Pattern Recognition Letters*, *33*(14), 1870–1881. doi:10.1016/j.patrec.2011.11.013

Chidananda, P., Srinivas, P., Manikantan, K., & Ramachandran, S. (2015). Entropy-cum-Hough-transform-based ear detection using ellipsoid particle swarm optimization. *Machine Vision and Applications*, *26*(2-3), 185–203. doi:10.100700138-015-0669-y

Cintas, C., Paschetta, C., Azevedo, S., Cesar, C., Cerqueira, S., Ramallo, V., ... Delrieux, C. (2017). Automatic ear detection and feature extraction using Geometric Morphometrics and convolutional neural networks. *IET Biometrics*, *6*(3), 211–223. doi:10.1049/iet-bmt.2016.0002

Emersic, Z., & Peer, P. (2015). Toolbox for ear biometric recognition evaluation, in *Proceedings of EUROCON 2015*, Salamanca,1-6.

Emersic, Z., Stepec, D., Struc, V., & Peer, P. (2017). Training Convolutional Neural Networks with Limited Training Data for Ear Recognition in the Wild. *Proceedings - 12th IEEE International Conference on Automatic Face and Gesture Recognition, FG 2017 - 1st International Workshop on Adaptive Shot Learning for Gesture Understanding and Production, ASL4GUP 2017, Biometrics in the Wild, Bwild 2017, Heteroge*, 987-994.

Hansley, E., Segundo, M., & Sarkar, S. (2017). Employing Fusion of Learned and Handcrafted Features for Unconstrained Ear Recognition. in IET Biometrics, 7(3), 215-223.

Nigam, A., & Gupta, P. (2015). Robust ear recognition using gradient ordinal relationship pattern. Lecture Notes in Computer Science (including subseries Lecture Notes in Artificial Intelligence and Lecture Notes in Bioinformatics), 9010, 617-632.

KEY TERMS AND DEFINITIONS

Biometrics: The use of human physical or behavioral characteristics for authentication purpose. This technology is mainly used in access control, forensic and surveillance activities.

Deep Learning: A subbranch of machine learning which inspires from the artificial neural network. It has eliminated the need to design handcrafted features as in deep learning features are automatically learned by the model from the data.

Ear Biometric: Ear is a unique biometric trait of human and possesses several advantages over the other biometrics traits. The ear is a useful biometric trait for forensic and surveillance, as it is non-intrusive and does not require user co-operation.

Enrolment: This is one of the biometric authentication stages. In this step, the unique features are extracted and stored in the database so that later they can be used to match the identity of an individual.

Intersection Over Union: A parameter to measure, how much predicted bounding boxes by the model are fit to ground truth bounding boxes.

Supervised Learning: In this learning, the model needs a labeled data for training. The model knows in advance the answer to the questions it must predict and tries to learn the relationship between input and output.

Transfer Learning: A method in deep learning in which a model which is a well trained on a large dataset, can be reused for the similar task with fewer data. With this method, it reduces the training time for the model as there is no need to train the model from scratch.

Unconstrained Environment (Wild): The images of the subjects are captured in the wild, which possesses variations like angle, scale, occlusion, light, background clutter, etc.

Unsupervised Learning: In this learning, the model does not require labeled data for training. The model learns the nature of data and does predictions.

Verification: A process of the biometrics-based authentication system, in which for a new query the unique features are extracted and matched against the database. If the match score is more than some threshold the user is authenticated as genuine otherwise an imposter.

Chapter 7
Multi-Modal Motion-Capture-Based Biometric Systems for Emergency Response and Patient Rehabilitation

Marina L. Gavrilova
University of Calgary, Canada

Ferdous Ahmed
University of Calgary, Canada

A. S. M. Hossain Bari
University of Calgary, Canada

Ruixuan Liu
University of Calgary, Canada

Tiantian Liu
University of Calgary, Canada

Yann Maret
University of Calgary, Canada

Brandon Kawah Sieu
University of Calgary, Canada

Tanuja Sudhakar
University of Calgary, Canada

ABSTRACT

This chapter outlines the current state of the art of Kinect sensor gait and activity authentication. It also focuses on emotional cues that could be observed from human body and posture. It presents a prototype of a system that combines recently developed behavioral gait and posture recognition methods for human emotion identification. A backbone of the system is Kinect sensor gait recognition, which explores the relationship between joint-relative angles and joint-relative distances through machine learning. The chapter then introduces a real-time gesture recognition system developed using Kinect sensor and trained with SVM classifier. Preliminary experimental results demonstrate accuracy and feasibility of using such systems in real-world scenarios. While gait and emotion from body movement has been researched in the context of standalone biometric security systems, they were never previously explored for physiotherapy rehabilitation and real-time patient feedback. The survey of recent progress and open problems in crucial areas of medical patient rehabilitation and rescue operations conclude this chapter.

DOI: 10.4018/978-1-5225-7525-2.ch007

INTRODUCTION

Rapid development of biometric technology has opened doors to a new class of fast and reliable identity management solution and has forever changed the research landscape (Jain et.al. 2008; Kisku et.al. 2013). Biometric relies on using observable unobtrusive traits (such as iris, fingerprint, face, voice or gait), to remotely authenticate a user. There has been a significant surge in the number of effective approaches to user authentication, discretionary access control to secure resources, and integration of biometric recognition algorithms with consumer products (such as cell phone with iris and/or fingerprint recognition, biometric door locks, industrial and military robots with voice recognition, self-driving cars with human silhouette identification, game consoles with gait recognition, etc.) (Jain et.al 2008; Kisku et.al 2013; Gavrilova & Monwar 2012). This chapter presents an overview of the most recent uses of Kinect sensor technology for biometric user identification and activity recognition, and presents an emerging methodology for emotion and intention recognition from body position and movement. The applications in real-world scenarios for medical diagnostics and search and rescue operations are discussed, along with numerous open problems in this domain.

Traditional definition of biometrics research scope usually includes recognizing someone's identity from the collected biometric data, which includes physiological, behavioural, soft, and social traits. Physiological features can be often collected visually (facial image, ear, iris etc.) or through some specialized devices, such as infrared sensors, remote temperature measuring devices, and so on. Behavioural characteristics include the way a person walks (gait), talks (voice) and writes (typing patterns, keystroke pressure). Soft biometrics include data which can be easily collected but not highly unique, such as age, gender, height or weight. One significant area where the development of biometric technologies can have a high tangible effect on society is emergency response and medicine. It is generating a lot of interest and getting traction in biometric research, as well as in related fields looking into human interaction, physiological studies, user profiling, pattern recognition, authorship identification and collective intelligence.

This book chapter outlines the current state of the art of Kinect sensor gait and activity authentication, while focusing on emotional cues that could be observed from human body and posture. It then presents a prototype of a new system which combines recently developed behavioral gait and posture recognition for human emotion identification. A backbone of the system is Kinect sensor gait recognition method, which explores the relationship between joint-relative angles and joint-relative distances through machine learning. The chapter then introduces a real-time gesture recognition system, developed using Kinect sensor and trained with SVM classifier, to recognize four emotions of the human body. Preliminary experimental results demonstrate accuracy and feasibility of using such systems in real-world scenarios. While gait and emotion from body movement has been researched in the context of standalone biometric security systems, they were never previously explored for physiotherapy rehabilitation and real-time patient feedback. Thus, the survey of recent progress and open problems in crucial areas of medical patient rehabilitation and rescue operations conclude this chapter.

BACKGROUND STUDY AND LITERATURE REVIEW

Rapid technological advancement has led to a new field of biometric security, in which emotional state of an individual can be predicted by analysing body movement patterns. Recognition of a person's emotional state and appropriate response is the foundation of all social interactions. As computing resources become more sophisticated, it is reasonable to treat them as social agents. Organizations such as financial services, healthcare, telecommunication, e-commerce and e-education are developing new approaches to recognize and predict human behaviour to create more secure biometric systems. Automatic recognition of emotion has many potential applications including but not limited to smart home services, patient behaviour monitoring, passenger behaviour profiling and so forth (Adey, 2009; Fragopanagos & Taylor, 2005; Gavrilova & Monwar, 2012). We start with the brief overview of emotion recognition in biometrics, before proceeding onto the sensing devices, datasets and recent methodologies to detect emotion from biometric traits, focusing specifically on emotions from motion recognition.

Emotion Recognition in Biometrics

Traditionally, emotions have been recognized from facial expression (Fasel & Luettin, 2003). However, it is also possible to recognize emotions from speech (Koolagudi & Rao, 2012), gaze (He et al., 2018), or body movements (Molchanov et al., 2015). A combination of multiple biometric modalities have been used to create more robust systems (Bänziger et al., 2009). In this chapter, we focus our attention on recognizing emotions from body posture and body movements.

Motion-based recognition system follows a general pipeline which starts with detection of the human body and subtraction of background information from every frame. In Dalal et al. (2006), researchers utilized the Histogram of Oriented Gradient (HOG) features for human silhouette detection. Another important contribution was made in Felzenszwalb et al. (2008), where geometric information extracted from the body was sued to create a human body model - the Deformable Partial Model (DPM). In the second part of the pipeline, a human posture is detected and tracked to reduce any redundant information. Initially, researchers predominantly used model-based approaches for posture detection, however in Siddiqui et al. (2010) the combination of gradient space similarity with the Monte Carlo approach was used to estimate a human posture. The final step is to build an appropriate representation of the extracted data and to apply a learning technique to map a body movement to a specific emotion.

Observing specific sections of the body, such as hands, can provide important cues to recognize a person's emotional state. It provides critical body language information, which can help determine whether a person is honest or insincere by observing the position of the hands (Molchanov et al., 2015). The pace of nodding the head is another indicator of a person's patience level (Pease & Pease 2004). Other attempts to understand emotion from body gestures were conducted using Dynamic Time Warping (DTW) and Hidden Naïve Bayes (HNB) to classify dynamic body gestures into anger, joy, pleasure or sadness (Castellano et al., 2007). A recent work by Seneca et al. (2016) used Artificial Neural Networks (ANN) to visualize human emotion continuously during a theatre performance. Most recently, Ahmed et al. (2018) applied score and rank-level fusion techniques to recognize human emotions from a body posture.

Deep Learning for Emotion Recognition

Deep learning techniques have drawn a significant attention in object recognition and information security domains due to their inherent capability to perform well in a case where domain-specific knowledge is required for feature representation. To reduce the limitations of classical methodologies, deep learning techniques such as convolution neural network (CNN) (Li & Chan, 2014) and multi-stage CNN (Newell et al., 2016) have been proposed. An early application of deep learning was introduced by Kosti et al.

(2017), who used two low-rank CNN to fuse background and movement information to recognize 26 emotions. Another approach was used in Fan et al. (2016) to recognize emotion from video sequences. A combination of Recurrent Neural Network (RNN) and Convolution Neural Network (CNN) was employed in this work to encode both structural and motion information for emotion recognition. Recently, researchers are focused on applying deep learning techniques for feature extraction of motion capture-based biometric systems. In Toshev and Szegedy (2014), a seven-layered Convolution Deep Neural Network to represent body joints from motion images were used. Similarly, Deep Neural Networks (DNN) were used to model large-scale movement of body parts such as the torso, hand, and leg (Neverova et al., 2013) with subsequent application of Recurrent Neural Network (RNN) and Long-short Term Memory (LSTM) for gesture recognition.

Overview of Motion Capture Devices

Existing motion capture technologies are centred around three main approaches: electromechanical, electromagnetic and optical tracking systems. Electromechanical systems require placing measuring devices at various points of the human body. This approach is restrictive in their weight and freedom of movement and thereby inhibiting the full range of movement. Electromagnetic sensors improve the range of motion to be captured, and are lighter and more comfortable than electromechanical devices. However, the wires attached to each sensor restrict the overall freedom of movement of the human subject. The least restrictive motion capture system is optical-based. Examples of optical-based motion capture systems are pre-calibrated multi-camera set-ups and depth camera-based image capturing devices. These devices allow full freedom of movement and the use of non-intrusive means for data acquisition. Microsoft Kinect is a low-cost depth-sensing camera technology that provides marker-less full-body tracking capability. It has been used for physical rehabilitation which requires accurate and appropriate tracking and feedback of performance. Such devices can be utilized to create an interactive game-based rehabilitation tool for balance training of adults with neurological injury (Lange & Chang 2011). They are also suitable for emotion recognition from a body posture or gait (Ahmed et al., 2018).

Datasets

With the advent of new motion-capturing devices, and the evolution of machine learning methods, there is a pressing need for high-quality datasets to test newly developed approaches to motion recognition. In 2018, a dataset named University of Idaho-Physical Rehabilitation Movement Data (UI-PRMD) (Vakanski et al., 2018), was created for the patients in physical rehabilitation programs. This publicly available dataset comprised of 10 healthy subjects who performed 10 repetitions of different physical therapy movements which were related to common exercises. The researchers then used Vicon Optical Tracker and Microsoft Kinect to capture a variety of motions. Majority of freely accessible datasets for emotion recognition contain acted expressions, but the most conventional way of collecting emotion data is to allow actors to immerse in a scene portraying an emotion. Although professionals are very good at mimicking various emotional states, they lack the spontaneity of a real-life emotion. Most of the authors predominantly focus only on a few basic emotions, such as sorrow, anger, or happiness. Fear, surprise, and disgust are other emotions that sometimes are more difficult to recognize. Some emotional states are named interchangeably, for example joy is often used instead of happiness.

One of the first publicly available body language databases was created to recognize 10 emotions by using RGB camera sensors to produce 206 video samples of different emotions (Gunes & Piccardi, 2005). A subset of LIRIS-ACCEDE video database contained RGB video data from 64 subjects for six different emotions (Baveye et al., 2015). Another work involved creation of a dataset extracted from the Microsoft Kinect v2 sensor (Saha et al., 2014). It comprised stimulating 10 subjects to emotions namely anger, fear, happiness, sorrow and relaxed state, from which 3D skeleton gesture data was gathered. A more recent emoFBVP database was created to recognize multi-modal emotion recognition and contained video recordings of the subject's face, body gesture, voice and physiological signals (Ranganathan et al., 2016). IN that research, 23 different emotions using audiovisual information were studied from 10 professionals, with each recording repeated up to six times. Lastly, another dataset called UTD-MHAD which contains four temporarily synchronized data modalities of RGB videos, depth videos, Kinect Camera and a wearable inertial sensor for 27 human actions is available (Chen et al., 2015). Table 1 presents a summary of some datasets for gesture-based expression recognition.

Research Challenges

As demonstrated by the recent research in this domain, gesture is an important and a versatile tool to communicate thoughts and emotions, but identification and detection of emotion-specific motion is a very challenging task. Specific motion cues are universally understood, for example nodding can be interpreted as a sign of confirmation. Turning away can be interpreted as a refusal and the expansion of nostrils can be a sign of an anger (Noroozi et al., 2018). But sometimes, interpretation of intent can also be subjective, and it may depend on the person making an observation. The emotion of a person may also depend on a person gender and/or culture (Pease & Pease, 2004). Research on automatic emotion recognition has mainly focused on face and voice as the primary source for emotion analysis. In fact, 95% of all the literature used the face as the only modality (De Gelder, 2009). Emotion analysis from posture and movement is still an unexplored domain with very high potential and numerous real-world applications.

Table 1. Summary of some of the publicly available datasets for gesture-based expression recognition

Reference	Name	Device	Body Parts	Modality	#Emotions	#Gestures	#Samples
[Gunes&Piccardi 2005]	FABO	Digital Camera	Face and Body	Visual	10	NA	206
[Gavrilescu 2015]	LIRIS-ACCEDE	Camera	Face and Upper Body	Visual	6	8	NA
[Saha et.al 2014]	NA	Kinect	Upper Body	Visual	5	NA	50
[Ranganathan et.al 2016]	emoFBVP	RGBD Camera	Face and Body	Visual, Audio	23	NA	1380
[Baltrušaitis et.al 2011]	GEMEP-FERA	Kinect	Upper Body	Visual	5	7	NA
[Fothergill et.al 2012]	MSRC-12	Kinect	Whole Body	Depth	NA	12	30

METHODOLOGY FOR EMOTION RECOGNITION FROM BODY MOVEMENT

Research Domain Overview

Emotional expression can be transmitted through verbal or non-verbal (gait, signature or facial expression) cues. Studies indicate that a human observer can easily recognize emotions expressed through body movement. Body movement can be retrieved very efficiently using surveillance cameras and wearable sensors. Development of inexpensive sensor technologies such as the Microsoft Kinect, which is a marker-free infrared motion capture system, enables continuous monitoring of three-dimensional body movement patterns. Fig. 1 shows some examples of depth images extracted using Kinect V2 sensor. The images show 2D color, 2D grey-scale, skeleton and a color depth-map (from left to right). In addition to different data streams that can be obtained from Kinect (RGB, depth, audio, etc.), it can also construct a 3D virtual skeleton from the human body and track it in real-time. Although skeletal data tracked by Kinect sensor is noisier than the traditional MOCAP data, the computationally inexpensive tracking of real-time skeleton contributed to recent spike in the action recognition methods that rely on the Kinect (Xia et al., 2012; Yang & Tian, 2014). More recently, new features based on Joint Relative Angle (JRA) and Joint Relative Distance (JRD) were introduced and trained using three classifiers –SVM, Decision trees, and KNN (Ahmed et al., 2016a; Ahmed et al., 2016b). For this reason, Kinect is an excellent choice to record human movement features and automatic recognition of emotion in a non-intrusive way. However, none of these works considered emotion recognition from the human body. The main research objective is thus to develop an automatic emotion recognition system from the analysis of body motion features acquired using Kinect sensor.

Figure 2 depicts the sequence of body skeleton joints, taken over the period of time, and corresponding to four distinct emotions: Happiness, Sadness, Anger and Fear. As it can be observed, while the body skeleton belongs to the same person in all the sequences, the body position at the corresponding time for different emotions is quite different (see, for example, body skeleton positions recorded at 10, 20 and 22 - the final image in the sequence). The question is how to determine which features characterize emotion expressed through body movements. In order to do it, we shall take a look at the various ways to discover knowledge about human perceptual and cognitive processes, responsible for emotional states (Kisku et al., 2013; Amir et al., 2016; Gavrilova et al., 2017). This acquired knowledge can be utilized to build a system for an automatic emotion recognition. This research attempts to answer the following essential questions:

Figure 1. Depth images of human gait captured using Kinect V2 sensor (Ahmed et al., 2015)

- What are the visual cues embedded in our daily movement that reflect specific emotional response?
- What is the extent of change in an individual's movement pattern with respect to a specific emotional response?
- How does one identify emotion specific spatially localized features and their individual impact for different emotional states?

One way to respond to such questions is to extract emotion specific features from human gait data based on already established frameworks and further analyze the effectiveness of such features by comparing performances exhibited by various other machine learning methods. One should keep in mind that visual cues present in body movement are very subtle and require significant processing for identification. Information available about the exact nature of encoding of biological and psychological traits from the visual motion patterns is quite insufficient. The emotional states of individuals are expressed using a distinct group of general movement features which are difficult to detect and generalize. For example, moving backwards might be indicative of fear, shame, and surprise. However, a slow drop of head and folded arm position reduces the chance of fear and surprise. Such information is often very contextual, making it difficult to build a robust system.

Figure 2. 3D Skeleton data for different walking emotion extracted using Kinect V2 sensor

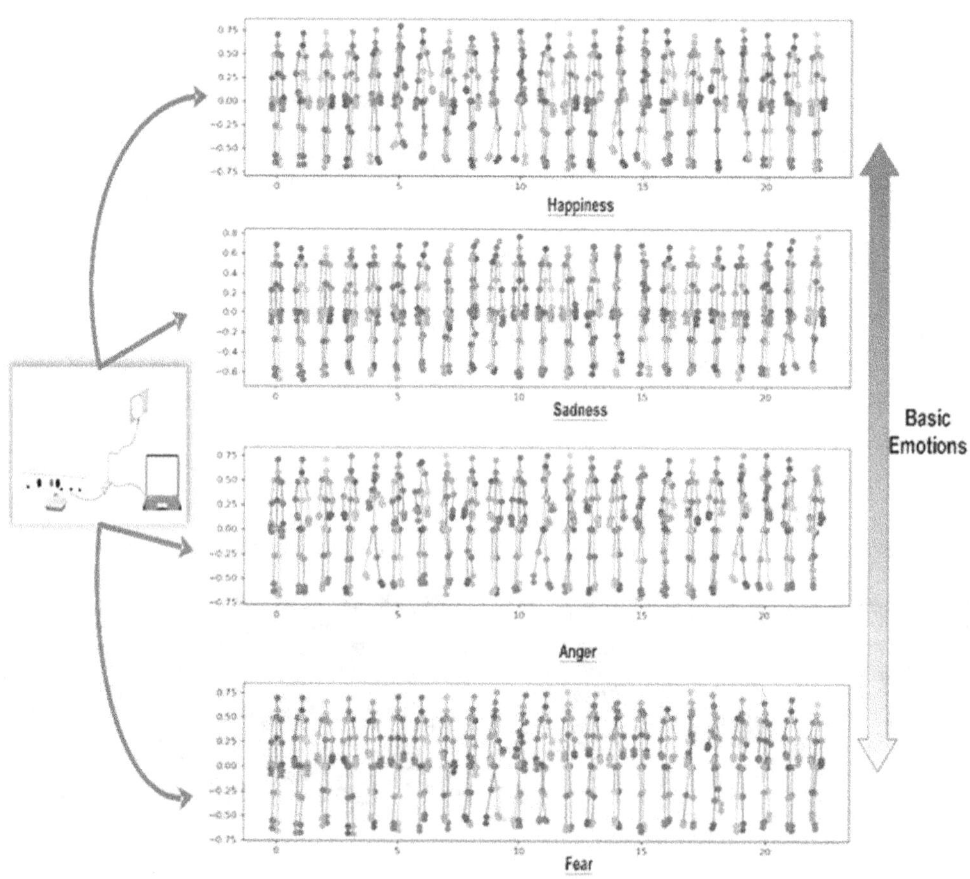

Emotion From Motion Recognition Systems

An emotion recognition system based on body movement can be modelled in a manner similar to an activity recognition system. However, the selected features must be consistent with physiological traits exhibited during a specific emotion. A reasonable approach here is to extract features based on an already established framework. For example, Rudolph Laban theorized a framework known as LMA (Laban Movement Analysis) to extract motion qualities (Laban & Ulmann, 2011). This framework has been used successfully to classify emotions during theatre performance (Aristidou et al., 2015). It incorporated ideas from anatomy, kinesiology and psychology to describe the human movement. The LMA framework has four components, each describing a quality of motion. These components are body, shape, effort and space. The body component generally comprises of structural information like the head inclination and flex angle. The shape component accounts for the change in shape for different emotional stimuli. The average distance of the joints from the base joint data can be used for the shape component. The effort component describes the intention and the dynamic attributes of the movement like kinetic energy and velocity of different joints of the body. Lastly, the space component captures information about how an individual utilizes the space around the body. Contraction index of the upper and lower body is used to measure space utilization during motion analysis of the human body. Figure 3 shows a conceptual flowchart of an emotion recognition system.

Once emotion specific geometric and kinematic features are extracted, these features are quantized to create a feature vector, which is subsequently used for training and testing purposes in classification task powered by the machine learning algorithms. Supervised machine learning algorithms such as Support Vector Machine (SVM), K-Nearest Neighbor (KNN), and Decision Trees (DTree) have already been used in similar applications for Kinect based action recognition (Ahmed et al., 2016), gait recognition (Ahmed et al., 2015), and in gender recognition from aesthetic preferences using an ensemble of classifiers trained by genetic algorithms (Azam & Gavrilova 2017a, 2017b). It is thus reasonable to assume

Figure 3. Conceptual flowchart of an emotion recognition system

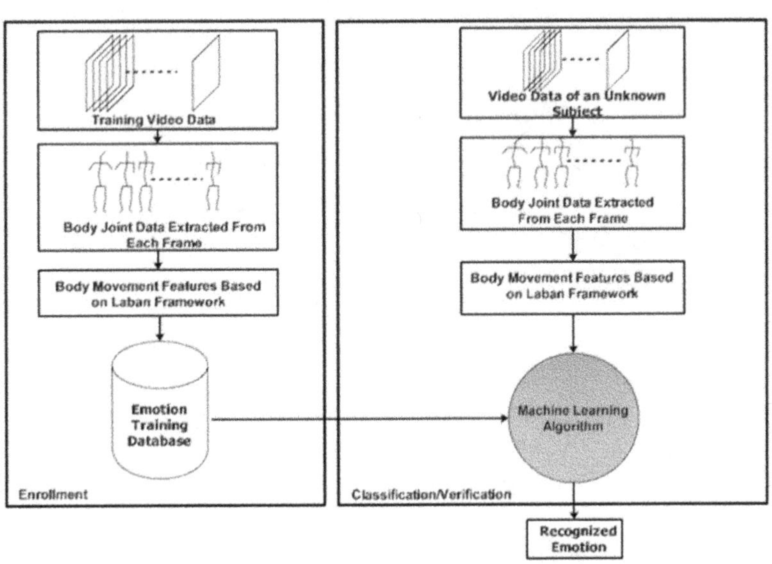

that these classifiers would produce similar performance for emotion recognition. Multiple expert systems can be trained in conjunction to create a more robust emotion recognition system. Information fusion techniques can be used to combine the scores by varying the individual contribution of the expert systems.

The first work of its kind has been recently reported in Maret et al (2018a), where research on how emotions could be identified and recognized from Kinect recordings of body joints (see Figure 4) was conducted. Their article describes a system capable of recognizing emotions such as friendly, distress, annoyed and neutral, based on body movement. Features extracted from 3D skeleton using Kinect sensor are classified by five machine learning techniques: KNN, SVM, Decision tree, Neural Network and Naive Bayes. A genetic algorithm is then invoked to find the best system parameters to achieve the higher recognition rate. The system achieved 98.96% average accuracy on the small scale experimental dataset recorded under controlled environment. The system integration with embedded hardware for emotion recognition in real time in an emergency response scenario is currently underway (Maret et al., 2018b).

Real-Time Gesture Recognition Methodology

As it was shown in the above discussion, many activity recognition approaches rely on machine learning for feature classification tasks. However, almost no research looked into usage of Kinect in real-time for tasks beyond gait and activity recognition. While there has been a surge in the number of action recognition systems developed based on Kinect sensor technology, recognizing emotion or performing crowd observation from the body movements or gestures has been rarely considered. Very recently, the first embedded hardware device that monitors the number of people inside the confined space using Kinect sensor was developed (Munier et al., 2017). This section outlines one of the first studies on the potential use of Kinect in situation awareness systems to provide additional cues in response to an emergency situation. It presents a new framework for using Kinect gait sensor to identify a person in distress, through

Figure 4. Kinect skeletal joints tracked for emotion recognition

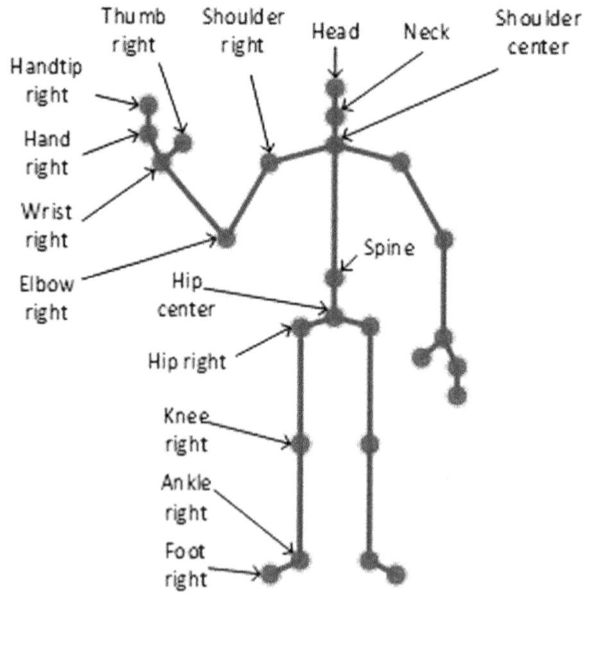

observation of a person's upper body movement. The classification relies on a commonly used supervised machine learning method: SVM and a framework to integrate Kinect based gesture recognition method with a real-time embedded system.

The overview of the methodology for the first prototype of a real-time Kinect emotion recognition system is presented below. Figure 5 depicts the flowchart of the proposed method for gesture recognition based on Kinect sensor capturing upper body motion. The tracking library, Body Skeleton Tracker (BST) is proposed to be used with OpenNI2 framework, which provides structured frames. Then, the BST returns the body joints corresponding to hand, elbow, shoulder (left and right), head and shoulder center, as shown in Figure 6. This process is performed using C library and the KEReco library developed specifically as an add-on to a Python script.

Kinect uses infrared sensor and depth image to detect and track a 3D object. The SDK also provides twenty-five 3-dimensional coordinates of the joint locations of the human body. Figure 7 shows the names of the joint locations extracted using Kinect. The coordinate system is transformed to hip center to eliminate the variability of the distance between the Kinect sensor and the subject. The skeleton joint data was recorded for each frame during the entire gait cycle. From the skeleton joint location data of each subject, geometric and kinematic features were collected using a proposed method.

In order to match the data from the BST, the tracking reference point has to be changed. Mathematical transformation have to be made to adjust the view, they are a translation from the top right to the middle of the camera field of view and two symmetry transformations on XZ and YZ axis. Equation 1 shows all the transformations applied.

$$\left(f_2 \, o \, f_1\right)\left(x, y, z\right) = \left(\begin{pmatrix} x \\ y \\ z \end{pmatrix} + \vec{v_1}\right)\begin{pmatrix} -1 \\ -1 \\ 1 \end{pmatrix} \tag{1}$$

A conversion from pixels to meters also has to be performed in order to match the data origin. The grabber process obtains every depth frame from Kinect and stores them in a temporary buffer. When a new frame is saved in the first buffer, any decoding thread can pick a depth frame. Then this thread will decode the depth frame and will return tracked joints. These tracker joints are stored in the second buffer which will be resynchronized if a recent frame is decoded faster than an older (see Figure 8). This process is explained below. The main issue of running a gesture recognition method on the embedded

Figure 5. Real-time gesture tracking system flowchart

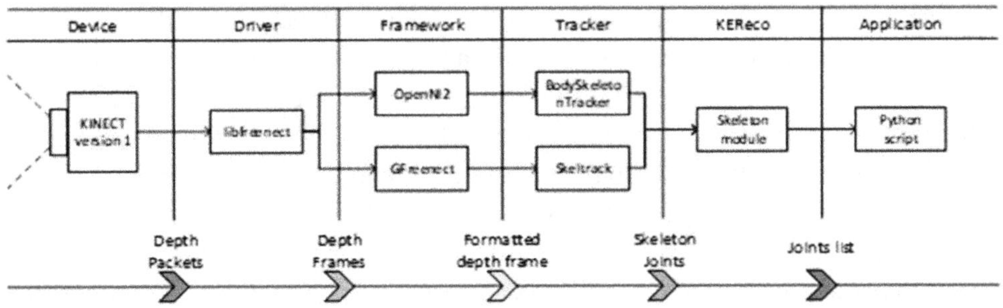

Figure 6. Extracted skeletal joints

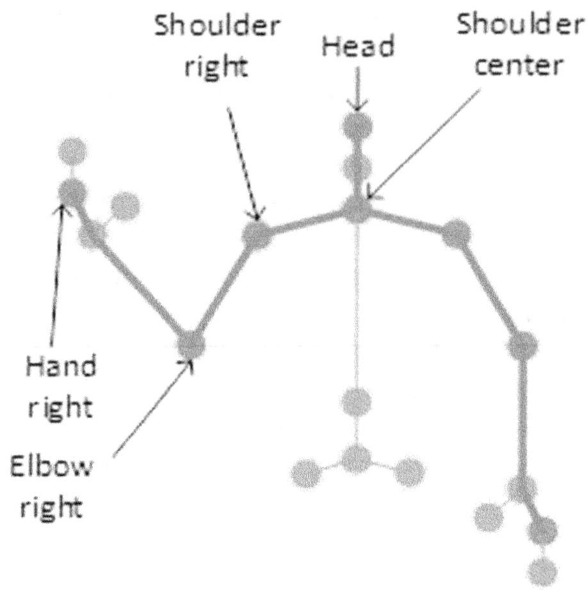

Figure 7. Extracted skeletal joints

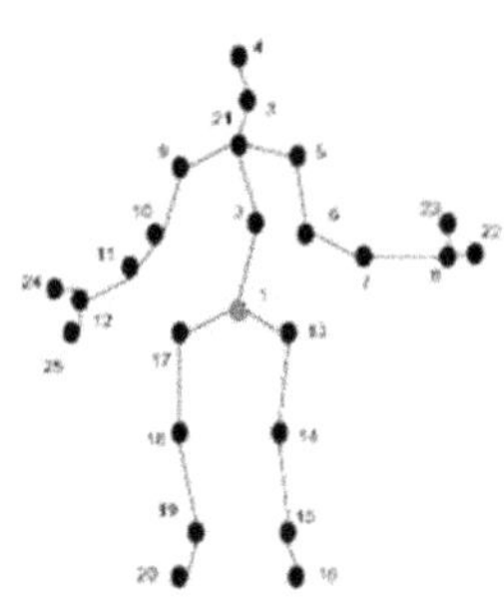

Right Knee (18)	Hip Center (1)	Right Elbow (10)
Right Ankle (19)	Spine Mid (2)	Right Wrist (11)
Right Foot (20)	Neck (3)	Right Hand (12)
Left Hand Tip (21)	Head (4)	Left Hip (13)
Left Hand Tip (22)	Left Shoulder (5)	Left Knee (14)
Left Thumb (23)	Left Elbow (6)	Left Ankle (15)
Right Hand Tip (24)	Left Wrist (7)	Left Foot (16)
Right Thumb (25)	Left Hand (8)	Right Hip (17)
	Right Shoulder (9)	

hardware is the skeleton decode time. Running on a single core takes up to three seconds to process all the 30 frames per second (FPS). At most 10FPS can be processed on a single core. The skeleton algorithm library was optimized to be run in real time. The optimization was done by using more threads to decode the skeleton over multiple cores. The optimization was made by increasing the number of threads that tracks the skeleton from the depth frame. The Kinect grabber thread stores frames inside a buffer where every other decoder threads can pick one of them randomly with the corresponding frame position (FID). A wake up condition activates all threads when depth frames are available. Once the decoder process is finished, all threads store skeleton's joints in a buffer where the frame will be resynchronized. If a thread has the next following frame ID, this thread can notify the callback. If all frame IDs are not consecutive, the thread will continue decoding the next frame. If the next thread can reorder correctly all the frames, then it will flush the buffer to the callback. Note that buffers are thread safe, which means that only a single thread can access the buffer at a time. If a thread is still decoding a first frame and the buffer becomes overflowed, a re-synchronization will be made. Finally, after resynchronization, all joints are sent to the application with a callback. The feature computation is performed similarly to

(Ahmed et al., 2016), which used JRD –Joint Relative Distance and JRA –Joint Relative Angle features. In order to recognize emotions based on arm gestures, Support Vector Machine (SVM) classifier was used. The best configuration to use within the SVM classifier was determined experimentally and used 75 features from 15 frames.

The frame buffer was used to hold at most 75 features before a new prediction was made by the classifier. After the classifier prediction, the buffer removed the oldest frame. The extracted features from the next frame would then be added to the buffer and a new prediction made.

Experimental Results

Some preliminary results obtained on a small scale sample dataset demonstrate that gestures are correctly recognized in real-time. The dataset consisting of eight subjects with four gestures was collected using Kinect version 2. The training set contained four gestures recorded at two distances: one collected two meters away from the Kinect and a second collected four meters away.

The experiments were run using two-fold cross-validation. This dataset was used to train the SVM classifier in order to predict confidence rate in a given gesture over other gestures. The main experiment is based on recognizing emotions from gestures in real time. Emotions to be recognized were: friendly, neutral, distress, neutral, annoyed and neutral. The neutral state was held for 5 second while others emotions for 10 seconds. The experiment was made using the SVM classifier and the Kinect V2 sensor. The Microsoft Kinect SDK 2 tracked all skeleton joints. From these joints features were extracted and then predicted by a classifier with the obtained confidence rate presented in Figure 9. These preliminary results demonstrate that the methodology is suitable for real-time body gesture recognition and the subsequent rescue operations. In order to avoid the transition effect when gestures changed, a threshold for gesture to be stable for 30 frames was set, which mitigated the issue. Another observation is that friendliness and annoyance were the two closest gestures, while all the other gestures were easily distinguishable. SVM confidence was above 85% for each gesture, computed according to the distance from the hyperplane.

Figure 10 shows a sequence of emotions from gestures, accurately recognized in 100% of cases in real-time for each frame. Note that since the developed hardware prototype is the first system of its kind, no comparators exist that could identify the four gestures based on sensor based recorded arm movement at the time. Thus, in order to validate our results, we extracted the AuC data for seven difference sce-

Figure 8. Extraction joints optimization

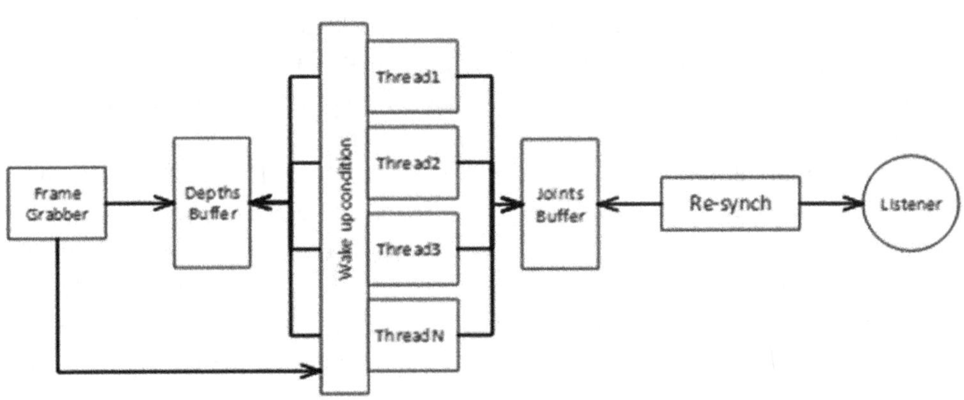

Figure 9. SVM continuoud gesture recognition on embedded system

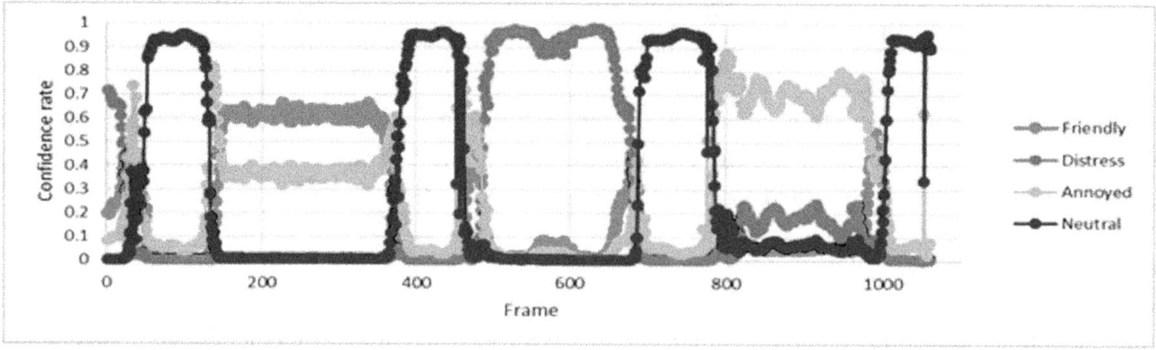

**For a more accurate representation see the electronic version.*

Figure 10. SVM real-time gesture recognition results

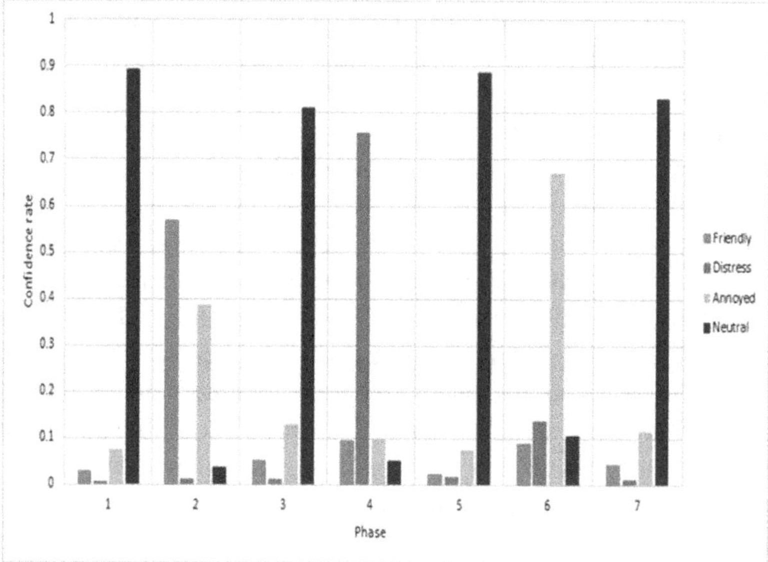

**For a more accurate representation see the electronic version.*

narios where the subject was asked to perform a sequence of arm gestures (Table 2 and Table 3) and the Confusion Matrix (Table 4 and Table 5). We performed experiments to compare method performance for PC (Personal Computer) and ES (Embedded System). The results demonstrated that both could predict the correct emotion with up to 97.7%. Note that the accuracy of the predicted class vs actual class was highest for PC for emotion of distress. This is the expected result, and very useful as the main intent of the system is to recognize a person who requires the assistance in emergency situations. The lowest recognition was for Friendly emotion (70.2% on PC, 61.7% on ES). The rest of the results were very high, with Distressed and Neutral emotions being recognized correctly in over 90% cases for both PC and ES, and Annoyance recognized with over 70% accuracy for ES and over 90% accuracy on PC. We conclude that for neutral emotion, the recognition rate is the same for PC and ES. For dynamic emotion recognition, the BodySkeletonTracker (BTS) performs slightly worse than the Microsoft SDK. We assume that this is due to the real-time tracking process implemented on the hardware. Further research is required to investigate this issue more comprehensively.

Table 2. Four emotions AUCurve for ES

		Area Under Curve			
		AUCFriendly	**AUCDistress**	**AUCAnnoyed**	**AUCNeutral**
Scenarios	1	0.479	0.216	5.564	89.74
	2	177.609	1.919	89.103	11.367
	3	1.128	0.728	6.258	116.884
	4	10.778	228.999	18.003	8.218
	5	0.975	0.89	6.275	110.858
	6	2.318	16.857	253.25	17.573
	7	1.432	1.251	6.751	119.564

Another observation is that the Microsoft skeleton is more stable when the same arm movement is of a certain duration, while the BST skeleton is less stable during transition. The BST system has to recalculate positions of joints during an arm movement and an emotion change. Despite that, the BST system provides a reliable real-time solution on an embedded board, which can be used in multiple real-world scenarios. The overall deviation of the embedded system prediction from the PC version is about 9%, which can be tolerated because the identified human emotion and the arm movement recognition still remains correct in all cases.

There are many application domains for the developed system. Emotion recognition systems can be used to enhance the decision-making capability of human operators. For security at an airport and surveillance systems, behavioural and other physiological traits can be monitored to capture true intention of a person [Adey 2009]. Assisted technologies can be developed to improve human-computer interactions. Intelligent avatars and other interactive dialog systems can improve user's experience while accessing smart communication devices (Faundez-Zanuy et al., 2013). Finally, emotion recognition system can also be used in rescue operations or in a patient observation scenario. Typically, an expert will observe emotions expressed through gestures or body movement to identify signature behaviour in the patients. An intelligent system capable of distinguishing emotional states of an individual can be useful for monitoring a patient and for making quick responses. The extension of this research will

Table 3. Four emotions AUCurve for ES

		Area Under Curve			
		AUCFriendly	**AUCDistress**	**AUCAnnoyed**	**AUCNeutral**
Scenarios	1	0.641	0.284	4.731	70.341
	2	140.861	3.025	100.381	15.731
	3	0.55	0.277	4.899	67.273
	4	23.468	176.992	22.352	6.186
	5	0.794	0.479	6.706	86.019
	6	21.768	30.831	150.526	18.873
	7	0.401	0.291	3.658	43.648

assist not only in establishment of new methodology in the domain of biometric security, but also in assisting new method integration in the health care domain, providing personalized healthcare targeting to specific patient needs.

DEVELOPED METHODOLOGY IN DISASTER RESPONSE AND MEDICAL APPLICATIONS

While the previous section discussed in detail how biometric sensors can capture and process biometric information from body movement, and recognize gestures in real-time, this section gives emphasis on the extension of this emerging research to the fields of medical patient treatment and disaster response. The prototype system presented in the previous section can be used in one of many emergency scenarios, as well as in multiple medical applications.

Although the scale and the methods may vary, technology has always played a crucial role in ensuring public safety. Within the domains of rescue operations and natural disaster recovery, biometric technology offers a promising future for first responders and victims alike. In life threatening rescue operations, technology needs to be accurate, portable, reliable and inexpensive. Microsoft's Kinect lends itself thoroughly to this effort by offering compact sensors at a comparatively low cost (Zhang et al., 2013). Rescue operations often take place in harsh environments which can be described as uncertain, dynamic, hazardous and lacking in resources (Wong et al., 2017). This is especially true in a case of natural disasters, where communication may be limited and unstable. Due to the possible number of victims following every crisis, manual labor alone can prove impractical when the affected area is substantial. With these constraints in mind, research into autonomous systems combined with Kinect sensors have recently gained traction (Al-Kaff et al., 2017; Osman et al., 2017; Wang et al., 2017). While the applica-

Table 4. Confusion matrix for Predicted Class for PC

		Predicted Class			
		Friendly	**Distress**	**Annoyed**	**Neutral**
Actual Class	Friendly	0.702	0.007	0.282	0.007
	Distress	0.004	0.977	0.008	0.009
	Annoyed	0.008	0.063	0.903	0.024
	Neutral	0.008	0.005	0.052	0.932

Table 5. Confusion matrix for Predicted Class for ES

		Predicted Class			
		Friendly	**Distress**	**Annoyed**	**Neutral**
Actual Class	Friendly	0.617	0.01	0.366	0.006
	Distress	0.029	0.945	0.02	0.005
	Annoyed	0.058	0.159	0.713	0.068
	Neutral	0.005	0.003	0.056	0.932

tion of the Kinect in real life scenarios can be limited due to occlusions, unfavorable lighting, noise and distance to a sensor, methodologies that seek to overcome those limitations are currently being actively developed. Some examples of such applications can be found below.

Natural Disasters

In the aftermath of a natural disaster, the natural environment can have noxious fumes, fires, exposed powerlines or unstable debris. Rescuer and victims' safety is a primary concern when navigating a disaster zone. Recent research has proposed the use of automatic approaches for providing aid and accessing the environment. By employing biometric identification algorithms, Kinect can be used to recognize and locate victims in distress. It was recently demonstrated that the Kinect RGB-D sensor can be used to construct a 3D model of an indoor environment, which may be too dangerous for humans to reach (Wang et al., 2017). Research has also been done to develop identification of persons and objects for Unmanned Aerial Vehicles (UAVs) during airborne rescue operations (Al-Kaff et al., 2017; Osman et al., 2017). These works are focused on decreasing human involvement in potentially dangerous situations, while remaining effective and robust in a limited environment. Usage of Kinect depth maps have been considered for disaster detection as well. Improved methodologies that handle occlusion of turbid waters for earthquake or tsunami detection have been tested in a lab setting, with experimental results outperforming current methods (Liu et al., 2017).

Refugee Camps

Large displacements of people are difficult to keep track of, particularly when the matter can evolve into a national security concern. It is very taxing for governments to deal with the maintenance of refugee camps or disaster centers, which often results in missing or unreported people. Biometric technology can reduce confusion because it does not require paperwork and has multi-modal systems to ensure accuracy. Microsoft's Kinect can employ gait recognition in refugee scenarios due to its speed, low cost and unobtrusive nature. Previous research supports the use of multiple Kinect sensors within the same system to increase coverage without a loss in accuracy (Banerjee et al., 2013). This would result in decreased queue times, more organized registration and accurate records. The non-pervasive gait identification additionally makes data collection more acceptable compared to other biometrics, such as traditional fingerprints for example.

Hostage Situations

There are many constraints to operations when hostages are involved. Technology must be compact, covert and silent to ensure victims safety. The Kinect sensors excel in this field because of its small size and ability to produce both RGB images and depth maps. Gesture recognition could see much usage in hostage situations, as the rescue team could communicate with the technology while being at a distance and without a sound. Morse code, for example, can be deciphered by the Kinect with relevant algorithms (Li et al., 2017). Extensive research into hand and finger tracking using the Kinect exists, though optimal performance occurs only in certain circumstances (Smari et al., 2016). Additional research involves the incorporation of the Kinect into radar biometric detection by simulation of Micro-Doppler signatures.

These signatures can deliver biometric details including heartbeats, blood vessels and breathing, with certain methods that can detect vital signals through walls (Amir et al., 2017).

Neurological Disease Treatment

As it was noted above, Kinect sensor is a light, a portable and a convenient device, which makes it useful not only for disaster and recovery, but also for medical providers and their patients (Roy et al., 2013; Stone & Skubic, 2013). In 2012, research team (Fernandez-Baena et al., 2012) compared the accuracy of optical motion capture professional system with reflective markers and Kinect motion capture in capturing joint angle data. They concluded that the precision of the Kinect is lower than that of an optical capture system, but its accuracy is sufficient for medical rehabilitation.

Some neurological diseases (such as stroke, Alzheimer's disease) and limb dysfunction caused by external trauma can seriously affect the patient's motor ability and quality of life. Intensive and repetitive training can help patients with physical rehabilitation and neural reorganization. Traditionally, physiotherapy and rehabilitation training is performed at a rehabilitation center of a hospital, and the patient needs to wear a tracking device. Several studies have evaluated the accuracy and feasibility of using Kinect for motion capture in rehabilitation systems. For example, in Chang et al (2011), researchers assessed the possibility of using Kinect-based system for rehabilitation of young adults with motor impairments. They used an A-B-A-B reversal replication design and reached a conclusion that Kinect-based systems enhance the therapeutic effect due to their interactive interface with audio and video feedback. However, the results were based on only two cases and more participants should to be used to validate the study. In Lange et al. (2011), researchers developed and evaluated an interactive game-based rehabilitation tool for balance training of adults with neurological injury. They demonstrated that the system was very well accepted by the patients.

Elderly Care and Rehabilitation

Elderly care and rehabilitation recently emerged as the two main directions in this field. In 2012, Stone and Skubic presented initial results of monitoring the gait of seven older adults continuously, using a Kinect sensor, along with a methodology for computing average gait parameters which may be predictive of future falls and adverse events in older adults (Stone & Skubic, 2012). In 2012, Zhang et al (2012) designed a method of extracting data from 2D images when a human subject was out of range of the Kinect's depth sensing, to enhance fall detection system capability. Soon after, researchers designed a Kinect-based system for in-home rehabilitation exercises monitoring and guidance (Zhao et al., 2014). The system implemented a set of correctness rules for each exercise and assessed the patient's movement in real-time to help the patient perform the exercise correctly. In addition, it provided an interactive feedback to patients for rehabilitation exercise monitoring applications.

Virtual and Augmented Reality for Medical Systems

Over the past decade, many researchers have focused on finding appropriate sensors and computing technology for motion capture to help develop ancillary medical systems (Chang et al., 2011; Fernandez-Baena et al., 2012). Virtual and augmented reality systems are being developed and tested to motivate patients. Use of such systems makes repetitive exercises more useful and interesting, and many of such systems are currently based on Kinect sensor. Kinect was intended as a game equipment, thus many studies combined its initial function with medical rehabilitation to design game-based rehabilitation system. In 2012, Gotsis and his team (Gotsis et al., 2012) designed an integrated hardware (consisting of a custom metal rig, six game tracks, a Microsoft Kinect, a laptop and a large screen) and software platform for rapid prototyping of virtual reality-based games for upper body exercise and rehabilitation. At the same time, Huang et al (2012) designed a smart glove for motion and finger angle extraction and integrating with Kinect for an upper extremities gaming application. However, such systems have to be more entertaining so that the patient could be better motivated. In 2013, a system known as "Kinect-o-Therapy" was developed, which incorporated serious games that process 3D input using the Microsoft Kinect motion sensor to build an effective rehabilitation system that gives feedback to the patient and the doctors (Roy et al., 2013). It used real-time motion capture data to animate characters in game, making rehabilitation more fun, engaging and rewarding. Since then, many researchers have focused on using Kinect in combination with motivating and fun recovery games, feedback systems with motion tracking, and medical systems for in-home use.

OPEN PROBLEMS AND FUTURE RESEARCH DIRECTIONS

As Kinect sensors emerged only recently as an active area of research with multiple applications, there are many problems and directions for the future work. Emotion recognition is an emerging area of research and has put forth several open-ended questions to the scientific community. One crucial question that needs to be answered is *whether body movement can be used as a standalone communication channel for emotional expression.* Research conducted up until now shows that it is beneficial to use body movement features in a multi-modal system alongside facial and voice features, as typically this results in a better performance for most practical applications (Gavrilova & Monwar, 2012). Biometric fusion can also in the future automatically recognize in real time if a patient is experiencing distress, and to inform health care provider on patient-specific recommendation of exercises or treatment.

Many spatially localized features can be extracted from the analysis of body movement. Finding out *the right subset of features and their individual contribution to classify emotion* is still an open problem (Maret et al., 2018a). Algorithms such as particle swarm optimization (PSO) and genetic algorithm (GA) can be used to generate the right subset of features along with the correct weight distribution to measure the individual impact of such features in emotion recognition. Finally, testing those methods on large-scale datasets in real-time will assist in providing definite answers to the open problems posted above.

One of the key issues is optimizing methods for *actual application outside of testing environments.* The constraints of size, speed, cost and accuracy have always been a limiting factor (Lange et al., 2011). Much of the proposed future work includes expanding the conditions in which Kinect can acceptably function. Since real disaster scenarios will not have favorable conditions, research must be done to address the innate weaknesses of the sensors or methodologies. This can include work into occlusion correction, lighting tolerance and noise filtering. Once usage of the Kinect is more universally applicable, mass adoption into rescue processes seems most logical.

In order to integrate the Kinect sensor in core medical applications, *a critical validation step is needed.* This will provide proper evaluation of the accuracy and latency of the action capture data obtained from Kinect for special diagnostic scenarios. In 2015, gait analysis of patients with Parkinson's disease (Bigy et al., 2015) was studied using healthy actors in a laboratory environment. Next year, the accuracy, precision, resolution, various measurement distances and angles of both Kinect 1 and Kinect 2 were investigated (Pohlmann et al., 2016). However, more studies to evaluate patient benefit in a controlled and randomized fashion should be performed in the future.

Another challenge in medical applications based on the Kinect is that repurposing a technology initially intended for young and healthy people into a system for elderly patients *requires new design strategies for the physiological and the psychological considerations* (Adey, 2009). First of all, medical rehabilitation games based on Kinect should be designed with high standards of captivating and long-term enjoyment potential while maintaining seamless methodologies of adaptation and monitoring of players' needs. But overexertion injuries in both individual movements and length of play are not encouraged. Also, user-friendly interfaces should be applied and wearable system should not be bulky. Thus, making existing medical systems more ergonomic is another open problem.

As the Kinect is an inexpensive device, some factors may affect its performance which may make it inadequate for some medical applications that require a high-level of accuracy. *In a multi-modal and multi-sensor scenario*, this issue should be carefully investigated. Over recent years, a new approach that advocates the use of multiple Kinect sensors was introduced to improve the data capturing accuracy. For example, (Beyl et al., 2013) discovered that the use of multiple Kinect sensors and careful calibration in the challenging uncontrollable environment during surgery could provide a median accuracy of 19.8 mm; however, the maximum error was around 100 mm, which was still too high to be practical. In addition, other sensors and equipment can be fused with Kinect to enhance quality and accuracy in the future through multi-modal and multi-sensor systems. Finally, *issues of data privacy* in the context of medical, virtual reality or search and rescue operations should not be overlooked.

CHAPTER SUMMARY

This book chapter outlined the current state of the art of Kinect gait and activity authentication, while focusing on emotional cues that could be observed from human body and posture. It presented a prototype of a new system which combined recently developed behavioral gait and posture recognition for human emotion identification. It then presented an overview of the first real-time gesture recognition system which could be used in emergency response or situation awareness systems, together with preliminary experimental results. The survey of recent progress and open problems in the areas of medical patient rehabilitation and rescue operations as linked to sensor based body movement concluded this chapter.

The chapter outlined some of the future directions of sensor based motion tracking research for a variety of application. The key open problems that should be addressed in the future research were also identified. They include introduction of new multi-Kinect and multi-sensor approaches for medical scenarios; development of the occlusion correction, lighting tolerance and a noise filtering methods; integration and testing of developed methodologies in real-time; careful assessment of the use of sensors for patient rehabilitation on a larger population sample and on different types of skeletomuscular conditions. In addition, it was pointed out that the multi-disciplinary team of experts should be involved in deploying new technology-enabled methodologies, and a special care and consideration should be given to privacy aspects of collecting and processing sensitive data.

ACKNOWLEDGMENT

Authors would like to thank NSERC DG program grant RT731064 and NSERC ENGAGE grant EGP522291 for partial support of this project, as well as all members of the Biometric Technologies laboratory at the University of Calgary and all collaborators for their valuable discussions during this chapter preparation.

REFERENCES

Adey, P. (2009). Facing airport security: Affect, biopolitics, and the preemptive securitization of the mobile body. *Proceedings of Environment and Planning: Society and Space*, *27*(2), 274–295.

Ahmed, F., Paul, P. P., & Gavrilova, M. (2015). DTW-based Kernel and Rank Level Fusion for 3D Gait Recognition using Kinect. *The Visual Computer*, *31*(6-8), 915–924. doi:10.100700371-015-1092-0

Ahmed, F., Paul, P. P., & Gavrilova, M. (2016). Joint-triplet motion image and local binary pattern for 3d action recognition using Kinect. In *Proceedings of 29th Int. Conference on Computer Animation and Social Agents* (pp. 111-119). Academic Press.

Ahmed, F., Paul, P. P., & Gavrilova, M. (2016). Adaptive Pooling of the Most Relevant Spatio-Temporal Features for Action Recognition. In *Proceedings of 2016 IEEE International Symposium on Multimedia (ISM 2016)* (pp. 177-180). IEEE. 10.1109/ISM.2016.0042

Ahmed, F., Sieu, B., & Gavrilova, M. (2018). Score and Rank-Level Fusion for Emotion Recognition Using Genetic Algorithm. In *Proceedings of 17th IEEE International Conference on Cognitive Informatics & Cognitive Computing* (pp. 46-53). IEEE.

Al-Kaff, A., Moreno, F. M., Escalera, A. D. L., & Armingol, J. M. (2017). Intelligent vehicle for search, rescue and transportation purposes. In *Proceedings of IEEE International Symposium on Safety, Security and Rescue Robotics (SSRR)* (pp. 110-115). IEEE. 10.1109/SSRR.2017.8088148

Amir, S., Waqar, A., Siddiqui, M. A., Majeed, M., Haseeb, H., & Mehmood, S. H. (2016). Kinect Controlled UGV. *Proceedings of Wireless Personal Communications*, *95*(2), 631–640. doi:10.100711277-016-3915-3

Aristidou, A., Charalambous, P., & Chrysanthou, Y. (2015). Emotion analysis and classification: Understanding the performers' emotions using the LMA entities. *Proceedings of Computer Graphics Forum*, *34*(6), 262–276. doi:10.1111/cgf.12598

Azam, S., & Gavrilova, M. (2017a). Person Identification Using Discriminative Visual Aesthetic, 30th Canadian Conference on Artificial Intelligence. In *Proceedings of AI'17 Springer, LNAI 10233* (pp. 15-26). Springer.

Azam, S., & Gavrilova, M. (2017b). Biometric Pattern Recognition from Social Media Aesthetics. *International Journal of Cognitive Informatics and Natural Intelligence, IGI Global*, *11*(3), 1–16. doi:10.4018/IJCINI.2017070101

Baltrušaitis, T., McDuff, D., Banda, N., Mahmoud, M., El Kaliouby, R., Robinson, P., & Picard, R. (2011). Real-time inference of mental states from facial expressions and upper body gestures. In *Proceedings of Automatic Face & Gesture Recognition (FG 2011), 2011 IEEE International Conference on, IEEE* (pp. 909–914). IEEE.

Banerjee, R., Sinha, A., & Chakravarty, K. (2013). Gait based people identification system using multiple switching Kinects. In *Proceedings of 13th International Conference on Intelligent Systems Design and Applications* (pp. 182-187). Academic Press.

Bänziger, T., Grandjean, D., & Scherer, K. (2009). Emotion recognition from expressions in face, voice, and body. In *Proceedings of the Multimodal Emotion Recognition Test (MERT) Emotion (Washington, D.C.)*, *9*(5), 691–698. doi:10.1037/a0017088 PMID:19803591

Baveye, Y., Dellandrea, E., Chamaret, C., & Chen, L. (2015). Liris-accede: A video database for affective content analysis. *IEEE Transactions on Affective Computing*, *6*(1), 43–55. doi:10.1109/TAFFC.2015.2396531

Beyl, T., Nicolai, P., Raczkowsky, J., Worn, H., Comparetti, M. D., & De Momi, E. (2013). Multi Kinect people detection for intuitive and safe human robot cooperation in the operating room. In *Proceedings of 16th International Conference on Advanced Robotics* (pp.1–6). Academic Press.

Bigy, A., Banitsas, K., Badii, A., & Cosmas, J. (2015). Recognition of postures and freezing of gait in Parkinson's disease patients using Microsoft Kinect sensor. In *Proceedings of 7th Annual International IEEE EMBS Conference on Neural Engineering* (pp. 731–734).IEEE.

Castellano, G., Villalba, S. D., & Camurri, A. (2007). Recognising human emotions from body movement and gesture dynamics. In *Proceedings of International Conference on Affective Computing and Intelligent Interaction* (pp. 71–82). Springer. 10.1007/978-3-540-74889-2_7

Chang, Y. J., Chen, S. F., & Huang, J. D. (2011). *A Kinect-based system for physical rehabilitation: A pilot study for young adults with motor disabilities.* Academic Press.

Chen, C., Jafari, R., & Kehtarnavaz, N. (2015). A multimodal dataset for human action recognition utilizing a depth camera and a wearable inertial sensor. In *Proceedings of IEEE International Conference on Image Processing (ICIP)* (pp. 168-172). IEEE.

Dalal, N., Triggs, B., & Schmid, C. (2006). Human detection using oriented histograms of flow and appearance. In *Proceedings of European Conference on Computer Vision* (pp. 428–441). Springer.

De Gelder, B. (2009). Why bodies? Twelve reasons for including bodily expressions in affective neuroscience. *Philosophical Transactions of the Royal Society of London. Series B, Biological Sciences*, *364*(1535), 3475–3484. doi:10.1098/rstb.2009.0190 PMID:19884142

Fan, Y., Lu, X., Li, D., & Liu, Y. (2016). Video-based emotion recognition using CNN-RNN and C3D hybrid networks. In *Proceedings of the 18th ACM International Conference on Multimodal Interaction* (pp. 445-450). ACM.

Fasel, B., & Luettin, J. (2003). Automatic facial expression analysis: A survey. *Pattern Recognition*, *36*(1), 259–275. doi:10.1016/S0031-3203(02)00052-3

Faundez-Zanuy, M., Hussain, A., Mekyska, J., Sesa-Nogueras, E., Monte-Moreno, E., Esposito, E., & Lopez-de-Ipiña, A. K. (2013). Biometric applications related to human beings, there is life beyond security. *Cognitive Computation*, *5*(1), 136–151. doi:10.100712559-012-9169-9

Felzenszwalb, P., McAllester, D., & Ramanan, D. (2008). A discriminatively trained, multiscale, deformable part model. In *Proceedings of CVPR* (pp. 1–8). IEEE.

Fernandez-Baena, A., Susín, A., & Lligadas, X. (2012). Biomechanical Validation of Upper-Body and Lower-Body Joint Movements of Kinect Motion Capture Data for Rehabilitation Treatments. In *Proceedings of 4th International Conference on Intelligent Networking and Collaborative Systems* (pp. 656-661). Academic Press.

Fothergill, S., Mentis, H., Kohli, P., & Nowozin, S. (2012). Instructing people for training gestural interactive systems. In *Proceedings of the SIGCHI Conference on Human Factors in Computing Systems* (pp. 1737–1746). ACM.

Fragopanagos, N., & Taylor, J. G. (2005). Emotion recognition in human–computer interaction. *Neural Networks*, *18*(4), 389–405. doi:10.1016/j.neunet.2005.03.006 PMID:15921887

Gavrilescu, M. (2015). Recognizing emotions from videos by studying facial expressions, body postures and hand gestures. In *Proceedings of Telecommunications Forum (TELFOR)* (pp. 720-723). IEEE.

Gavrilova, M., Ahmed, F., Azam, S., Paul, P., Sultana, M., Rahman, W., & Zohra, F. T. (2016). Emerging Trends in Security System Design using the Concept of Social Behavioral Biometrics. Information Fusion for Cyber-Security Analytics, 691, 229-252.

Gavrilova, M., & Monwar, M. (2012). *Multimodal Biometrics and Intelligent Image Processing for Security Systems*. IGI Global.

Gavrilova, M., Wang, Y., Ahmed, F., & Paul, P. P. (2017). Kinect Sensor Gesture and Activity Recognition. *IEEE Consumer Electronics Magazine*, *7*(1), 88–94. doi:10.1109/MCE.2017.2755498

Gotsis, M., Tasse, A., Sweden, M., Lympouridis, V., Poulos, I., Thin, A., ... Jordan-Marsh, M. (2012). Mixed reality game prototypes for upper body exercise and rehabilitation. In *Proceedings of IEEE Virtual Reality Workshops (VRW)* (pp. 181-182). Costa Mesa, CA: IEEE.

Gunes, H., & Piccardi, M. (2005). Affect recognition from face and body: early fusion vs. late fusion. In *Proceedings of 2005 IEEE International Conference on Systems, Man and Cybernetics* (vol. 4, pp. 3437–3443). IEEE. 10.1109/ICSMC.2005.1571679

He, H., She, Y., Xiahou, J., Yao, J., Li, J., Hong, Q., & Ji, Y. (2018). Real-Time Eye-Gaze Based Interaction for Human Intention Prediction and Emotion Analysis. *Proceedings of Computer Graphics International*, *2018*, 185–194.

Huang, M. C., Xu, W., Su, Y., Lange, B., Chang, C. Y., & Sarrafzadeh, M. (2012). Smartglove for upper extremities rehabilitative gaming assessment. In *Proceedings of 5th International Conference on Pervasive Technologies Related to Assistive Environments, PETRA '12* (pp. 20:1–20:4). ACM.

Jain, A., Flynn, P., & Ross, A. (2008). Handbook of Biometrics: Eds. doi:10.1007/978-0-387-71041-9

Kisku, D. R., Gupta, P., & Sing, J. K. (2013). *Advances in biometrics for secure human authentication and recognition*. CRC Press. doi:10.1201/b16247

Koolagudi, S., & Rao, K. (2012). Emotion recognition from speech: A review. *Proceedings of International Journal of Speech Technology, 15*(2), 99–117. doi:10.100710772-011-9125-1

Kosti, R., Alvarez, J. M., Recasens, A., & Lapedriza, A. (2017). Emotion recognition in context. *The IEEE Conference on Computer Vision and Pattern Recognition (CVPR)*. doi: 10.1109/CVPR.2017.212

Laban, R., & Ullmann, L. (2011). *The Mastery of Movement* (4th ed.). Binsted, UK: Dance Books Ltd.

Lange, B., Chang, C. Y., Suma, E., Newman, B., Rizzo, A. S., & Bolas, M. (2011). Development and evaluation of low-cost game-based balance rehabilitation tool using the Microsoft Kinect sensor. In *Proceedings of Engineering in Medicine and Biology Society, EMBC, 2011 Annual International Conference of the IEEE* (pp. 1831-1834). IEEE.

Lee, Y. S., & Chung, W. Y. (2012). Visual sensor based abnormal event detection with moving shadow removal in home healthcare applications. *Sensors (Basel), 12*(1), 573–584. doi:10.3390120100573 PMID:22368486

Li, R., Nguyen, M., & Yan, W. Q. (2017). Computer Input of Morse Codes Using Finger Gesture Recognition. In *Proceedings of International Conference on Digital Image Computing: Techniques and Applications (DICTA)* (pp. 1-8). Academic Press.

Li, S., & Chan, A. B. (2014). 3d human pose estimation from monocular images with deep convolutional neural network. In *Proceedings of Asian Conference on Computer Vision* (pp. 332–347). Springer.

Liu, H., Zhang, Y., Li, Y., Zhou, Q., Tadoh, R., Uemura, T., ... Serikawa, S. (2017). Depth Map Reconstruction for Underwater Kinect Camera Using Inpainting and Local Image Mode Filtering. *IEEE Access: Practical Innovations, Open Solutions, 5*, 7115–7122. doi:10.1109/ACCESS.2017.2690455

Maret, Y., Oberson, D., & Gavrilova, M. (2018a). Identifying an Emotional State from Body Movements Using Genetic-based Algorithms. In *Proceedings of 17th International Conference on Artificial Intelligence and Soft Computing* (vol. 10841, pp. 474-485). Springer. 10.1007/978-3-319-91253-0_44

Maret, Y., Oberson, D., & Gavrilova, M. (2018b). Real-Time Embedded System for Gesture Recognition. *Proceedings of IEEE Conference on Systems, Man and Cybernatics.*

Molchanov, P., Gupta, S., Kim, K., & Kautz, J. (2015). Hand gesture recognition with 3d convolutional neural networks. In *Proceedings of the IEEE Conference on Computer Vision and Pattern Recognition Workshops* (pp. 1–7). IEEE.

Munier, S., Arora, R., Hesling, C., Li, J., Francis, J., Shelton, C., ... Berges, M. (2017). Real-time fine-grained occupancy estimation using depth sensors on ARM embedded platforms. *Proceedings of the 23rd IEEE Real Time and Embedded Technology and Application Symposium.*

Neverova, N., Wolf, C., Paci, G., Sommavilla, G., Taylor, G., & Nebout, F. (2013). A multi-scale approach to gesture detection and recognition. In *Proceedings of the IEEE International Conference on Computer Vision Workshops* (pp. 484–491). IEEE.

Newell, A., Yang, K., & Deng, J. (2016). Stacked hourglass networks for human pose estimation. In *Proceedings of European Conference on Computer Vision* (pp. 483–499). Springer.

Noroozi, F., Corneanu, C., Kamińska, D., Sapiński, T., Escalera, S., & Anbarjafari, G. (2018). *Survey on emotional body gesture recognition. IEEE Transactions on Affective Computing.*

Osman, H. I., Hashim, F. H., Zaki, W. M. D. W., & Huddin, A. B. (2017). Entryway detection algorithm using Kinect depth camera for UAV application. In *Proceedings of IEEE 8th Control and System Graduate Research Colloquium (ICSGRC)* (pp. 77-80). IEEE.

Pease, B., & Pease, A. (2004). *The definitive book of body language*. Bantam.

Pöhlmann, S., Harkness, E., Taylor, C., & Astley, S. M. (2016). Evaluation of Kinect 3D Sensor for Healthcare Imaging. *Journal of Medical and Biological Engineering*, *36*(6), 857–870. doi:10.100740846-016-0184-2 PMID:28111534

Ranganathan, H., Chakraborty, S., & Panchanathan, S. (2016). Multimodal emotion recognition using deep learning architectures. In *Proceedings of 2016 IEEE Winter Conference on Applications of Computer Vision (WACV)* (pp. 1–9). IEEE.

Roy, A. K., Soni, Y., & Dubey, S. (2013). Enhancing effectiveness of motor rehabilitation using Kinect motion sensing technology. In *Proceedings of 2013 IEEE Global Humanitarian Technology Conference: South Asia Satellite (GHTC-SAS)* (pp. 298-304). IEEE.

Saha, S., Datta, S., Konar, A., & Janarthanan, R. (2014). A study on emotion recognition from body gestures using kinect sensor. In *Proceedings of Communications and Signal Processing (ICCSP), 2014 International Conference on, IEEE* (pp. 056–060). IEEE.

Senecal, S., Cuel, L., Aristidou, A., & Magnenat-Thalmann, N. (2016). Continuous body emotion recognition system during theater performances. *Proceedings of Computer Animation and Virtual Worlds*, *27*(3-4), 311–320. doi:10.1002/cav.1714

Siddiqui, M., & Medioni, G. (2010). *Human pose estimation from a single view point, real-time range sensor. In CVPRW* (pp. 1–8). IEEE.

Smari, K., & Bouhlel, M. (2016). Gesture recognition system and finger tracking with kinect Steps. In *Proceedings of 7th International Conference on Sciences of Electronics, Technologies of Information and Telecommunications (SETIT)* (pp. 544-548). Academic Press.

Stone, E., & Skubic, M. (2012). Passive, in-home gait measurement using an inexpensive depth camera. In *Proceedings of Initial results: International Conference on Pervasive Computing Technologies for Healthcare* (pp. 183-186). IEEE.

Toshev, A., & Szegedy, C. (2014). Deeppose: Human pose estimation via deep neural networks. In *Proceedings of the IEEE Conference on Computer Vision and Pattern Recognition* (pp. 1653–1660). IEEE.

Vakanski, A., Jun, H., Paul, D., & Baker, R. R. (2018). *A Data Set of Human Body Movements for Physical Rehabilitation Exercises*. Academic Press. doi:10.3390/data3010002

Wang, H., Zhang, C., Song, Y., & Pang, B. (2017). Information-fusion based robot simultaneous localization and mapping adapted to search and rescue cluttered environment. In *Proceedings of 18th International Conference on Advanced Robotics (ICAR)* (pp. 511-517). Academic Press. 10.1109/ICAR.2017.8023658

Webster, D., & Celik, O. (2014). Systematic review of Kinect applications in elderly care and stroke rehabilitation. *Journal of Neuroengineering and Rehabilitation*, *11*(1), 1–24. doi:10.1186/1743-0003-11-108 PMID:24996956

Wong, C., Yang, E., Yan, X. T., & Gu, D. (2017). An overview of robotics and autonomous systems for harsh environments. In *Proceedings of 23rd International Conference on Automation and Computing (ICAC)* (pp. 1-8). Academic Press.

Xia, L., Chen, C., & Aggarwal, J. (2012). View invariant human action recognition using histograms of 3d join. In *Proceedings of IEEE Conference CVPR* (pp. 20–27). IEEE.

Yang, X., & Tian, Y. (2014). Effective 3d action recognition using eigenjoints. *Journal of Visual Communication and Image Representation*, *25*(1), 2–11. doi:10.1016/j.jvcir.2013.03.001

Zhang, C., Tian, Y., & Capezuti, E. (2012). Article. In Lecture Notes in Computer Science: Vol. 7382. Privacy preserving automatic fall detection for elderly using RGBD cameras. Springer-Verlag.

Zhang, Z., Liu, A., Gao, Z., & Liu, C. (2013). A Kinect-based 3D sensing and human action recognition solution for urban search and rescue environments. In *Proceedings of IEEE RO-MAN* (pp. 344-345). IEEE. 10.1109/ROMAN.2013.6628488

Zhao, W., Feng, H., Lun, R., Espy, D. D., & Reinthal, M. A. (2014). A Kinect-based rehabilitation exercise monitoring and guidance system. In *Proceedings of 2014 IEEE 5th International Conference on Software Engineering and Service Science* (pp. 762-765). IEEE.

Chapter 8
Pre-Clinical ASD Screening Using Multi-Biometrics-Based Systems

Tanu Wadhera
Dr. B. R. Ambedkar National Institute of Technology Jalandhar, India

Deepti Kakkar
Dr. B. R. Ambedkar National Institute of Technology Jalandhar, India

Gurjot Kaur
Dr. B. R. Ambedkar National Institute of Technology Jalandhar, India

Vasudha Menia
Dr. B. R. Ambedkar National Institute of Technology Jalandhar, India

ABSTRACT

Autism Spectrum Disorder (ASD) screening is still an ongoing process due to few objective and effective screening approaches. The consideration of facial patterns, fingerprints, and other metrics such as footprints and ear size can well explain the ASD phenotypes. These physical anomalies provide a simpler and more objective approach to screen the disorder through rather than considering the complex biological factors. Moreover, it is very easy to acquire these metrics as compared to those lengthier and restrictive procedures. The screening engine in which both the biometrics are integrated has better and reliable outcomes as compared to singular approaches. The objective of this chapter is to present an ASD screening system based on the combination of biometrics of face and fingers. The novelty of the chapter is that the classifier-based matching and fusion of the modalities has been proposed. Hence, multi-biometric-based pre-clinical system has the potential to follow ASD-affected individuals continuously, objectively, and periodically.

DOI: 10.4018/978-1-5225-7525-2.ch008

INTRODUCTION

Biometrics is widely used in pattern recognition based on the physiological or behavioral features such as fingerprint, DNA, earlobe geometry, iris pattern, gait and facial pattern (Jain, Ross & Prabhakar, 2004). The biometric based identification system relies on the features that are unique and free from temporal variations (Jain, Flynn, & Ross, 2007). These are widely used in surveillance, personalized identification, screening, interaction, encryption, and verification. The screening and monitoring of healthcare diseases can be done using the biometric traits of the individuals (El-Latif, Hossain, & Wang, 2017). In healthcare systems, biometrics help in detecting ocular pathologies (Trokielewicz, Czajka, & Maciejewicz, 2017), genetic disorders and neurodevelopment disorders such as ASD (Kazemi et al, 2017) and in assessing emotions and finding the effect of music on emotions (Alhussein, 2016). In this chapter, the biometric system for screening the Autism Spectrum Disorder (ASD) affected individuals has been utilized.

ASD represents a spectrum of etiologically and clinically non-uniform neuro-developmental disorders. It majorly comprises of autism, Pervasive Development Disorder Not Otherwise Specified (PDD-NOS), and Asperger's syndrome (AS) disorders. These disorders have been merged under a single umbrella and their clinical symptoms should accomplish the diagnostic criteria mentioned in DSM-5 (Diagnostic and Statistical Manual-V; Tanu & Kakkar, 2018). ASD affects the social interaction and communication ability of the individuals and leads to restricted & repetitive behavior (Tanu & Kakkar, 2018). It is diagnosed by targeting the behavioral symptoms, outcome oriented patterns such as social skill, clinical course of disorder such as regression, neurological abnormalities and physical modalities (Obafemi-Ajayi et al., 2015). The behavioral symptoms include: social interaction, sensory processing, aggression, attention deficit and motor movements/gait patterns. The neurological abnormalities include: brain patterns, seizures, deafness and cognitive ability. The physical identity includes the facial morphology and different modalities viz. fingerprints, footprints, ear size and iris. These identities are advantageous due to ease of acquiring, analysis and non-invasiveness & have detected various developmental disorders such as Down syndrome and Williams syndrome (Gilani et al., 2015). Diagnosis of ASD is a challenging task since the behavioral and the developmental traits of the subject have to be carefully observed by an expert to detect disorders. Recent years, have observed a shift in the detection of ASD towards the physical modalities viz. facial morphology and dermatoglyphics which act as assistive tools for the detection and are solely based on behavioral observations. The detection through these biometric based approaches is relatively simple and does not involve the iterations associated with detection.

Based on the recent advances and usage, the dermatoglyphics (finger prints) and face morphology studies have been considered in this chapter. Dermatoglyphics is a scientific process of taking impressions or patterns of fingerprints, palm and feet for the analysis of disorders. The patterns are hereditarily decided due to which these are highly variable and no two people or even the twins have the same pattern. It is strongly related to genetics and hence, this is highly utilized for diagnostic purposes. The dermal ridges start developing in the 13th week (Okajima, 1975) of intrauterine life and completely develop up to the 21st week. From then onwards, ridges remain same and consequently can be considered as fossils of a particular time of pre-birth development (Fatjó-Vilas, Gourion, Campanera, Mouaffak, Levy-Rueff, Navarro & Fañanás, 2008; Fearon, Lane Airie, Scannell, McGowan, Byrne & Waddington, 2001). So, this has a great role in identification of an individual and diagnosis of hereditary disorders (Turek, 1990). Though, the interest in dermatoglyphics research sustained from the earliest starting point of the twentieth century, but it has now gone into a stage where it is drawing an incredible number of researchers from all portions of biology and medicine. Moreover, it has many advantages over other diagnostic techniques such as unlike most human traits, dermal ridges are not affected with growing years and environmental conditions, and detailed structure of individual ridges is extremely variable (Prabhu, Issrani, Mathur, Mishra, & Sinha, 2014; Kaur, & Batra, 2013). There is a strong correlation between ridge differentiation and limb embryogenesis as various developmental disorders are reflected in the ridges (Arrieta, Martinez, Criado, Simon, Salazar, & Lostao, 1990). So the ridge pattern is impaired by certain abnormalities that mainly include genetic disorders.

The other modality used in ASD diagnosis is face morphology in which the specific facial features are measured by considering various face parameters. For example, the small & flat nose, small mouth puts an individual into Down syndrome category and facial asymmetry, reduced facial height and enhanced mouth breadth puts individuals into ASD category (Aldridge et al., 2011; Hammond et al., 2008). The development of the brain and face goes alongside and hence, any condition that affects the brain's development could manifest itself in the face of a person (DeMyer, Zeman, & Palmer, 1964). The facial morphology of children with ASD, hence, could be subtly different than the Typically Developing children (TD). In face morphology, the specific facial features are measured by considering various facial landmarks and detecting abnormalities as compared to the normal facial structure. The detection through facial structure could prompt a minimal effort screening and easy analysis device as compared to rest of the biometrics (due to the availability of highly developed and precise 3D modeling softwares) for the subjects with ASD. Hence, it could likewise help pinpoint hereditary reasons for the various patterns of the numerous disorders. As the analysis is done through images, it follows a non intrusive measurement approach. Moreover, the assessment is detailed as it includes the minute facial patterns. Facial morphology has been widely used in forensic science for example, in the investigation of missing people, identifying individuals, in the recognition and classification of multiple ethnic groups and also in the medical environment for planning treatments and diagnosis (Thomas, Claes, Shaweesh, & Clement, 2011).

There are other metrics such as eye gaze patterns, ear size, footprints, prominent forehead (Manouilenko, Eriksson, Humble, & Bejerot, 2014; Ozgen, Hellemann, De Jong, Beemer, & van Engeland, 2013). The gaze pattern abnormalities shown by ASD individuals have also been found in schizophrenia individuals (Papagiannopoulou, Chitty, Hermens, Hickie, & Lagopoulos, 2014). The overlapping of the characteristics of gaze modality is the reason that it is not possible to distinguish ASD individuals from other disordered individuals. The ear shape, ear lobe, & footprints have also been found to be associated with the traits of ASD (Manouilenko, Eriksson, Humble, & Bejerot, 2014). But, these changes in different features of the ear also exist in the people belonging to different ethnic groups and hence, it cannot act as a prominent screening factor for ASD. This is the reason that individual biometrics provide a little help in screening ASD and modalities other than face patterns and fingerprints have not been considered in the domain of this chapter. There are few other limitations of uni-metric based systems and some of them are listed below (El-Latif, Hossain, & Wang, 2017):

1. Lacks in distinguishing the individuals properly.
2. More susceptible to noise and can have unacceptable error rates.
3. Less reliable, distorted & inaccurate as uni-biometric systems are based on single sources.
4. Less universal, poor compatibility and cannot be deployed for meaningful screening of the individuals.
5. Age effects and other abnormalities can lead to poor distinguishable metrics such as faded or underdeveloped fingerprint ridges at old age and poor iris images in case of pathological eye conditions.
6. More prone to the spoof attacks, data imitation and inter-subject similarities.

Many experimental studies have shown that multi-biometric systems have improved accuracy (approx. 99.5%), less distorted and not vulnerable to spoof over the uni-biometric systems (Taha & Norrozila, 2015). Based on these advantages, the current chapter is providing a platform for fusing the facial and fingerprint modalities for better screening of ASD. The rest of the chapter is organized into following different subsections: the first section provides a literature survey of the biometric modalities used in ASD screening. The second section highlights the gaps in the uni-metric based diagnosis. In the next section, the multi-biometric based system proposal and its methodology have been framed in detail. Finally, the discussion has been provided along with the future work of metric based screening approaches.

BACKGROUND AND LITERATURE SURVEY

The biometrics application has been found in a number of fields that include: security systems, identification and verification of persons. In the medical field, the different metrics such as body dimensions or blood type are also utilized for employing metric based healthcare monitoring systems. In the first phase of the literature, introduction to dermatoglyphics and its role in ASD screening has been surveyed. The second phase provides the knowledge about face morphology.

Dermatoglyphics

Dermatoglyphics is study of prints (fingers) and its physical composition is partitioned into factorial and stochastic components (Graham, Raz, Hel-Or, & Nevo, 2010; Lajus, Graham, & Kozhara, 2003). The factorial part is the deviation among subjects that can be additionally subdivided into hereditary, ecological and ontogenetic segments. The stochastic segment represents the variation inside people and instability in the subjects or any developmental issue (De Bruin, Graham, Louwerse and Huizink, 2014). Fingerprint ridge patterns are divided into three basic types, i.e. Arches, Loops, and Whorls, as shown in Figure 1, adopted from (Marera, Oyieko, & Agumba, 2015), and computed by measuring frequency of ridges in corresponding pattern and disposition of tri-radii (a point where three sets parallel ridges intersect).

The different types of fingerprint patterns are explained as:

1. **Arches:** It is framed by the progression of pretty much parallel edges, which cross the pattern territory and is curved proximally. It is one of the least difficult patterns found on the fingers. The arch design can be plain/simple (Schaumann, & Alter, 2012) and tented as shown in Figure 2 (a, b).
2. **Loops:** It refers to the progression of ridges that enter the pattern zone on one side of the digit (finger number), twist suddenly, and then leaves the pattern zone on a similar side. They might be large or small, tailor short, vertically or horizontally arranged, plain loop or twofold loop. The loop design can be ulnar and radial type as shown in Figure 3(a,b), adopted from (Schaumann & Alter, 2012).
3. **Whorls:** It is a sort of ridge with at least two tri-radii. One tri-span is on radial and the other on the ulnar side. The different sub-types of whorl are shown in Figure 4 (a,b,c,d), adopted from (Brazelle, & Brazelle, 2018).

Figure 1. Basic Patterns in Dermatoglyphics

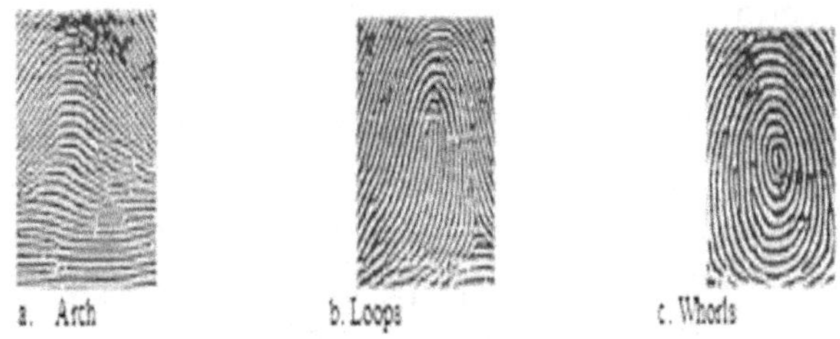

Figure 2. Types of Arch

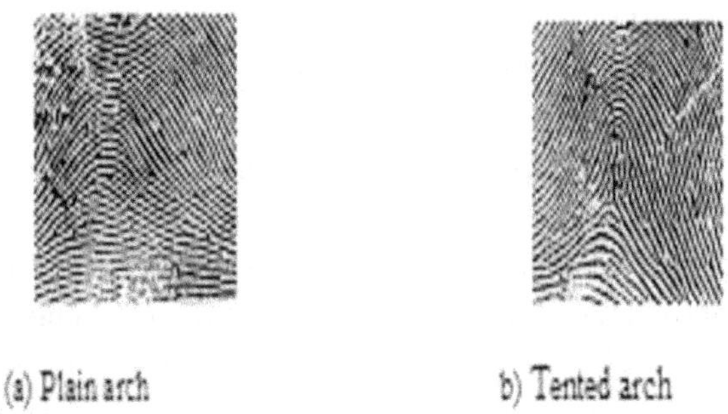

Figure 3. Types of loops

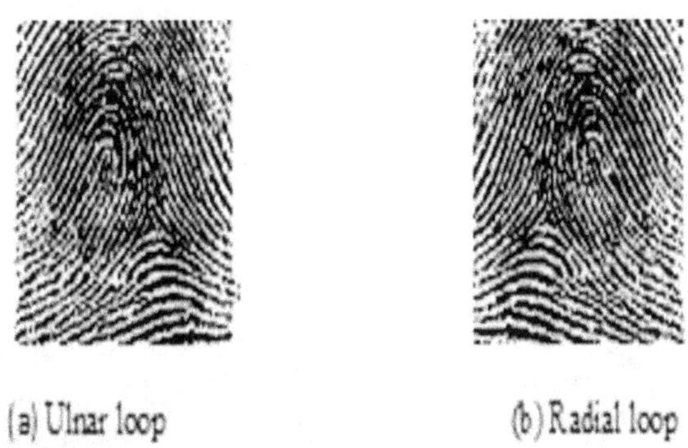

Figure 4. Types of whorls

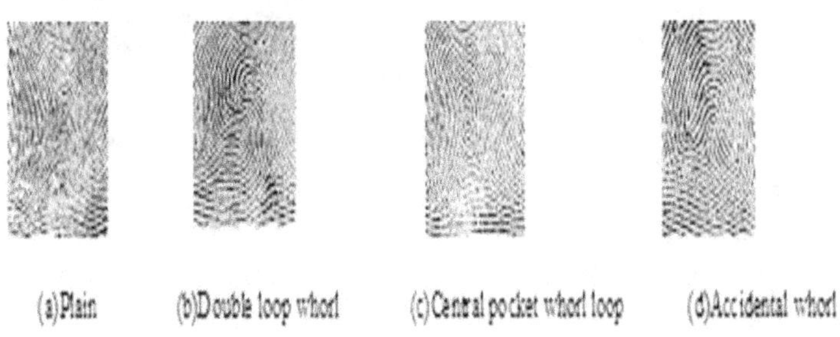

The fingerprint based screening systems extract the above mentioned features/patterns by making use of different minutiae extraction techniques (Bansal, Sehgal, & Bedi, 2011) and diagnosed different disorders such as Schizophrenia (Fananas, Mora & Bertranpetit, 1990), Klinefelter syndrome (Forbes, 1964), Turner syndrome (Reed, Reichmann, & Palmer 1977), Trisome18, Trisome13 and Trisome21 (Schaumann & Alter, 1976), and Kanner's syndrome. Not only these medical disorders, but dermatoglyphics is also linked with neurodevelopment disorder such as ASD. This disorder has very wide spectrum and covers a large range of neurodevelopmental symptoms under single umbrella. It is advantageous to screen ASD using dermatoglyphics as children show variant behavioral traits at the time of diagnosis and it becomes difficult to control them as some of the diagnostic techniques are painful, time-consuming, not easily available and very costly. Hence, dermatoglyphics comes out with an easier way for screening the disorder. Exploring the literature one can discover a number of strategies for recording the dermatoglyphics. The difference lies in their prerequisites for gear, time and experience, and in the idea of the prints created. The different techniques that can be utilized, as shown in Figure 5, adopted from (Gupta & Gupta 2013).

The study of dermatoglyphics in ASD screening has been briefly described in the tabular form in Table 1.

The dermatoglyphics dataset has been analyzed for ridge pattern and ridge count to compare ASD and control subjects using parameters such as arch, whorl and loop. The frequencies of the major patterns on the finger of ASD and control subjects are computed for comparison (Hartin& Barry, 1979, Kazemi, Fayyazi-Bordbar, & Mahdavi-Shahri, 2017). The table summary has provided an overview of the importance of fingerprints in scanning the disorder.

Facial Morphology

Facial morphology is a non-invasive and time saving method that ensures minimum discomfort (physically or mentally) to the subjects. The connection of facial asymmetry with the development of the brain

Figure 5. Methods of Recording Dermatoglyphics (a) Inkless Method (b) Transparent Adhesive Tape Method

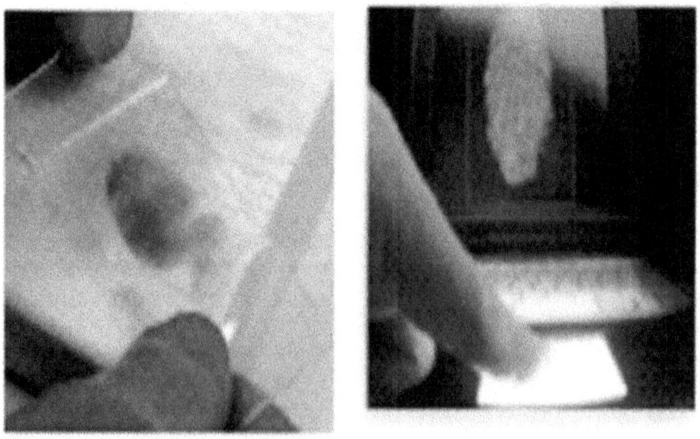

(a) (b)

Table 1. ASD Screening Using Dermatoglyphics

Reference	Assessment Criteria		Modality	Findings
	ASD Dataset	Mean Age (yrs)		
Hartin & Barry, 1979	32 subjects; using ink method	7	Fingerprints	(i) Difference in dermal pattern and ridge line disruption. (ii) No difference in ridge count.
Wolman, Campbell, Marchi, Deutsch, & Gershon, 1990	95 subjects; using inkless Faurot method	-	Fingerprints	(i) Consistency among hands, ATD angle values, total ridge counts, incidence. (ii) Effects on the ridge count of arches, double loops, and whorls are larger than control.
Arrieta, Martinez, Criado, Simon, Salazar, & Lostao 1990	60 subjects; using ink method		Fingerprints	(i) Higher frequency of transitional radial loops and a lower frequency of dicentric whorls. (ii) Total finger ridge count and radial count are lower than control.
Stošljević, & Adamović, 2013	182 subjects; using Futronic-FS80	-	Fingerprints	(i) Difference in the distribution of both arch and whorl pattern between both subjects.
Okoh Peter, Oghenemavwe Loveday & Yorkum Leyira, 2013	20 subjects; using a digital scanner	-	Fingerprints	• (i) Frequency distribution of the digital patterns of ASD and control subject was: arch (49.5%, 25.5%), loop (3.5%, 8.5%). • (ii) Mean values of the a–b ridge counts of ASD female and male are less than controls.
Kazemi et al., 2017	104 subjects; using a scanner	10	Fingerprints	(i) Higher count of loops on their right and left thumbs and in their index fingers. (ii) Lower in ridge counts for the right, left thumb and in the index finger observe when compared with the controls.

in ASD is explicitly described by Hammond et al. (2008) and hence, these atypicalities can easily screen ASD and its sub-types. The facial measurement can be done using three methodologies (Dindaroğlu, Kutlu, Duran, Görgülü, & Aslan, E, 2016), explained as:

1. **Manual/Direct-Measurement:** Morphologic points are marked on the face and then the distances between them are measured manually using digital millimeter sliding caliper.
2. **2D-Photography:** The images are captured using the professional cameras. For example Canon EOS 40D and calibrated using software such as View Box 4.0. These are the snapshots of the subjects from different directions and require attention during acquisition only. These are less complex than 3D image acquisition.
3. **3D-Stereophotogrammetry:** These images are acquired using 3dMDflex systems & measurement of different facial parameters is obtained using 3dMD Patient software. It offers various advantages such as high speed for acquiring data, non-invasive, user friendly, accurate and reliable. The advanced 3D photogrammetric gadgets and spearheading improvements in morphometric investigation techniques now empower more delicate surface shape examinations of facial morphology (Hammond et al., 2004) and hence, have become the 'gold standard' in facial measurement.

Among these, 3D-stereophotogrammetry offers various advantages such as high speed for acquiring data, non-invasive, user friendly, accuracy and reliability. After capturing the 3D photographs, the first step is to determine the significant facial landmarks and to collect their coordinate's data using the 3dMD patient software program (Farkas, 1994). The different facial landmarks from the images of the faces of sample population are shown in Figure 6. These landmarks are termed biologically significant (Farkas, 1994).

In order to evaluate facial differences, the linear, Euclidean, hamming or geodesic distances are measured between the considered landmarks. These features have been computed and compared with ASD individuals and the summary of the studies have been provided in Table 2.

The tabular summary shows that the studies have recognized ASD based on different facial phenotypes observed. The studies have opted to classify the ASD from TD using facial landmarks. All the dataset is self acquired. This is one of the reasons for limited studies in metric based ASD diagnosis. One of the approaches has used clustering algorithm to recognize the subgroups of ASD boys that showed distinct clinical and behavioral patterns and optimized the algorithm performance using a Cluster Validity Analysis Platform (CVAP) tool based on MATLAB (Obafemi-Ajayi et al., 2015). The Gradient based Feature Selection (GEFS) algorithm (Hus, Gotham, & Lord, 2014) is also used to demarcate the masculine and feminine ASD features and rearrange the landmarks in iterations that enhances the classification accuracy. The connection of facial asymmetry with the development of the brain in ASD is explicitly described by Hammond et al. (2008). The results obtained demonstrated remarkable facial asymmetry in boys with ASD, particularly in the supra and periorbital regions anterior to the frontal pole of the right hemisphere of the brain. In one study, two hypotheses were investigated - first hypotheses tested the facial distinctions between the ASD affected adolescent boys of Caucasian ethnicity and the unaffected boys (Aldridge et al., 2011). The second hypothesis tested whether there could be certain subgroups within ASD based on facial phenotypes. The results obtained demonstrated remarkable facial asymmetry in boys with ASD, particularly in the supra and periorbital regions anterior to the frontal pole of the right hemisphere of the brain.

In another study, Autism Spectrum Quotient (AQ) scores were calculated based on the associated masculinity and femininity with facial morphology and the voices of the subjects (Tan et al., 2015). The results emphasized that the voices of the higher AQ men showed enhanced masculinity as compared to the men with lower AQ scores. The 23 significant facial landmarks from the facial images of the sample population have been used to detect ASD (Gilani et al., 2015) and minimal redundancy maximal rel-

Figure 6. Different Significant Landmarks of Facial Morphology

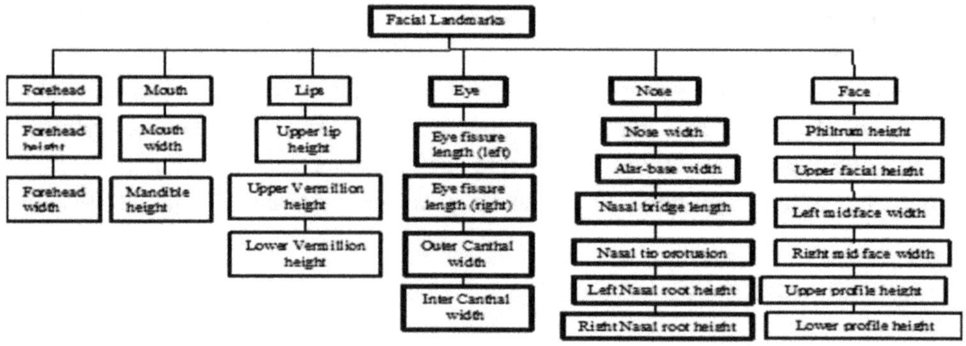

Table 2. ASD Screening with Face Morphology

Reference	Techniques Used	Modality	Assessment of Subjects		Findings in ASD
			No. of ASD Subjects	Age (Mean) in Years	
Hammond et al., 2008	Dense surface-modeling and pattern matching algorithms to distinguish faces.	Facial asymmetry.	3D face images (72 boys and 41 unaffected mothers) were captured using photogrammetric devices	9.2	(i) Remarkable facial asymmetry (supra and per orbital regions) in ASD boys. (ii) Mothers (Non-ASD) of children with ASD also show higher facial asymmetry.
Aldridge et al., 2011	3-D Stereophotogrammetry using 3dMD cranial system	Facial morphology	Face images of 65 boys were acquired using 3dMD cranial System	10.14	Subgroup1 characterized 75% ASD traits, subgroup 2 showed 20% & remaining ASD subjects showed 62%.
Tan et al., 2015	Capturing photos, distributing questionnaires and recording voices.	Facial morphology	Black & white images of 76 children were acquired through digital Canon camera	17-19	(i) No significant effect on AQ scores for male facial features & female voices. (ii) Less feminine facial features for female & more masculine voices for male.
Whitehouse et al., 2015	3D photos and GEFS algorithm	Facial morphology	3dMD face system captured photographs of 97 males and 86 females	Males- 22.81 Females 22.91	(i) Established connection b/w prenatal testosterone levels & facial morphology. (ii) Higher testosterone amount resulted in masculinity enhancement in both genders.
Obafemi-Ajayi et al., 2015	3D stereo-photogrammetric images and varied cluster analysis has been done for classification.	Facial morphology	Facial images of 62 boys were acquired through 3dMD face system.	8-12 years	(i) Facial features categorized ASD with unique subgroups of cognitive impairment and regression of language. (ii) Obtained accuracy of classification by SVM= 95.16%, MLP=93.55%, RF= 91.94%.
Gilani et al., 2015	3dMDface system acquired facial images and features were extracted using gender classification and scoring algorithm.	Facial morphology	Facial images of Low-AQ males (33) & females(33, High-AQ males(25) & females(25).	22.81	(i) Four distances varied significantly between high & low-AQ groups. (ii) ASD males showed less masculine features and ASD females showed less femininity as compared to TD individuals.
Diana Weiting et al., 2017	3D photogrammetry to analyze important facial landmarks.	Facial morphology	3D facial images of 54 boys and 20 girls with ASD were acquired through 3dMD face system.	3-12	(i) Both boys and girls with ASD showed enhanced masculinity. (ii) ASD boys with greater social-communication difficulties showed typical facial features, but no association between gender scores and restricted and repetitive behavior.

ASD: Autism Spectrum Disorder; AQ: Autism Quotient; MLP:Multi Layer Perceptron; RF: Random Forest; SVM: Support Vector Machine; TD: Typically Developed evance algorithm demarcated the male and female facial features. The four out of six parameters showed significant difference between the two subgroups. Recently, hypermasculinisation and connection of typical facial morphology with social and communication skills in ASD were investigated (Diana et al., 2017). The results showed that the facial area did not vary significantly amongst the two genders for the subjects with and without ASD, however, boys with ASD showed less gender scores and more masculinity than the control boys while the girls with ASD showed less femininity than the control girls.

This survey shows that face morphology and dermatoglyphics are providing a better platform for understanding the physical anomalies based picture of ASD and its association with different sub-types. It could be analyzed through rigorous literature review that the biometrics can provide the necessary assistance in ASD screening. The relation between these two modalities is a potential approach that can open new and easy ways of screening ASD.

GAPS IN EXISTING TECHNIQUES

The gaps in the literature survey conducted in the above section and other prevailing techniques for detection of ASD have been framed in this section. The number of screening techniques is growing rapidly and so is the ASD affected individuals. Even though, the techniques are growing rapidly, but lacks at some points due to which ASD screening is still an ongoing process. Some of the gaps are listed below:

1. The subjective and theoretical nature of screening scales makes it difficult for the early detection of ASD. It requires dedicated time and observation of the subjects. The techniques need to be more objective.
2. The complexity of the disorder increases with age that may lead to greater chances of co-existence of co-morbidities. The parameters associating the disorder with its sub-types are required for better identification.
3. Use of neuro-imaging techniques frequently for the detection may get very costly for the average population. Also, these techniques pose a number of physical restrictions, sedation and a lot of circuitry making it very tough for the individual under observation.
4. The invasive nature of detection methods may lead to discomfort of the subjects.
5. The single biometric sensor may get vulnerable to noise. The measured parameter may be contorted because of flawed measurement conditions.
6. Unimodal biometrics may present undesired results due to inter- class similarities. For example- for identical twins, face recognition may not be able to present correct results and other biometrics approach like fingerprints can provide the necessary help.
7. The dermatoglyphics process only computes the frequencies of the patterns and not realizing the features by proper extraction techniques. There is a need to extract the features of the fingerprints for more accurate and reliable classification of the disorder.
8. The frequency distribution of finger pattern is overlapping among some of the disorders. This problem can lead in the direction of wrong diagnosis. Therefore, the frequency range needs to be well defined so that overlapping can be avoided.

There are multiple gaps linked to the prevailing detection techniques. Hence, there is a need for some detection procedure which does not cause any kind of discomfort to the subjects and save them from the frequent visits to the doctor for diagnosis. An assistive tool needs to be designed to provide the necessary and sufficient information at the early stage to start with the training or treatment process. A multi-biometric based approach can be very useful and reliable in this context as the reliability increases when multiple parameters are considered. Moreover, it does not involve discomfort and multiple sittings just for the screening at initial stages of the disorder.

MULTI-BIOMETRIC SYSTEMS

As per the advantages of multi-biometric systems over uni-metric systems, the present chapter is also utilizing multi-biometric system. This integration of information from different metrics has better performance (Kumar, & Zhang, 2006) that can be utilized in real-world applications such as template protection, remote authentication. The application has extended to the healthcare domain to enhance the safety of healthcare monitoring units, identity management of patients as well as doctors and detect patient's emotions, and ocular pathologies. In the healthcare field, fused biometrics are majorly used in identity management, security and record keeping as manual patient tagging is full of errors and vulnerabilities that can cost one's life. For example, in Texas, a biometric based system addressing frauds in Medicaid system has been in operation since 2004. Also, in South Africa a pilot fingerprint based identification system has been launched in the health care sector to tackle diseases especially HIV/AIDS epidemic. The application of metrics is shifting towards the disease diagnosis to make this process more accurate.

The multi-biometric systems need to merge different individual modalities on a single platform. In order to combine two metrics on a single platform, a fusion key is required. The different levels of fusion (Jain, Nandakumar, Ross, 2005, Ross & Jain, 2004), in obtaining multi-biometric systems, are shown in Figure 7.

The fusion levels are decided corresponding to the stage at which fusion is required, that is, pre-matching or post-matching fusion. The sensor and feature level fusions take place at pre-matching stages and score, decision and rank level fusions take place in post-matching stages.

- **Score-Level Fusion:** A score fusion approach is adopted to classify the data in the output layer and is shown in Figure 8. The feature vectors are processed, individual matching score is found and then the scores are fused together to find a score for better classification. It is classified into two different approaches on the basis of match score processing such as feature vector classification and combination. This fusion type has lesser complexity, more availability & more accuracy, so it is widely used and preferred as compared to other techniques (Aravinth, & Valarmathy, 2016; Jain, Nandakumar, & Ross, 2005). It is utilized in a number of Artificial Intelligence algorithms such as genetic and evolutionary computation systems (Alford et al., 2011).
- **Decision-Level Fusion:** At the decision level, the biometric features are extracted and classified into either accepted or rejected category and then outputs of different biometric classifier's are fused together to finally classify the metric. The boolean result of individual metrics is combined to provide the final decision. Due to the usage of individual modalities results, it is also called abstract level fusion. The fusion of modalities uses different algorithms such as fuzzy k-means, fuzzy vector quantization and radial basis function (Chatzis, Bors, & Pitas, 1999). It is a least informative type fusion as the decision is based on individual modalities. The fusion approach is shown in Figure 9.

Figure 7. Fusion Levels in Multi-biometric Systems

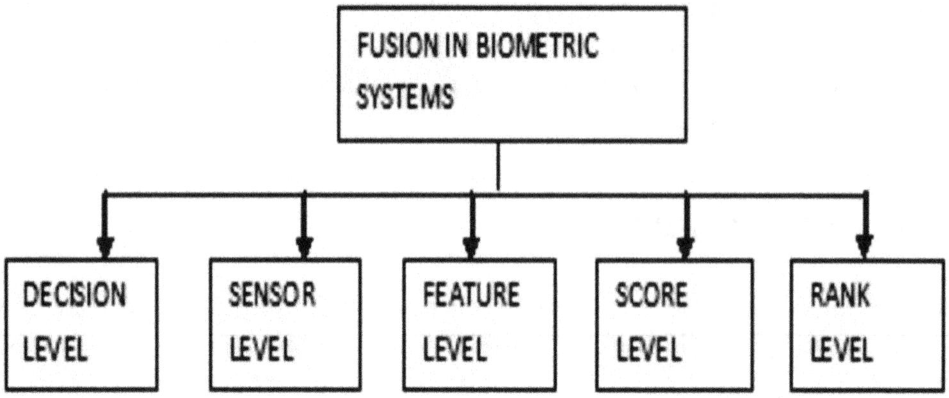

Figure 8. Score-Level Fusion

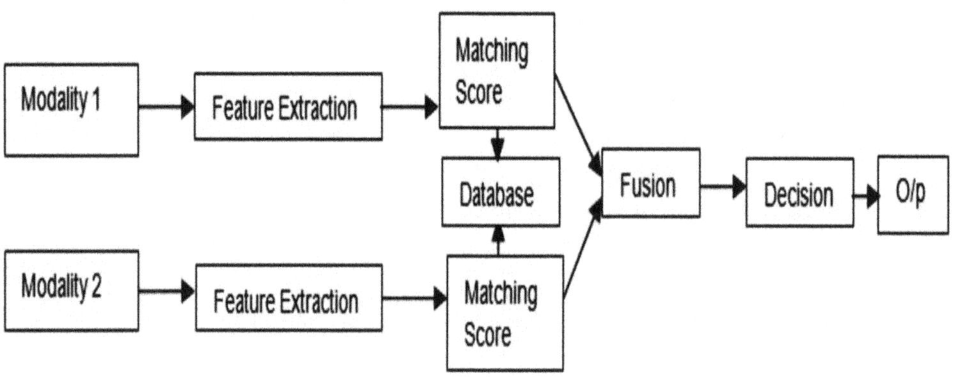

Figure 9. Decision-Level Fusion

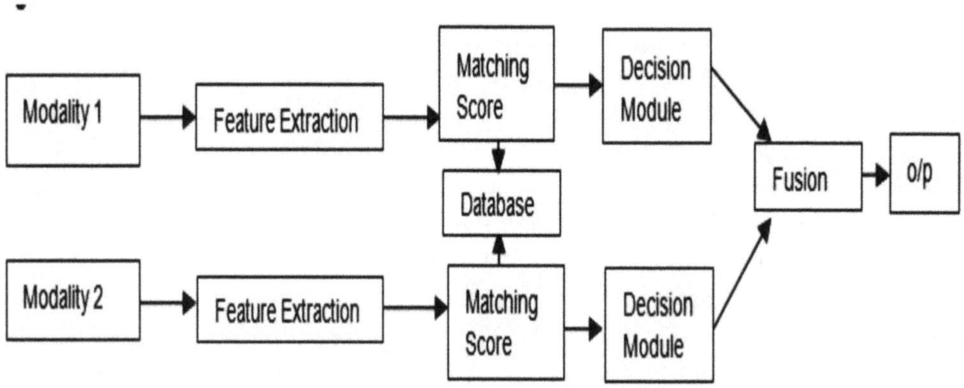

- **Feature-Level Fusion:** The features contain most of the biometric traits information and the fusion of these can provide better and accurate results only if the different modalities are very much similar to each other (Ross, & Govindarajan, 2005). The feature fusion, as shown in Figure 10, involves feature vectors obtained from single biometric modality collected from different sensors or entities or generated from multiple modalities. It creates a set of important features by correlating the features extracted using different algorithms to improve the accuracy of the system. For example, in case of fingerprints, feature fusion is performed in the minutiae and vein patterns and shape of metrics to form feature vector.
- **Sensor Level Fusion:** Two or more samples of the similar modality are scanned by the sensor to fuse them more accurately before feature extraction. These modalities are combined in raw form and the features are extracted from the combined modality (Mwaura, Mwangi, & Otieno, 2017). The data from multiple scans are fused together to form 2D or 3D scans of metrics to enhance accuracy. It can be employed in image based face recognition systems such as fusing infrared and visible face images and hence, also called data or image level fusion. This type of fusion, as shown in Figure 11, has been utilized in fingerprinting mosaicking that is required in touchless fingerprint systems (Choi, Choi, & Kim, 2010), iris based recognition systems (Ghouti, & Bahjat, 2009) and audio-visual systems (Froba, Rothe, & Kublbeck, 2000).
- **Rank Level Fusion:** The rank level fusion, as shown in Figure 12, combines the modality ranks and decides on the basis of new rank generated after fusion. It is widely used in identifying the systems on the basis of their ranks. The fusion of ranks can be done using different methods such as highest rank, logistic regression and borda count methodology (Mwaura, Mwangi, & Otieno, 2017). The disadvantage of using this fusion is that it supposes that all matching methods perform equally, which is not possible in a real time scenario. This type of fusion needs more exploration as it is not utilized extensively in the literature.

A hybrid-level fusion can also be attained by combining the different levels of fusion such as feature and decision level or score and feature level. One of the study has reported that more security can be attained by doing integration at the fusion stages such as score and decision level fusion (Marfella, Marasco, & Sansone, 2012). The preliminary studies that have employed multi-biometric systems have

Figure 10. Feature-Level Fusion

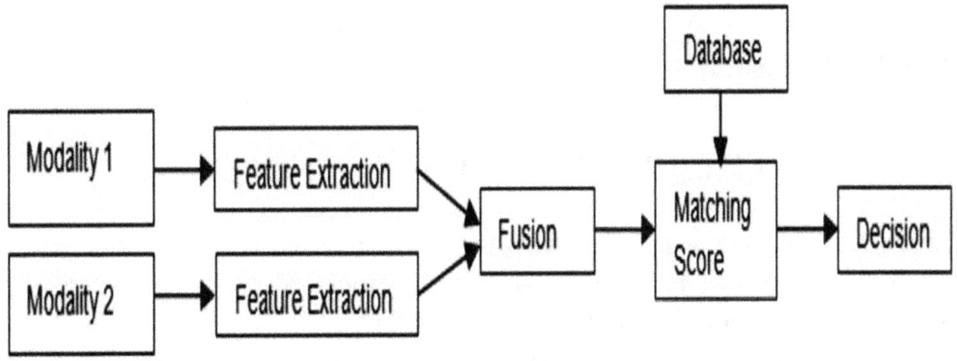

Figure 11. Sensor-Level Fusion

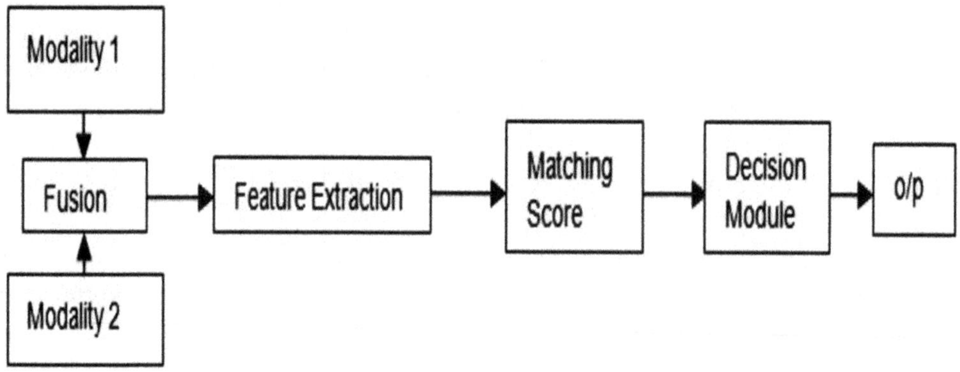

Figure 12. Rank-Level Fusion

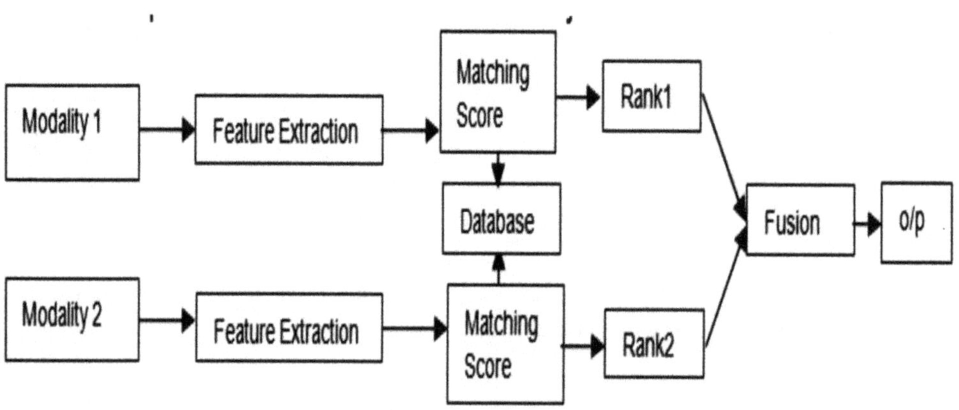

been reviewed in the tabular form in Table 3. This summary clearly shows that biometric systems are not only widely utilized in verification and identification systems but also in the healthcare domain.

There are various other datasets utilized in multi-biometric systems such as IIT-Delhi dataset providing palmprints and FKP (Hossain, & Chetty, 2011), ITK dataset containing images of palm-dorsa-vein-pattern (Ojansivu, & Heikkilä, 2008), CASIA-Iris-Distance dataset comprising eyes and facial images (Svoboda, Bronstein, & Drahansky, 2015). The dataset are not widely available for ASD and needs to be captured before carrying out the screening process. The proposed methodology has also explained the proper procedure for acquiring the required set of metrics.

PROPOSED APPROACH AND METHODOLOGY

As per the above discussion, it is clear that multi-biometric approaches have more accuracy, security and reliability as compared to other singular approaches. The face morphology and dermatoglyphics provide easy, non-invasive and more informative platform for screening ASD individuals. These metrics based strategies can solve many problems related to other diagnostic approaches and provide solutions to the

Table 3. Multi-biometric Based Systems

Reference	Modality	Dataset	Type of Fusion	Application
Kumar, Hanmandlu, Sanghvi, & Gupta, 2010	Palmprints and Hand vein	Self-acquired	Decision Level	In recognizing the persons with an accuracy of 99%.
Kumar, Hanmandlu, Sharma & Gupta, 2011	Ear and Iris	IIT Delhi	Rank and Decision Level	In metric based recognition system (FAR-0.09% and FRR-1.5%).
Hanmandlu, Grover, Gureja, & Gupta, 2011	Face and Ear	IIT Kanpur dataset	Score Level	A Recognition system with an accuracy (95.53%), FRR (5.55%) and FAR (3.40%).
Meraoumia, Chitroub, & Bouridane, 2012	Fingerprint, FingerKnucle Print (FKP)	Hongkong PloyU dataset	Sensor Level	In person recognition systems.
Long, & Hanh, 2012	Fingerprint and Face	Publicly available DB4FVC2004 dataset	Feature level	In authentication systems with False Acceptance Rate (FAR) of 4.95% and False Rejection rate (FRR) of 1.12%.
El-Latif, Hossain, & Wang, 2017	Iris and Face	CASIA-Iris-Thousand and NVIE dataset	Score Level & Normalization	In monitoring health care & detecting any fraud. Investigating the patient's mood & ocular pathologies.
Hashemian, & Pourghassem, 2017	Facial Images and EEG metrics	Self-acquired using BIMEC EEG machine	Decision Level	Proposed structure effectively distinguished ASD from TD individuals by interpreting EEG signals related to different facial expressions.
Wu, José, Nurnberger, & Torres, 2018	Physiological movement data	Acquired using electromagnetic sensors	---	Provided clinical and objective behavioral metric for ASD diagnosis.

ASD: Autism Spectrum Disorder; EEG: Electroencephalography; FAR: False Acceptance Rate; FRR: False Rejection Rate; IIT- Indian Institute of Technology; NIST: National Institute of Standards and Technology; NVIE: Natural Visible and Infrared Facial Expression Database; TD: Typically Developing

existing gaps in the literature. The high-tech approach based on the biometrics can set new standards for capturing and assessing behavior traits and treating ASD individuals. This technology based solution can describe the inside body picture of ASD affected individuals.

The proposed system is using the multi-biometric approach by fusing the face and fingerprints in order to screen ASD with more accuracy. The fusion technique used in the system is the classifier based score-level fusion. The novelty of the proposed approach is that it is using machine learning based classifiers for finding the matching score as well as for classifying the fused score. The proposed multi-biometric system, as shown in Figure 13, is providing an integrated approach to screen ASD and non-ASD individuals. The face and finger metrics for testing can be acquired from the subjects using following methodology:

Finger Prints and Feature Extractors

The metric dataset required in the proposed methodology is not publicly available but needs to be acquired on its own. In order to capture the data, the procedure has been explained for both the metrics used in the proposed system. There are a couple of prerequisites that needs to be considered before capturing the fingerprints of ASD individuals. These are:

Figure 13. Multi-biometric System for ASD Diagnosis

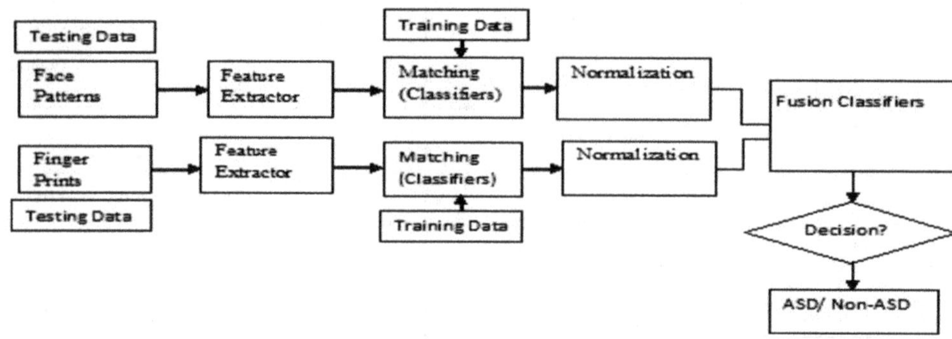

- There needs to be coordination between ASD and control subjects on the factors viz. IQ level, age, sex, socioeconomic level and geographic circumstances.
- Every single ASD subject should be diagnosed for autism, not any control subject.

The quantitative and qualitative samples that are collected from more than 30 subjects are compared for each single variable in ASD and control subjects. The statistical procedures can be carried out for comparing the frequency of patterns in ASD and control individuals using independence G-test with Yates correction, the t-test with the arcsine transformation of qualitative values, and Student's *t*-test for quantitative (Okoh Peter *et al*, 2013). Fingerprint patterns have cores and tri-radii and the classification is based on the number of tri-radii present and ATD-angles that are measured with the Image J software program or PASS software (Kazemi, Fayyazi-Bordbar, & Mahdavi-Shahri, 2017). Other dermatoglyphic characteristics are described quantitatively either by ridge counting or by measuring distances/ angles between specified points that can be determined by the number of lines in center of print pattern. In order to determine the ridge structure and to form feature vectors from fingerprints, the gray scale based minutiae extraction algorithm can be utilized (Bansal, Sehgal, & Bedi, 2011). The ridge line following and fuzzy based algorithms resolves the complex fingerprints to a point pattern that can be easily matched and processed (Gao et al., 2010). The different algorithms such as scale invariant feature transform algorithm (Mwaura, Mwangi, & Otieno, 2017), principal component analysis, ridge thinning or machine learning based classifiers such as neural networks (Lim, Jiang, & Yau, 2002) or fuzzy networks (Sagar, Ngo, & Foo, 1995) can also be utilized for minutiae extraction. The image enhancement and segmentation techniques (such as Gabor-filters) can be utilized before to improve the image and for extraction.

Face Patterns and Feature Extractors

The prerequisites that need to be considered before acquiring the facial images of ASD individuals are:

- The expert system has to be designed considering the ethnic non-variability as the variability may result in the inaccurate results.
- The factors viz. IQ level, age, sex, socioeconomic level and geographic circumstances need to be matched while recruiting the non-ASD group.

Out of different facial imaging systems the 3-D Stereophotogrammetry is more advanced and better option. The 3-d facial images captured are given to 3dMD cranial system (Aldridge, Boyadjiev, Capone, DeLeon, & Richtsmeier, 2005) to determine 3-D coordinates of facial landmarks under study. Using this system, the images are acquired by projecting the random light samples onto the subjects. Four modular units capture a 360 (full view of the head to create a 3D model with the help of three digital machines installed cameras present within each unit. The resulting images need not to be superimposed to get a single composite picture as required in dense surface modeling due to the use of multiple cameras. It constructs more precise and accurate 3-D models that can be visualized through 3dMD Patient software on the desktop computer (Weinberg, Scott, Neiswanger, Brandon, & Marazita, 2004). The significant facial landmarks were determined on the acquired 3-D images and their coordinate's data was collected using the 3dMD patient software program (Farkas, 1994). The linear and geodesic distances (curved 3 D distances) can be computed between the considered landmarks and a distance based feature vector can be constructed. The feature vector is not limited to distances, but other features can also be examined using algorithms such as Local Gabor XOR Pattern (LGXP; Ali, & Abdalla, 2011). The computed linear and geodesic distances of ASD subjects and controls can be compared by student t- tests, f-tests and chi-square tests through correlation coefficient.

The feature vectors obtained from both the modalities are given to machine learning based classifiers for further processing of the modalities.

Machine Learning Classifiers

In this proposal, the machine learning based classifiers have been utilized for matching the features of the modality and also for integrating them to provide a final score. These automate the model development process without any requirement of probabilistic assumptions or manual thresholding. The different machine learning based classifiers used for finding the behavioral differences in ASD individuals are: Linear Discriminant Analysis, Decision-tree, Random-Tree, Logistic Regression, K-Nearest Neighbor (KNN) and Support Vector Classification (SVC). The tree-based algorithms are less efficient as they need twice the features compared to others and SVC has been found very relying classifier as it depends on an extreme set of examples for developing a model (Duda, Ma, Haber, & Wall, 2017).

The SVC classifier is mainly utilized for binary decisive purposes. In this system, it is used for matching the pattern of the testing image with the training image. The classifier generally classifies the data into two classes, but here one-vs-all method is used for classifying into different classes according to the number of the individuals. On the basis of this separation SVC will generate a matching score for all the individuals. The number of the SVC for matching depends upon the number of samples/subjects recruited. For example, if there are 15 individuals so they need to be classified into 15 classes and 15 SVCs are required.

Normalization

After generating the matching scores, normalization is the first step before score level fusion. The extracted features are from the different domains and normalization can bring them on a common platform. Since, the data or score distribution of individual matching classifiers need not be of the same type, on the same scale or location. There are different ways that can be adopted for normalizing the scores (Jain, Nandakumar, & Ross, 2005; Scheirer, Rocha, Micheals, & Boult, 2011). For example, the fingerprint modality produces feature vector in the range [1, 100] and the face modality in [1, 2000] and combining these modalities in their raw form is not possible. Therefore, normalization is the pre-requisite step prior to score-level fusion. The scores obtained from both the machine learning classifiers are normalized to reduce redundancy and other anomalies and to bring data on a common platform. This step brings the scores of both biometrics on the similar scale for efficient fusion and better accuracy. It gives better recognition performance than others because it has feasible & practical information which is more robust to noise in the data. The commonly used normalization techniques are minimum-maximum, Z-score, the likelihood-ratio, hyperbolic tangent and various others (Abate, Nappi, Riccio, & Sabatino, 2007). The minimum-maximum and Z-score methods are simple but very much sensitive to the outliers present in the score matrix. In comparison, the hyperbolic tangent is more efficient, but it involves a large number of parameters and the most promising technique is likelihood ratios (Nandakumar, Chen, Dass, & Jain, 2008). This step brings the metric-score values in the [0, 1] interval to perform the score-level fusion. The choice of normalization technique depends upon the type of biometric, outliers and final result output. The ill-chosen methodology can lead to very poor recognition rate and performance.

Fusion Classifiers

After the generation of similar score of both the modalities, the normalized scores are fused together using classifier based score-level fusion. This is the last fusion step where the two normalized vectors are fused. The fusion is the core requirement for combining the multiple biometrics into a single approach. The classifier employed for fusion purpose can be a single classifier or group of classifiers. The system proposed here is aimed to work with the combination of different classifiers as the combination can provide better efficiency. The choice of classifiers can be fuzzy based, Naïve Bayes, neural learning networks or various others according to the requirement. The classifier combination can preserve the efficiency of the system by retaining the advantages of the different classifiers and refining their individual disadvantages. The combination can authenticate the modalities with more reliability and accuracy. For example, the scores of each modality are fed to a fuzzy based classifier. This classifier can convert the scores into fuzzy linguistics such as high, low, or medium by using if-then rules. There are different membership functions for defining these rules such as triangular, max-membership, centroid or weighted average method. The rule can be defined as: IF (Score 1 is Low) AND (Score 2 is Low), THEN individual is non-ASD. After defining these rules for obtaining scores, the de-fuzzifier is used to convert rule's decision into crisp values. These crisp values indicate whether the classifier identified the individual as ASD or non-ASD. In case, a combination of classifiers is used, then the performance is determined by computing the correlation between the classifier and desired/threshold outputs.

There are other ways of fusing the scores in the score-level fusion such as transformation based (sum-rule) or density-based (likelihood ratio-test) but these methods fail in distinguishing the unknown features accurately. On the other hand, the classifier based approach can provide better accuracy (Horng et al., 2009) and do not ignore the farther distances in inter-class and closer in intra-class (El-Latif, Hossain, & Wang, 2017). However, this approach requires a very large and high quality data.

ASD or Non-ASD

The final fused scores obtained from the classifiers are fed to the decision module. This module will compare the value with the threshold value to categorize the individual into ASD or non-ASD group.

The performance of the multi-biometric system can be computed by calculating different parameters such as equal error rate, false acceptance rate, accuracy or false rejection rate.

Evaluation Metrics

The performance of the proposed screening system can be computed by employing evaluation metrics comprising False Acceptance Rate (FAR), False Rejection Rate (FRR), Equal Error Rate (EER), accuracy and Receiver Operating Characteristic (ROC) curves (Aravinth, & Valarmathy, 2016; Ross, Nandakumar, & Jain, 2006). These evaluation metrics can demonstrate the effectiveness of the proposed system and can help in a comparative analysis with the other existing multi-biometric based systems.

- **False Acceptance Rate (FAR):** This measures the likelihood that the biometric system will incorrectly diagnose non-ASD and puts them in ASD category. It is defined as the ratio of the number of false diagnoses (non-ASDs as ASDs) by the total number of diagnosis attempts.
- **False Rejection Rate (FRR):** This measures the likelihood that the biometric system will incorrectly diagnose the ASD individuals and puts them in non-ASD category. It is defined as the ratio of the number of false diagnoses (ASDs as non-ASDs) by the total number of diagnosis attempts. It is defined as the ratio of the number of false rejections in the biometric system by the total attempts.
- **Equal Error Rate (EER):** This compares the performance of system with another system by computing whether FAR and FRR rates are equal or have a minimum distance. The lower value of EER indicates that the system is better.
- **Genuine Acceptance Rate (GAR):** It is the ratio of the number of correctly diagnosed individuals genuine by the total number of cases provided to the biometric system. It is the complement of FRR such that GAR=1-FRR.

In this manner the proposed system's performance can be evaluated by computing the accepted quality metrics.

The usage of metrics in healthcare was initiated by the applications such as in access control, enhancing patient security, and record storage. The other usages include: iris based recognition technology is used in MethaDoseprogram for the treatment of heroin addicted citizens (Marohn, 2006). Similarly, a recognition and healthcare monitoring system based on iris, thermal and visible facial modalities have been proposed for achieving better performance and accuracy (). In ASD, the biometrics wristbands measures heart rate, galvanic skin response and 3D limb movements to diagnose the disorder easily. A gas discharge visualization method has been used to acquire biometrics of ASD individuals and reported that psycho-emotional levels were significantly different in ASD as compared to controls (Kostyuk, Rajnarayanan, Isokpehi, & Cohly, 2010). Recently, one of the charity centers has planned to employ biometric technology to understand ASD. The system proposed in this section is providing a new direction for the use of biometrics in the healthcare field. It provides a diagnosis angle of the biometrics such that these modalities can act as a baseline in understanding and screening ASD.

DISCUSSION

ASD affected individuals, who are admitted to hospitals or other health care centers, remain marooned there for longer periods of time because of lack of expertise platforms that can understand their behavior and find a suitable intervention methodology. The tracing of different physical modalities is an efficient and easy approach for targeting disorder technologically and objectively. The present chapter has proposed a system based on the fusion of the biometrics for screening ASD. The system is novel in itself as the matching and fusion of metrics is done using hybrid classifiers. This system has the potential to provide a cutting-edge technology and capability to assist the clinicians/neurologists in screening ASD.

Detection of ASD through prevailing techniques takes a few months as the behavioral traits of the subject are carefully studied before concluding whether the subject has ASD or not. The severity of the disorder may increase with age, hence, there is a need for early screening through simple assistive tools. One such assistive tool could be designed by considering the physical anomalies of the affected individuals such as fingerprints and facial patterns are more informative and have greater variations in ASD subjects as compared to other modalities such as ear shape, feet pattern, head circumference (Manouilenko, Eriksson, Humble, & Bejerot, 2014). Moreover, detection through facial morphology and fingerprints is a reliable and an easy approach. The most subtle face patterns could be demarcated between ASD and TD population (Diana et al., 2017) and the approach is valid for each age group. The direct detection through images, hence, provides an easier and less time taking detection procedure (Angkustsiri et al., 2011). It could be analyzed through rigorous literature review that the facial morphology can provide the necessary assistance for ASD detection.

In dermatoglyphics, dermal palmer ridges are exceedingly valuable in biological studies. Their prominent variable factors are not duplicated in other individuals, even in monozygotic twins or even in a similar individual from one area to another. What makes dermatoglyphics essential as markers for diseases and traits is the way these marker develop at a specific period in the fetus. Irregularities in the development procedure are responsible for distorting the arrangement of dermal ridges that might happen because of action in abnormal genes, chromosomal aberrations and even from poisoning. Therefore, dermatoglyphics is helpful in making the first step towards the diagnosis of various medical and neurological disorders.

The biometrics individually, seems to be very helpful in screening ASD. The screening system can be more accurate, reliable and efficient if a multi-biometric based system is designed (El-Latif, Hossain, & Wang, 2017). The collection of the metrics from several modalities and finding their features are important requirements for carrying out the successful fusion of the modalities. The proposed screening system is combining the metrics (facial and finger prints) that contains wide variations and feature vector. The machine learning based algorithms for matching and combining the score together have the potential to enhance the performance and efficiency of these systems. The comparative analysis of the literature has provided better options for feature extraction, normalization and classification. The proposed system, if implemented, has the potential to screen ASD pre-clinically and assist clinicians in conducting detection at initial stages.

The sensory differences and external stimuli factors such as smell or noise can lead to stress and unpredictable mood swings in ASD individuals. The metrics based system can screen ASD as well as can be extended to provide care to affected individuals who are extremely reactive to these external factors. These systems can be developed and spread throughout the world, without restricting or torturing ASD individuals, to learn and treat their behavioral traits even before they can develop fully. Hence, this could be the beginning of developing smart, and assistive high-technology based tools for following and screening individuals with ASD by measuring their biological signs.

FUTURE STUDIES

More work needs to be done on enhancing the efficiency of multi-biometric systems, for better screening of ASD, by using different classifiers such as deep learning based classifiers and feature extraction techniques. The systems need to be spread worldwide to provide a screening facility remotely and anytime. The dermatoglyphics frequency range should be treated as features so that metric-based system could be trained to detect disorder easily. These systems need to be enhanced for assisting the tracing of other neuro-developmental disorder viz. Down syndrome. In future, the work needs to be devoted more towards the detection of emotions, mood and anxiety co-morbidities using multi-biometric systems because these external factors have major proportion leading to ASD. Also, the ethnic variability may result in the inaccurate results so the integrated expert system should be designed considering the ethnic non-variability.

REFERENCES

Abate, A. F., Nappi, M., Riccio, D., & Sabatino, G. (2007). 2D and 3D face recognition: A survey. *Pattern Recognition Letters, 28*(14), 1885–1906. doi:10.1016/j.patrec.2006.12.018

Aldridge, K., Boyadjiev, S. A., Capone, G. T., DeLeon, V. B., & Richtsmeier, J. T. (2005). Precision and error of three-dimensional phenotypic measures acquired from 3dMD photogrammetric images. *American Journal of Medical Genetics. Part A, 138*(3), 247–253. doi:10.1002/ajmg.a.30959 PMID:16158436

Aldridge, K., George, I. D., Cole, K. K., Austin, J. R., Takahashi, T. N., Duan, Y., & Miles, J. H. (2011). Facial phenotypes in subgroups of prepubertal boys with autism spectrum disorders are correlated with clinical phenotypes. *Molecular Autism, 2*(1), 15. doi:10.1186/2040-2392-2-15 PMID:21999758

Alford, A., Hansen, C., Dozier, G., Bryant, K., Kelly, J., Abegaz, T., . . . Woodard, D. L. (2011, June). GEC-based multi-biometric fusion. In *Evolutionary Computation (CEC), 2011 IEEE Congress on* (pp. 2071-2074). IEEE. 10.1109/CEC.2011.5949870

Alhussein, M. (2016). Automatic facial emotion recognition using weber local descriptor for e-Healthcare system. *Cluster Computing, 19*(1), 99–108. doi:10.100710586-016-0535-3

Ali, H. S., & Abdalla, M. I. (2011). Score-level fusion for efficient multimodal person identification using face and speech. *International Journal of Computer Science and Information Security, 9*(4), 48.

Angkustsiri, K., Krakowiak, P., Moghaddam, B., Wardinsky, T., Gardner, J., Kalamkarian, N., ... Hansen, R. L. (2011). Minor physical anomalies in children with autism spectrum disorders. *Autism, 15*(6), 746–760. doi:10.1177/1362361310397620 PMID:21610186

Aravinth, J., & Valarmathy, S. (2016). Multi classifier-based score level fusion of multi-modal biometric recognition and its application to remote biometrics authentication. *Imaging Science Journal*, *64*(1), 1–14. doi:10.1080/13682199.2015.1104067

Arrieta, M. I., Martinez, B., Criado, B., Simon, A., Salazar, L., & Lostao, C. M. (1990). Dermatoglyphic analysis of autistic Basque children. *American Journal of Medical Genetics. Part A*, *35*(1), 1–9. doi:10.1002/ajmg.1320350102 PMID:2301457

Bansal, R., Sehgal, P., & Bedi, P. (2011). *Minutiae extraction from fingerprint images-a review.* arXiv preprint arXiv:1201.1422.

Brazelle, M., & Brazelle, S. (2018). Hand Determination of Whorl Patterns Using Axis Slant. *Journal of Forensic Identification*, *68*(1), 77.

Chatzis, V., Bors, A. G., & Pitas, I. (1999). Multimodal decision-level fusion for person authentication. *IEEE Transactions on Systems, Man, and Cybernetics-Part A: Systems and Humans, 29*(6), 674-680.

Choi, H., Choi, K., & Kim, J. (2010). Mosaicing touchless and mirror-reflected fingerprint images. *IEEE Transactions on Information Forensics and Security*, *5*(1), 52–61. doi:10.1109/TIFS.2009.2038758

De Bruin, E. I., Graham, J. H., Louwerse, A., & Huizink, A. C. (2014). Mild dermatoglyphic deviations in adolescents with autism spectrum disorders and average intellectual abilities as compared to typically developing boys. *Autism Research and Treatment*. PMID:25478224

DeMyer, W., Zeman, W., & Palmer, C. G. (1964). The face predicts the brain: Diagnostic significance of median facial anomalies for holoprosencephaly (arhinencephaly). *Pediatrics*, *34*(2), 256–263. PMID:14211086

Dindaroğlu, F., Kutlu, P., Duran, G. S., Görgülü, S., & Aslan, E. (2016). Accuracy and reliability of 3D stereophotogrammetry: A comparison to direct anthropometry and 2D photogrammetry. *The Angle Orthodontist*, *86*(3), 487–494. doi:10.2319/041415-244.1 PMID:26267357

Duda, M., Ma, R., Haber, N., & Wall, D. P. (2017). Use of machine learning for behavioral distinction of autism and ADHD. *Translational Psychiatry*, *6*(2), e732. doi:10.1038/tp.2015.221 PMID:26859815

El-Latif, A. A. A., Hossain, M. S., & Wang, N. (2017). Score level multibiometrics fusion approach for healthcare. *Cluster Computing*, 1–12.

Fananas, L., Moral, P., & Bertranpetit, J. (1990). Quantitative dermatoglyphics in schizophrenia: Study of family history subgroups. *Human Biology*, 421–427. PMID:2373511

Farkas, L. (1994). Anthropometry of the head and face in clinical practice. *Anthropometry of the Head and Face*, *2*, 71–77.

Fatjó-Vilas, M., Gourion, D., Campanera, S., Mouaffak, F., Levy-Rueff, M., Navarro, M. E., & Fañanás, L. (2008). New evidences of gene and environment interactions affecting prenatal neurodevelopment in schizophrenia-spectrum disorders: A family dermatoglyphic study. *Schizophrenia Research*, *103*(1), 209–217. doi:10.1016/j.schres.2008.04.015 PMID:18583099

Fearon, P., Lane, A., Airie, M., Scannell, J., McGowan, A., Byrne, M., & Waddington, J. (2001). Is reduced dermatoglyphic a–b ridge count a reliable marker of developmental impairment in schizophrenia? *Schizophrenia Research*, *50*(3), 151–157. doi:10.1016/S0920-9964(00)00089-X PMID:11439235

Forbes, A. P. (1964). Fingerprints and palm prints (dermatoglyphics) and palmar-flexion creases in gonadal dysgenesis, pseudohypoparathyroidism and Klinefelter's syndrome. *The New England Journal of Medicine*, *270*(24), 1268–1277. doi:10.1056/NEJM196406112702402 PMID:14133664

Froba, B., Rothe, C., & Kublbeck, C. (2000). Evaluation of sensor calibration in a biometric person recognition framework based on sensor fusion. In *Automatic Face and Gesture Recognition, 2000. Proceedings. Fourth IEEE International Conference on* (pp. 512-517). IEEE.

Gao, X., Chen, X., Cao, J., Deng, Z., Liu, C., & Feng, J. (2010, September). A novel method of fingerprint minutiae extraction based on Gabor phase. In *Image Processing (ICIP), 2010 17th IEEE International Conference on* (pp. 3077-3080). IEEE.

Ghouti, L., & Bahjat, A. A. (2009, December). Iris fusion for multibiometric systems. In *Signal Processing and Information Technology (ISSPIT), 2009 IEEE International Symposium on* (pp. 248-253). IEEE. 10.1109/ISSPIT.2009.5407577

Gilani, S. Z., Tan, D. W., Russell-Smith, S. N., Maybery, M. T., Mian, A., Eastwood, P. R., ... Whitehouse, A. J. (2015). Sexually dimorphic facial features vary according to level of autistic-like traits in the general population. *Journal of Neurodevelopmental Disorders*, *7*(1), 14. doi:10.118611689-015-9109-6 PMID:25901187

Graham, J. H., Raz, S., Hel-Or, H., & Nevo, E. (2010). Fluctuating asymmetry: Methods, theory, and applications. *Symmetry*, *2*(2), 466–540. doi:10.3390ym2020466

Gupta, R. K., & Gupta, A. K. (2013). New, easy and effective method to take dermatoglyphic prints. *National Journal of Medical Research*, *3*(1), 45–47.

Hammond, P., Forster-Gibson, C., Chudley, A. E., Allanson, J. E., Hutton, T. J., Farrell, S. A., ... Lewis, M. E. S. (2008). Face–brain asymmetry in autism spectrum disorders. *Molecular Psychiatry*, *13*(6), 614–623. doi:10.1038/mp.2008.18 PMID:18317467

Hammond, P., Hutton, T. J., Allanson, J. E., Campbell, L. E., Hennekam, R., Holden, S., ... Murphy, K. C. (2004). 3D analysis of facial morphology. *American Journal of Medical Genetics. Part A*, *126*(4), 339–348. doi:10.1002/ajmg.a.20665 PMID:15098232

Hanmandlu, M., Grover, J., Gureja, A., & Gupta, H. M. (2011). Score level fusion of multimodal biometrics using triangular norms. *Pattern Recognition Letters*, *32*(14), 1843–1850. doi:10.1016/j.patrec.2011.06.029

Hartin, P. J., & Barry, R. J. (1979). A comparative dermatoglyphic study of autistic, retarded, and normal children. *Journal of Autism and Developmental Disorders*, *9*(3), 233–246. doi:10.1007/BF01531738 PMID:489510

Hashemian, M., & Pourghassem, H. (2017). Decision-Level Fusion-Based Structure of Autism Diagnosis Using Interpretation of EEG Signals Related to Facial Expression Modes. *Neurophysiology*, *49*(1), 59–71. doi:10.100711062-017-9630-1

Horng, S. J., Chen, Y. H., Run, R. S., Chen, R. J., Lai, J. L., & Sentosal, K. O. (2009, December). An improved score level fusion in multimodal biometric systems. In *Parallel and Distributed Computing, Applications and Technologies, 2009 International Conference on* (pp. 239-246). IEEE.

Hossain, E., & Chetty, G. (2011, October). Multimodal face-gait fusion for biometric person authentication. In *Embedded and Ubiquitous Computing (EUC), 2011 IFIP 9th International Conference on* (pp. 332-337). IEEE.

Hus, V., Gotham, K., & Lord, C. (2014). Standardizing ADOS domain scores: Separating severity of social affect and restricted and repetitive behaviors. *Journal of Autism and Developmental Disorders, 44*(10), 2400–2412. doi:10.100710803-012-1719-1 PMID:23143131

Jain, A., Nandakumar, K., & Ross, A. (2005). Score normalization in multimodal biometric systems. *Pattern Recognition, 38*(12), 2270–2285. doi:10.1016/j.patcog.2005.01.012

Jain, A. K., Flynn, P., & Ross, A. A. (Eds.). (2007). *Handbook of biometrics*. Springer Science & Business Media.

Jain, A. K., Ross, A., & Prabhakar, S. (2004). An introduction to biometric recognition. *IEEE Transactions on Circuits and Systems for Video Technology, 14*(1), 4–20. doi:10.1109/TCSVT.2003.818349

Kaur, J., & Batra, A. P. S. (2013). Role of dermatoglyphics in medical disorders. *Indian Journal of Fundamental and Applied Life Sciences, 3*(3), 536.

Kazemi, M., Fayyazi-Bordbar, M. R., & Mahdavi-Shahri, N. (2017). Comparative dermatoglyphic study between autistic patients and normal people in Iran. *Iranian Journal of Medical Sciences, 42*(4), 392. PMID:28761206

Kostyuk, N., Rajnarayanan, R. V., Isokpehi, R. D., & Cohly, H. H. (2010). Autism from a biometric perspective. *International Journal of Environmental Research and Public Health, 7*(5), 1984–1995. doi:10.3390/ijerph7051984 PMID:20623006

Kumar, A., Hanmandlu, M., Sanghvi, H., & Gupta, H. M. (2010, September). Decision level biometric fusion using Ant Colony Optimization. In *Image Processing (ICIP), 2010 17th IEEE International Conference on* (pp. 3105-3108). IEEE.

Kumar, A., Hanmandlu, M., Sharma, V., & Gupta, H. M. (2011, December). Rank based hybrid multimodal fusion using PSO. In *International Conference on Swarm, Evolutionary, and Memetic Computing* (pp. 217-224). Springer. 10.1007/978-3-642-27172-4_27

Kumar, A., & Zhang, D. (2006). Personal recognition using hand shape and texture. *IEEE Transactions on Image Processing, 15*(8), 2454–2461. doi:10.1109/TIP.2006.875214 PMID:16900698

Lajus, D. L., Graham, J. H., & Kozhara, A. V. (2003). Developmental instability and the stochastic component of total phenotypic variance. *Developmental instability: causes and consequences*, 343-363.

Lim, E., Jiang, X., & Yau, W. (2002). Fingerprint quality and validity analysis. In *Image Processing. 2002. Proceedings. 2002 International Conference on* (Vol. 1, pp. I-I). IEEE.

Long, T. B., & Hanh, T. (2012, September). Multimodal biometric person authentication using fingerprint, face features. In *Pacific Rim International Conference on Artificial Intelligence* (pp. 613-624). Springer.

Manouilenko, I., Eriksson, J. M., Humble, M. B., & Bejerot, S. (2014). Minor physical anomalies in adults with autism spectrum disorder and healthy controls. *Autism Research and Treatment*. PMID:24782925

Marera, D. O., Oyieko, W., & Agumba, G. (2015). Variation in Dermatoglyphic patterns among diabetics in Western Uganda population. *African Journal of Science and Research*, *7*(3), 20–25.

Marfella, L., Marasco, E., & Sansone, C. (2012, September). Liveness-based fusion approaches in multibiometrics. In *Biometric Measurements and Systems for Security and Medical Applications (BIOMS), 2012 IEEE Workshop on* (pp. 1-7). IEEE. 10.1109/BIOMS.2012.6345779

Marohn, D. (2006). Biometrics in healthcare. *Biometric Technology Today*, *14*(9), 9–11. doi:10.1016/S0969-4765(06)70592-6

Meraoumia, A., Chitroub, S., & Bouridane, A. (2012, June). Multimodal biometric person recognition system based on fingerprint & finger-knuckle-print using correlation filter classifier. In *Communications (ICC), 2012 IEEE International Conference on* (pp. 820-824). IEEE.

Mwaura, G. W., Mwangi, W., & Otieno, C. (2017). Multimodal Biometric System:-Fusion Of Face And Fingerprint Biometrics At Match Score Fusion Level. *International Journal of Scientific & Technology Research*, *6*(4), 41–49.

Nandakumar, K., Chen, Y., Dass, S. C., & Jain, A. K. (2008). Likelihood ratio-based biometric score fusion. *IEEE Transactions on Pattern Analysis and Machine Intelligence*, *30*(2), 342–347. doi:10.1109/TPAMI.2007.70796 PMID:18084063

Obafemi-Ajayi, T., Miles, J. H., Takahashi, T. N., Qi, W., Aldridge, K., Zhang, M., ... Duan, Y. (2015). Facial structure analysis separates autism spectrum disorders into meaningful clinical subgroups. *Journal of Autism and Developmental Disorders*, *45*(5), 1302–1317. doi:10.100710803-014-2290-8 PMID:25351828

Ojansivu, V., & Heikkilä, J. (2008, July). Blur insensitive texture classification using local phase quantization. In *International conference on image and signal processing* (pp. 236-243). Springer. 10.1007/978-3-540-69905-7_27

Okoh Peter, D., Oghenemavwe Loveday, E., & Yorkum Leyira, K. (2013). *Dermatoglyphic Patterns of Autistic Children in Nigeria*. Academic Press.

Ozgen, H., Hellemann, G. S., De Jonge, M. V., Beemer, F. A., & van Engeland, H. (2013). Predictive value of morphological features in patients with autism versus normal controls. *Journal of Autism and Developmental Disorders*, *43*(1), 147–155. doi:10.100710803-012-1554-4 PMID:22669539

Papagiannopoulou, E. A., Chitty, K. M., Hermens, D. F., Hickie, I. B., & Lagopoulos, J. (2014). A systematic review and meta-analysis of eye-tracking studies in children with autism spectrum disorders. *Social Neuroscience*, *9*(6), 610–632. PMID:24988218

Prabhu, N., Issrani, R., Mathur, S., Mishra, G., & Sinha, S. (2014). Dermatoglyphics in health and oral diseases-A review. *JSM Dent*, *2*(4), 1044.

Reed, T., Reichmann, A., & Palmer, C. G. (1977). Dermatoglyphic differences between 45, X and other chromosomal abnormalities of Turner syndrome. *Human Genetics*, *36*(1), 13–23. doi:10.1007/BF00390431 PMID:858621

Ross, A., & Jain, A. K. (2004). Multimodal biometrics: An overview. *Proceedings of the 12th European Signal Processing Conference, 1221*-1224.

Ross, A. A., & Govindarajan, R. (2005, March). Feature level fusion of hand and face biometrics. In *Biometric Technology for Human Identification II* (Vol. 5779, pp. 196–205). International Society for Optics and Photonics. doi:10.1117/12.606093

Ross, A. A., Nandakumar, K., & Jain, A. K. (2006). *Handbook of multibiometrics* (Vol. 6). Springer Science & Business Media.

Sagar, V. K., Ngo, D. B. L., & Foo, K. C. K. (1995, October). Fuzzy feature selection for fingerprint identification. In *Security Technology, 1995. Proceedings. Institute of Electrical and Electronics Engineers 29th Annual 1995 International Carnahan Conference on* (pp. 85-90). IEEE. 10.1109/CCST.1995.524738

Schaumann, B., & Alter, M. (1976). Embryogenesis and genetics of epidermal ridges. In *Dermatoglyphics in Medical Disorders* (pp. 1–11). Berlin: Springer. doi:10.1007/978-3-642-51620-7_1

Schaumann, B., & Alter, M. (2012). *Dermatoglyphics in medical disorders*. Springer Science & Business Media.

Scheirer, W. J., Rocha, A., Micheals, R. J., & Boult, T. E. (2011). Meta-recognition: The theory and practice of recognition score analysis. *IEEE Transactions on Pattern Analysis and Machine Intelligence*, *33*(8), 1689–1695. doi:10.1109/TPAMI.2011.54 PMID:21422483

Stošljević, M., & Adamović, M. (2013). Dermatoglyphic characteristics of digito-palmar complex in autistic boys in Serbia. *Vojnosanitetski Pregled. Military-Medical and Pharmaceutical Review*, *70*(4), 386–390. doi:10.2298/VSP1304386S PMID:23700943

Taha, K. E., & Norrozila, S. (2015). A survey of multi-biometrics and fusion levels. *Indian Journal of Science and Technology*, *8*(32). doi:10.17485/ijst/2015/v8i32/91488

Tan, D. W., Russell-Smith, S. N., Simons, J. M., Maybery, M. T., Leung, D., Ng, H. L., & Whitehouse, A. J. (2015). Perceived gender ratings for high and low scorers on the Autism-Spectrum Quotient consistent with the extreme male brain account of autism. *PLoS One*, *10*(7), e0131780. doi:10.1371/journal.pone.0131780 PMID:26186689

Tanu & Kakkar, D. (2018). Diagnostic Assessment Techniques and Non-invasive Biomarkers for Autism Spectrum Disorder. *International Journal of E-Health and Medical Communications*.

Thomas, C. D. L., Claes, P. D., Shaweesh, A. I., & Clement, J. G. (2011). Three-dimensional facial shape archetypes for identification and diagnosis. In Biological Shape Analysis (pp. 131-144). Academic Press.

Trokielewicz, M., Czajka, A., & Maciejewicz, P. (2017). Implications of ocular pathologies for iris recognition reliability. *Image and Vision Computing*, *58*, 158–167. doi:10.1016/j.imavis.2016.08.001

Turek, S. (1990). Dermatoglyphics and schizophrenia: Analysis of quantitative traits. *Collegium Antropologicum, 14*(1), 137–150.

Weinberg, S. M., Scott, N. M., Neiswanger, K., Brandon, C. A., & Marazita, M. L. (2004). Digital three-dimensional photogrammetry: Evaluation of anthropometric precision and accuracy using a Genex 3D camera system. *The Cleft Palate-Craniofacial Journal, 41*(5), 507–518. doi:10.1597/03-066.1 PMID:15352857

Whitehouse, A. J., Gilani, S. Z., Shafait, F., Mian, A., Tan, D. W., Maybery, M. T., ... Eastwood, P. (2015). Prenatal testosterone exposure is related to sexually dimorphic facial morphology in adulthood. *Proc. R. Soc. B, 282*(1816), 20151351.

Wolman, S. R., Campbell, M., Marchi, M. L., Deutsch, S. I., & Gershon, T. D. (1990). Dermatoglyphic study in autistic children and controls. *Journal of the American Academy of Child and Adolescent Psychiatry, 29*(6), 878–884. doi:10.1097/00004583-199011000-00006 PMID:2273014

Wu, D., José, J. V., Nurnberger, J. I., & Torres, E. B. (2018). A biomarker characterizing neurodevelopment with applications in autism. *Scientific Reports, 8*(1), 614. doi:10.103841598-017-18902-w PMID:29330487

Chapter 9 Digitization of Paper Electrocardiogram: A Review

Priyanka Gautam
Dr. B. R. Ambedkar National Institute of Technology Jalandhar, India

Ramesh Kumar Sunkaria
Dr. B. R. Ambedkar National Institute of Technology Jalandhar, India

Lakhan Dev Sharma
MLV Textile and Engineering College, India

ABSTRACT

In order to diagnose a possible cardiac disorder, ECG (electrocardiogram) signals are usually recorded on standard grid papers in hospitals. Many efforts have been made to advance the technology in order to improve the diagnosis and management of cardiovascular disease. There is a need to convert the existing ECG records into digital forms, as it is the most efficient method to store and analyze ECG attributes for clinical uses. The main purpose of this chapter is to review the existing algorithms for digital conversion of paper ECG. It discusses the various challenges and a systematic study on different methods that have been used so far to convert paper ECG records into digitized form so that they can be retrieved efficiently. Initial challenge involved in the digitization process is gridline removal. In this process, information of ECG signal is also removed. None of the existing methods provide flawless gridline removal. The paper ECG used in hospitals differs in shape, size, formats, so the main challenge in digitization process is to achieve a worldwide ECG format.

DOI: 10.4018/978-1-5225-7525-2.ch009

INTRODUCTION

Technologies that are used to compute human behavior or physical features are known as Biometrics such as iris, face, fingerprints, retina, hand geometry, voice or signatures and using such measures to detect and recognize individuals. Biometric act as an acrimonious alternative for identity verification such as passports, driving licenses, ID cards or PIN numbers in authentication. No need to bring any type of additional ID documents by the user because biometrics uses unique physical traits for authentication purpose. Biometric credentials cannot be forgotten, lost, easily cloned or guessed like other traditional authorization systems such as passwords, ID cards or personal identification numbers (PINs). Usual biometric systems involve a nomination and verification/identification phase. In enrollment process "live samples" of a person's biometric is acquired for identification purpose, followed by processing process and after that stored as templates. In verification process captured biometric samples are matched to the enrolled template, which is stored so as to verify/identify user identity (Lumini et al., 2017, Merone et al., 2017, Hejazi et al., 2016).

First fundamental fingerprint recognition system was presented in early 20 century. In search of new biometric modalities, research community spent too much energy to get any behavioral or physical trait which assure the conditions of universality, varying with time, discriminate between the population, readily collectable and difficult to cheat/reproduce. On the basis of above criteria several distinctive symptoms are identified, such as physiological (e.g. fingerprint, face, and iris), behavioral (e.g. signature, voice, gait), soft (e.g. gender, height and ethnicity) and medico-chemical (e.g. ECG, DNA) (Lumini et al., 2017, Sidek et al., 2014).

Unlike traditional biometric modalities, for authentication purpose the heart is likely a most safe for biometric modality since it is confined in the structure of the body, making it difficult for the opportunists to forge, change, replicate, mimic etc. The electric signal produced by the heart can be captured in a non-invasive way from the body's surface using ECG sensors. Most of the available devices for ECG acquisition in developing and underdeveloped countries use Paper-ECG. To create a country wide database for citizen identification/ verification, it is required to develop techniques for digitization of these ECG's available in paper form (i.e. Paper-ECG) (Islam et al., 2017, Gutta et al., 2016).

Human Heart

The Human Heart is a muscular organ having four chambers that are located just to the left of the mid-plane of the thoracic cavity. The superior two chambers, i.e. atria are divided by a wall like structure known as interatrial septum. The inferior two chambers, i.e. ventricles are divided by a similar structure known as the interventricular septum. The relaxing or filling phase of a cardiac chamber is known as diastole; the contracting or pumping phase is known as systole.

Impure blood flows into the right atrium through the veins with lack of oxygen, these veins known as superior and inferior vena cava. The right atrium collects impure blood from head, neck, chest and arms through the superior vena cava. It collects impure blood from the remainder of the trunk and the legs through the inferior vena cava. During the pumping phase of atrial, blood flows through the tricuspid valve from right atrium to the right ventricle. During ventricle contraction, the right ventricle is pumped out the impure blood to the lungs through the pulmonary valve for purification (oxygenation). Purified blood from the lungs is received by left atrium, which is passed to the left ventricle during atrial systole through the mitral valve. The most important and the largest part of cardiac chamber are known as the left ventricle. The strongest contraction is provided by the left ventricle among all the cardiac chambers, as it has to pump out the purified or oxygenated blood via aortic valve and aorta opposite to the pressure of the rest of the vascular system of the whole body.

Special pacemaker cells released from sino-atrial (SA) node which controls the Heart Rate whereas SA node is located at the junction of the right atrium and superior vena cava. The autonomous and central nervous systems generates the impulses so that they can deliver neurotransmitters acetylcholine (for vagal stimulation, resulting reduced heart rate) or epinephrine (for sympathetic stimulation, resulting increased heart rate) to control the firing rate of SA node. The normal heart rate of human being is near 70bpm. Heart rate is slower during sleep, but bradycardia is a type of disorder which indicates abnormal slow heart rate below 60bpm. During powerful exercise or athletic activity instantaneous increment in heart rate occurs that reaches a value 200bpm. Tachycardia is a type of disorder in which constant high heart rate occurs due to illness, diseases or cardiac abnormalities (Rangayyan, 2002, Guyton et al., 2006, Najarian et al., 2012, Iaizzo, 2005). Figure 1 shows the diagram of human heart.

Electrocardiogram

Electrical activity of the heart is recorded in the form of a waveform known as an electrocardiogram (ECG), which was instigated by Willem Einthoven in 1895 as a tool to diagnose the possible cardiac disorder. It shows the working of the heart, whether it is working normal or abnormal (Barold S. S., 2003, Sajjan, 2013). Recording of ECG is performed by placing electrodes on the skin. Electrodes are used to take bioelectric signal from the surface of the body before it transfer to amplifier for recording or displaying purpose. Ionic conduction in tissues is converted into electronic conduction with the help of

Figure 1. Diagram of Human Heart (Sharma et al., 2017)

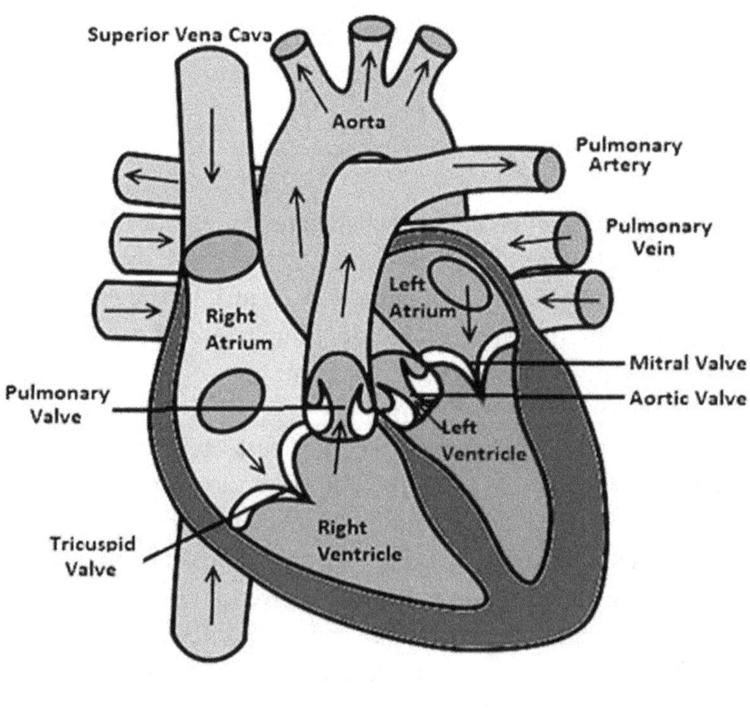

electrodes which is mandatory for measurements. There are two types of electrodes - surface electrodes and deep-seated electrodes, which usually used to take bioelectric signals. When surface electrodes are placed on the surface of the body, then they take potential difference from surface of tissue without harming the live tissue, whereas deep-seated electrodes showing the internal electric potential difference of tissue or cell. They used to monitor cardiac health (Khandpur, 2003, Williams et al., 2011).

Electrical System of Heart

The electrical system of the heart generates a sequence of waves in cardiac cycle. Initially firing started from the SA node. Slow moving contraction (depolarisation) of atria occurs to pump blood into the ventricles. This results as a P wave in the ECG. P wave is slow and low amplitude wave due to slow contraction with duration of about 60-80 ms and amplitude of about 0.1-0.2 mV. Due to excitation of wave atrio-ventricular (AV) node faces a propagation delay, which results an iso-electric segment after the P wave of about 60-80ms in the ECG, called as the PR segment. Rapid depolarization (contraction) of the ventricles occurs to pump blood out, this results a QRS complex wave of the ECG. QRS wave is a sharp biphasic or triphasic wave having amplitude of about 1mV and duration of about 80ms. An iso-electric segment after the QRS complex of about 100-120 ms called as ST segment. It represents the time duration between the end of the depolarization of ventricles and start of the T wave. T wave is the result of repolarization or recovery period of the ventricles, having amplitude of 0.1-0.3 mV and duration of 120-160ms. U wave is also the part of the ECG, but it is rarely seen may be due to repolarization of the papillary muscles (Rangayyan, 2002, Thaler, 2015, Hampton, 2013). Figure 2 shows the waveform of ECG (electrocardiogram).

Figure 2. ECG waveform

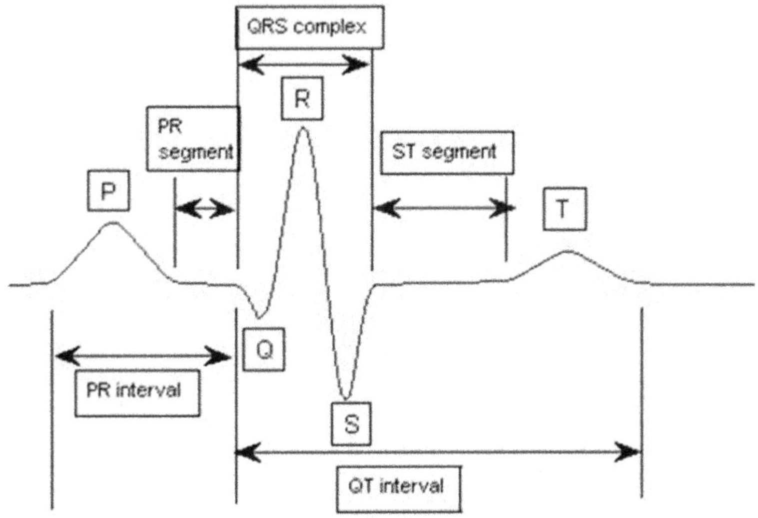

Requirement of Digitization of Paper-ECG

There is a need to convert the existing Paper-ECG records into digital forms, as it is the most efficient method to store and analyze ECG attributes (quality/features etc.) for clinical uses. The manual calculation of Paper-ECG is cumbersome as analyses of morphological parameters are imprecise and time consuming (Waits et al. 2016, Bhullar et al. 1993). ECG signals are a precious source of information; therefore retrieval of past ECG signals from graph paper is still needed (Karsikas et al. 2007). When sharing and diagnosis of ECG signals are required, in that case digitization of Paper-ECG signals becomes cost effective (Deepak et al. 2012). In the age of Computer Technology, formation of digital datasets is necessary for storing, processing or analyzing of information relating to any area. A fully computerized Paper-ECG information extraction and analyzing system is greatly acceptable today because of its wider application (Mitra et al. 2004). The advantages of fully computerized Paper-ECG information extraction and analyzing system includes – pliability of data processing, decrement of noise, management of data, improved precision and various additional potential (Mitra et al. 2003). One of the main advantages of digitization of Paper-ECG signal is a recovery of signal information from old books, journals, laboratory records, clinical charts and similar non-electronic media (Widman et al. 1991). The extensive goal of digitization of Paper-ECG is to encourage fast transmission, advancement in the research field and better analysis to improve patient care (Waits et al. 2016). Figure 3 shows image of 12-lead ECG waveform.

Figure 3. Image of 12-Lead ECG waveform (O'keefe et al., 2009)

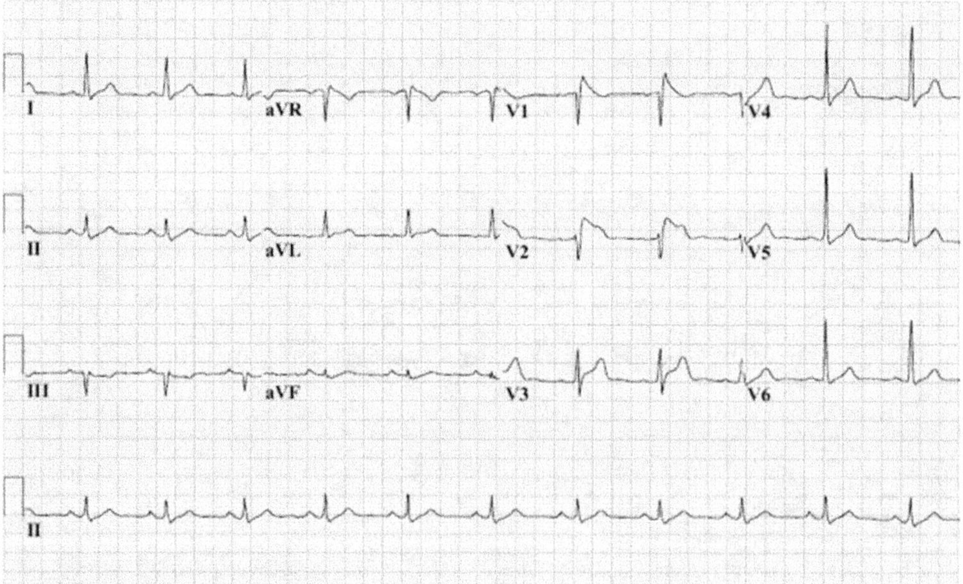

LITERATURE REVIEW

There are various digitization algorithms as discussed (Bhullar et al. 1993, Mitra et al. 2004, Widman et al. 1991, Zang et al. 1987, Wang et al. 1996, Leung et al. 1998, Badilini et al. 2005, Chebil et al. 2008, Ravichandran et al. 2013). Basic steps to digitize ECG are same, but some variation occurs, such as: type of scanner, resolution, techniques to remove grid lines, validation parameters, etc. The basic steps of the algorithm are scanning process, grid removal, linking algorithm, post processing and data retrieval. In scanning process simply Paper-ECG is scanned with the help of a scanner. Resolution of the scanner depends on the requirement. Keeping in mind both storage requirements as well as accuracy of retrieval of Paper-ECG, resolution of the scanner is set (Bhullar et al. 1993, Mitra et al. 2004, Widman et al. 1991, Wang et al. 1996, Chebil et al. 2008, Ravichandran et al. 2013). In the second process gridline removal is performed with various techniques. In this process image processing technique is used to remove the background of Paper-ECG image in such a way that, all things are removed except ECG waveform (Mitra et al. 2004, Zang et al. 1987, Wang et al. 1996, Badilini et al. 2005, Chebil et al. 2008, Ravichandran et al. 2013). Linking algorithm is an algorithm in which 2D image received after grid removal is converted into 1D signal. Various techniques are used to convert image to signal. 1D signal means at every time of instance get its corresponding voltage level (Bhullar et al. 1993, Mitra et al. 2004, Widman et al. 1991, Zang et al. 1987, Wang et al. 1996, Chebil et al. 2008, Ravichandran et al. 2013). The next step is post processing of the signals collected after conversion. In this task, some filtering is applied according to necessity (Bhullar et al. 1993, Widman et al. 1991, Wang et al. 1996, Ravichandran et al. 2013). Retrieval of information of ECG signal has executed in the last step (Ravichandran et al. 2013). Figure 4 shows some basic steps involved in conversion of paper-ECG into digital signal for all algorithms.

Zang et al. (1987) this paper presents an algorithm for converting Paper-ECG into 1D signal for storage on floppy disk. The entire process is executed with the help of the M75 image system which is made by the IIS Corporation with a VAX11/730 as its host computer. Paper- ECG is scanned with the help of Dage 68 TV camera and forming an image having a size of 512x512 with 8-bits per pixel of grey scale. Initially pre-processing of an image is performed to eliminate noise. In this process select a proper value of grey threshold to separate ECG signal image from noise. The locations of R-peaks are very important in analysis of ECG signal and they are often difficult to detect. So, the next step is to detect R-peaks with

Figure 4. Basic steps for digitization of Paper-ECG

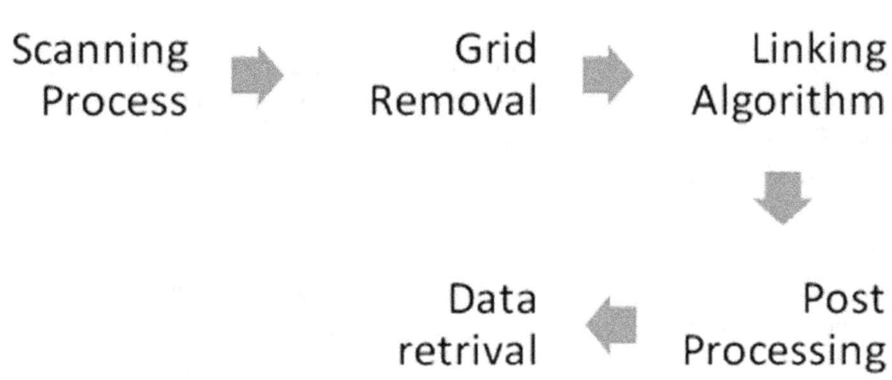

the help of Template Matching. After detecting R-peaks grid line removal performs with the help of gradient transformation. Gradient transformation performs edge detection it is applied to the Y-axis of the image which converts solid grid line into many isolated points these are easily distinguished from ECG waveform. Calculate mean spacing of these isolated grid lines which can be used as a ruler to measure heart rate, QRS complex width and RR interval. Eight Neighborhood tracing is used to get continuous and singled valued ECG waveform, this method provide continuous ECG waveform everywhere except for QRS complex where discontinuity may occur. Next step is to perform reconstruction of the QRS complex. Finally digitization of Paper-ECG is well performed.

Widman et al. (1991) this paper presents accuracy of the semi automated technique for converting Paper-ECG waveform into digital signal using an optical scanner. This paper used one non-mechanical and three mechanical ECG recorders. Analog signal received from one of the chart recorder was routed to DT2821-F A-to-D converter. ECG waveform is converted into bitmap image using Hewlett Packard ScanJet optical scanner with 300 dpi resolution. Image having size 8.5x11 inch occupied 1.05 megabytes disk storage. Standard lossless compression programs are used to compress bitmap images to 20-50 kilobytes to save storage space. There are four steps to perform digitization of Paper-ECG. First step, correction of each image performed at the pixel level to remove unused grid lines and stray marks. A line finding algorithm is used to find the contour of ECG waveform. Second, identify the upper and lower contour of ECG waveform in the bitmap image. Third, this algorithm temporary indicates "thin line" of the ECG wave by connecting midpoints of a series of line segments. Last step, temporary thin line is corrected by the user using a mouse-based interface which allows regenerate thin line at user selected point.

Bhullar et al. (1993) this paper presents a measurement of dispersion of the QT interval of the ECG. If dispersion of the QT interval of the ECG is increased it means the risk of cardiac arrhythmia is also increased. It presents a personal computer based technique for converting Paper-ECG into digital signals and automatic measurement of dispersion of the QT interval. Here 12-lead ECGs are simultaneously recorded on paper at standard speed 25mm/s using Hewlett Packard or Siemens AB electrocardiographs. Scanning of these recorded ECGs is performed with the help of flatbed scanner (Hewlett Packard ScanJet Plus) with resolution 300dpi. These recorded ECGs are passed through scanner to produce a pictorial file image in commercial .PCX format. To divide the image into .PCX subfiles corresponding to individual leads a graphical editor is used. Preprocessing of the scanned image is performed to remove noise and fill in the gaps. Thinning of image performed using Zhang-Suen algorithm to recompense for finite line width in original ECG recording. An algorithm is used to extract digital information from the thinned image. After that high pass and low pass digital filtering done at 0.01Hz and 100Hz. Automatic analysis used in this paper was designed to detect QT onset, QT offset and RR interval from QRST waveform for validation purpose also.

Lawson et al. (1995) this paper presents a method which was developed to digitize scanned Paper-ECG waveform for analysis by using Hewlett Packard (HP) program and to collect this information into clinical database. Paper-ECG is scanned with 200 dpi resolution using 8-bit greyscaling into two half page TIFF graphic files. Grayscale threshold is used to differentiate between ECG waveform and gridlines which is decided by the user. This threshold value is used by the software to separate the contour of the waveform from the surrounding image using skeletonization operation. This operation provides a set of coordinate pairs which define line segments centered over the ECG tracing. Linear interpolation is used to connect these points. During the tracing process, the user has the ability to adding and deleting points in order to modify the computer's skeleton. These traces are sampled at 500 Hz and converted into a proprietary binary storage format of Hewlett-Packard. For analysis these ECG files are transferred to a HP Pagewriter XLi digital electrocardiograph.

Wang et al. (1996) this paper presents designed a prototype system to convert Paper-ECG records into digital signals so that they can be either retrieved efficiently as needed or examine by signal processing algorithm or transmitted through computer networks for clinical uses. Paper-ECG records are scanned using a flatbed scanner with 300 dpi resolution. To fit requirement of display screen maximum size of scanned ECG image restricted to 640x480 pixels. Concept of masking is used in heuristic filtering (pre-filtering) algorithm so that to remove gridlines and random noise and to fill possible gaps in ECG image which introduce during scanning and digitization process. Two types of masks are applied to ECG images that are horizontal and vertical mask. These two masks are iteratively applied to the ECG image to remove random noise and grid lines. A thinning algorithm is used to obtain the contour of ECG waveform. Here distance approach thinning is used to obtain a skeleton of ECG waveform. Post filtering algorithm is used to extract ECG data by tracing the skeleton of ECG signal which are then stored in ASCII format.

Chia et al. (1996) this paper presents a precise and automatic system to extract features of the Paper-ECG records. The scanning is performed with the help of an HP flatbed color scanner (ScanJet 4k) which offer true color (24bits) scanning with 600 dpi resolution. 12 lead Paper-ECG records are scanned using color a scanner with 300 dpi resolution. The system is designed in such a way that the gridlines are removed automatically and modified run-length coding technique is used to compress resulting image in real time with no loss of information. At the same time, scanned image of 12 lead ECGs is segmented automatically so that to isolate the 12 individual leads and rhythm strip. The signal recovery algorithm is applied to each segmented image to generate a series of ASCII values which representing the actual ECG signal. Thinning is used to recover lineal structure of data without destroying its connectivity. Data values are then extracted from the contour of the ECG, filtering is applied to remove high frequency noise introduce by scanning, thinning and extraction process.

Leung et al. (1998) this paper discusses an oversampling technique which is known as Delta- Sigma Modulation, which is used for digitization of Electro-physiological signals such as ECG. This modulation technique uses a very simple circuitry which consists of the operational amplifier and flip-flop. This modulation technique uses oversampling as well as feedback integration also. The advantage of this technique is that the circuitry used in this is simple for both modulation and demodulation. This paper implements this technique using Matlab software program as well as experimental setup also. Matlab program is used for simulation of Delta- Sigma Modulator. A sampled ECG signal is used as an input for the modulator. Reconstructed signal is obtained by passing a digital output signal of modulator to a digital 2nd order Butterworth low pass filter. Experimental setup of modulator is designed using quad operational amplifier LF444 and a CMOS D flip-flop (4013). ECG signal is an input for the modulator which is provided by a function generator from YOKOGAWA. A digital bit stream is obtained as the output of modulator which is then passed into a 1st order low pass filter to obtain the reconstructed signal. For further analysis of the reconstructed signal WINDAQ computer acquisition tools is used to record signals from circuit to computer disk. Data obtain from this tool is imported to Matlab for further analysis.

Kao et al. (2001) this paper presents a morphological method to extract information from Paper-ECG charts using an optical scanner and a computer. This paper describes the 3 different types of background of Paper-ECG: uniform background, colored grid background and black grid background. Here two parallel systems are used for digitization. In the first system, ECGs are recorded on 3 -channel electrocardiograph in which 12 leads are recorded in a group. In the second system, ECGs are recorded on analog magnetic tape and computer system. Scanning of Paper-ECG records having a black grid background is performed using optical scanner with 300 dpi resolution. Extraction of images of the ECG wave that having a uniform or colored background grid, is performed using color and area filter. Morphological operation is used to remove background with black grid. Noisy spots and characters are removed by using a background noise filter. If the ECG waveform without gridlines having broken points then for connecting the waveform linear interpolation is used. At last, thinning algorithm is used to obtain ECG signal.

Lobodzinski et al. (2003) this paper presents an optical ECG waveform recognition (OEWR) technique for extraction of information of ECG waveform from Paper-ECG records. Hardcopy ECGs are printed on special red grid paper (type M1737) from original database available on Hewlett-Packard Xli cardiograph. These Paper-ECGs are scanned in 24-bit RGB mode using an Epson 1640 optical scanner with resolution of 600 dpi. Size of scanned images of Paper-ECG is about 100Mbytes. There are three main steps of OEWR technique. In First step, pre-filtering of ECG raster performs which involves removal of skew errors, pixel classification, cropping and scaling of image. Removal of background gridlines from ECG image are not performed in OEWR technique because gridlines and ECG waveform are overlapped. If removal of gridlines is performed it means some information of ECG signal is also removed due to which accuracy of OEWR technique is reduced. Instead of this it uses a statistical filter and an adaptive line tracking algorithm. Second steps involve linking and filtering of waveform segments. In last step data is extracted from ECG waveform.

Mitra et al. (2004) this paper presents an automatic data extraction system which converts Paper-ECG record into digital time database. Scanning of Paper-ECG performed with a flatbed scanner (HP ScanJet 4C model) with high resolution (1200 dpi) and 256 gray tone spectral resolutions. With the help of histogram analysis a suitable threshold is selected to convert gray scale image into a binary image. Gridlines of Paper-ECGs are eliminate after thresholding operation, but some dotted portions may exist in some cases which are removed by using a Run Length Smearing algorithm. Thinning algorithm is used to obtain contour of ECG waveform, especially to avoid repetition of unnecessary data in the data set of database. Co-ordinate of each black pixel is extracted because the ECG waveform is represented by the black pixel in the image and these values are serialized according to coordinate system of the chart. Sorting of raw data (co-ordinate) according to their abscissa value is performed with the help of sorting routine. Sorted data obtained from sorting routine is imported to the regeneration domain of the system to check the pattern of captured signal with original waveform. Now sampling period of reconstructed signal is calculated. After calculation of sampling period Discrete Fourier Transform was performed for each data to observe phase and frequency response properties of every ECG signal.

Badilini et al. (2005) this paper presents a computer application known as ECGScan which is developed for converting Paper-ECG records into digital ECG signal. This application opens image files in various graphic file formats which involve Microsoft Windows bitmap, JPEG, TIF, PNG, and graphics interchange. User interaction is introduced in this application that uses the concept of the active rectangle in which user selects the rectangular portion of a loaded ECG image with the help of computer mouse. A grid detector used in this application which is able to estimate grid angle (i.e. presence of tilt effect) and to remove nonrepetitive elements (like marks and strains). Three different types of mode are used to detect ECG grid in this application that are exact, range and manual. The user has an option to choose either digitization of single lead at a time or digitization of group of simultaneous lead or digitization of full Paper-ECG record. This paper presents digitization of Paper-ECG record based on active contour and dynamic programming. Removal of gridlines for digitization is not required in this algorithm; if any gaps are occurring in ECG waveform then they are automatically interpolated during the digitization process. Anchor points in image are manually set by the user. Finally digitized ECG signals saved in various formats which may be Extensible Markup Language, binary or American Standard Code for Information Interchange.

Morales et al. (2005) this paper presents a method which is used for digitization and synchronization of the ECG signal for use in vectorcardiography. Conversion of Paper-ECG printout into image files by using a conventional computer scanner. ECG records are divided into 4 sections of 6 leads each so that meeting the synchronization conditions. Automatically digitization of the 24 signals is performed in developed computer software. These sections are manually selected by the user if selection of all sections is completed then unwanted data is filter-out by computer software. Finally, the determination of the isoelectric line level is performed manually for all the recordings and pixel scale is applied to the digitized image. This application produces 24 digitized signals. Electrocardiographic signals must be recorded simultaneously for derivation of VCG. Therefore, synchronization of these digitized signals is necessary for the derivation of VCG from ECG data. Efficiency of synchronization process is determined by the figure of merit which is obtained by Dower analysis. A transformation matrix is proposed by the Dower which is known as "Frank XYZ Transfer matrix" which is used for retrieval of ECG from VCG XYZ leads. The Inverse Dower Technique is also available in which VCG XYZ leads are reconstructed form ECG leads. Synchronization of ECG signals is performed by cross correlation of continuous signals.

Karsikas et al. (2007) this paper presents a method for the digitization of Paper-ECG record and systematical comparison of nine ECG morphological parameters. The basic idea of this experimental setup was to develop a hardcopy of ECG recording and digitize it back to create effects of original ECG records. There are five different cases are considered in this experiment. In the first three cases, original 12-lead digital ECG signals are analyzed so that to measure the effect of the alignment of grouped channels and non-simultaneous sampling between these channel groups. In the last two cases, analysis of digitized Paper-ECGs is performed through simultaneous and non-simultaneous data. Five cases to measure the morphological parameters are: case D, case DA, case DANs, case P and case PANs. Scanning of Paper-ECG records are performed using an optical flatbed scanner Epson Perfection 1250 (Seiko EPSON Corp, UK), with resolution of 600 dpi in black and white mode. Raster images are the output of a scanner which is act as an input to Un-Scan-It Graph Digitizing System 6.0 (Silk Scientific Corporation, Utah, USA) for tracking and digitization of the contour of ECG waveform in the images. The output of the system is stored in ASCII format.

Chebil et al. (2008) this paper presents a method for digitization and extraction of Paper-ECG record using neighborhood and median approach as well as selection of appropriate image resolution during scanning. Scanning is also an important step during the digitization of Paper-ECG record. During the scanning process three main factors should be considered these are: resolution of image, bit depth and format of output file. The scanning is performed in this paper with the help of Epson scanner which is type of perfection 2480 photo. All the digitization process performs in this paper with the help of MATLAB software. This software supports various types of image file formats these are: BMP, JPEG and TIFF file format. Scanner has a wide range of PPI from 100 to 4800 and 8-bit grayscale is available. This paper selects BMP file format as a format of image. Image thresholding algorithm is used to extract ECG waveform from gridline background. Due to image thresholding algorithm some unwanted isolated black pixel present in the image, which need to be removed. Transformation of extracted data into its digital form is done in three steps. First, zero voltage level is located by using referencing pulse. Second, column wise median approach is applied to get single valued black pixel. A median approach is preferred over mean, because it prevents the effect of unwanted isolated pixels. Neighborhood approach is used to remove remaining unwanted data. Resulted data is obtain from this stage is represent in the form of x-y coordinates of ECG waveform.

Ravichandran et al. (2013) this paper presents a MATLAB-based tool for conversion of hardcopy of the ECG into digital ECG signal. Scanning of Paper-ECG is performed so that to convert Paper-ECG into 8-bit grayscale digital image. Resolution of the image is scaled down to 300 dpi so that to reduce computational costs without compromising quality of the reconstructed signal. Skew correction is used to correct the angular orientation of Paper-ECG. This tool uses a graphical user interface (GUI) for skew correction. Binary thresholding operation is used to remove gridlines from a background of ECG image and change of grayscale images into a binary image. Histogram of image is used to determine the threshold. Median filtering and interpolation are used to remove the "salt and pepper" noise introduce in the background of ECG and to remove discontinuities occur in ECG signal due to removal of major gridlines. Now column-wise pixel scan is use to find the skeleton of ECG waveform based on black pixel or zero location. Upper and lower contour of ECG signal is obtained for each column since the printed form of ECG signal has some thickness along with it. X-axis and y-axis scales are used to obtain the time-voltage pair for mean of the contour of the ECG signal. Extraction of patient information from the Paper-ECG record is done with the help of template-based optical character recognition (OCR) so that to interface the patient information into the medical record.

Table 1. Summary of conversion algorithms

First Author	Year	Name of Scanner	Image Specification	Keywords	Validation Parameter	Software or System	Result	Number of Samples
Zang et al.	1987	Dage 68 TV camera	size of 512x512 with 8-bits per pixel of grey scale	Histogram filtering, Template Matching, Gradient transformation, Eight Neighborhood tracing	Comparison of original ECG image and their reconstructed counterparts	M75 image system	n/a	11
Widman et al.	1991	Hewlett Packard ScanJet optical scanner	size 8.5x11 inch, 300dpi resolution	line finding algorithm, thinning	RR interval and QRS complex (correlation coefficient, RMS error)	MS-DOS Turbo C 2.0	Digital and digitized signal are "identical"	261 QRS complex and 207 RR intervals
Bhullar et al.	1993	flat bed scanner	.PCX format, 300 dpi	Preprocessing, thinning, high- and low- pass filtering	RR and QT intervals (correlation coefficient, Bland–Altman, sign tests)	Unspecified graphical editor	Correlation within ~1%	116(series1), 112(series2)
Lawson et al.	1995	n/a	200 dpi, 8-bit grayscale, TIFF graphic file	Grayscale threshold, skeletonization operation, Linear interpolation	Qualitatively (compare ECG-generated matrices from scanned Vs original ECGs, as well as automated measurements of scanned ECGs vs manual reading)	Hewlett Packard (HP) program	Accurately preserves essential features	84 (compare with original ECGs) and 107 (compare with hand measurements)
Wang et al.	1996	Flatbed scanner	300 dpi resolution, size 640x480 pixels	Binarization, thinning, pre- and post- filtering (tracing)	Qualitatively	Windows, GUI	n/a	1
Chia et al.	1996	HP flatbed color scanner	300 dpi resolution, 24 bits	run-length coding, Signal recovery algorithm, Thinning,	RMS error, cross correlation,	PC system (P90 processor, 161 MByte RAM, 1.2 GByte hard disk, super VGA card with, 1024x768 resolution display).	Value of cross correlation is 0.99	12

continued on following page

Table 1. Continued

First Author	Year	Name of Scanner	Image Specification	Keywords	Validation Parameter	Software or System	Result	Number of Samples
Leung et al.	1998	n/a	n/a	Delta- Sigma Modulator, flip-flop, operational amplifier, Butterworth LPF	PRD (percent root mean square difference), PAD (percent mean absolute difference)	Matlab, Experimental setup	23% PRD and 27% PAD	1
Kao et al.	2001	Optical scanner	300 dpi resolution	Morphological operation, linear interpolation, thinning	Correlation, APRD (average percent root-mean-square (rms) Difference)	Computer analysis system	97% correlation and 5.25% APRD	12
Lobodzinski et al.	2003	Epson 1640 optical scanner	600 dpi resolution, 24-bit, size 100 Mbytes	OEWR (ECG waveform recognition), waveform segment linking and filtering	Direct comparison (difference plotting)	Windows 2000 OS, MATLAB, Java, C++ and C subroutines	Sample-by-sample differences of 1–3 pixels (4–12 uV)	240
Mitra et al.	2004	flatbed scanner	1200 dpi resolution and 256 gray tone spectral resolution	Binarization, background separation, thinning, data extraction, sorting	Harmonics of V4, V6: normal vs. ischemic vs. infarcted (t test)	Self-developed	Accuracy of ~98.4%	60
Badilini et al.	2005	n/a	n/a	Grid detector, skew correction, waveform detection, anchor	Direct comparison (RMS), QT intervals (Bland–Altman)	Self-developed, C++ based	Precision within 5–6 ms	60
Morales et al.	2005	Conventional computer scanner	n/a	Vectorcardiography, synchronization, Dower analysis	NRMS error (normalized RMS error)	Self developed	13.49% (max NRMS), 9.85% (min NRMS)	12
Karsikas et al.	2007	optical flatbed scanner Epson Perfection 1250	600 dpi resolution	morphological parameters, Signal contour extraction, alignment	T-amplitude, T-area, and derived parameters: loop and non-dipolar (Bland–Altman, Median abs. Deviation, Mann–Whitney, RMS error)	SPSS; "Un-Scan-It Graph Digitizing System 6.0"	Mean absolute deviation = +5.85% to −4.86%; relative median error for T wave amplitude = −1.7 ± 5.4%, T wave area = 18.6 ± 21.1%; RMS err: 10–12%	30
Chebil et al.	2008	Epson scanner	PPI range (100 to 4800), 8-bit grayscale	Scanning, threshold algorithm, neighborhood and median approach,	Relative error percentage (PR interval, QRS duration, QT interval, heart rate), RMS (root mean square error)	MATLAB	Relative error (0% min and 2.5% max), RMS error (3% in worst case), PPI range (400 to 600)	5
Ravichandran et al.	2013	n/a	8-bit grayscale, 300 dpi resolution	Grid detection, skew correction, GUI, column-wise pixel scanning, optical character recognition	Direct comparison; PR, QRS, RR, QT, QTc intervals (kappa statistic, signal correlation match)	MATLAB version 7.10	Kappa stat >85%; signal fit 75–80%; correlation 85–90%; intra-observer and inter-observer correlations 80–100%	10

DISCUSSION

This section summarizes the numerous existing algorithms by providing various information's such as the type of the scanner used, image specifications, keywords, validation parameter, used software, results and the number of samples. There are various types of scanner is used for the scanning process, mostly optical and flatbed scanner is used. Initially scanning process is started with a TV camera (Zhang et al., 1987). Some of them also uses a color scanner (Chia et al., 1996, Lobodzinski et al., 2003) which provide true color scanning with 24-bit RGB mode. Specification of images based on size, resolution and format of the images. Mostly images describe in 8-bit grayscale, but some of them in 24-bit RGB also. Complexity is reduced by using 8-bit grayscale representation rather than 24-bit RGB representation of images. The images are mostly scanned with a resolution of 300 dpi, but some of them scan with 200, 600 and 1200 dpi resolution. Image resolution selected in such a way that they give better retrieval of ECG as well as reduces the computational cost also. Chebil et al. (2008) discusses about the selection of image resolution. It compares RMS error for 150, 300, 400 and 600 PPI values. It is observed that the RMS error reduces with increase in PPI value. The error is quite comparable if the range of PPI values is between 400 and 600, therefore the image resolution should be selected in this range during scanning. Most of the papers not defined the format of images, but two of them used PCX and TIFF image format. There are various types of image formats are available such as Windows Bitmap (BMP), Joint Photographic Experts Group (JPEG), Tagged Image File Format (TIFF), Graphics Interchange Format (GIF) and Windows Paintbrush (PCX). Format of images also depends on the software associated with scanners. Chebil et al. (2008) observed that the software associated with the Epson scanner can support only three file formats such as BMP, JPEG and TIFF. It is observed that the quality of the BMP and TIFF file formats are better than JPEG, but their size of files is larger. JPEG is not preferred because it uses a lossy compression technique where as others are lossless compression technique. TIFF is also not preferred because it uses LZW compression which is not supported by MATLAB. So this paper chooses BMP file format for better results. Keywords are provided so that the reader easily gets the main topics of the paper. By go through these keywords reader easily understand the paper's whole algorithm and work. Keywords already studied in the literature review section. Validation parameters are used to validate the existing algorithm. There are various validation parameters such as the correlation coefficient, RMS error, cross correlation, PRD (percent root mean square difference), PAD (percent mean absolute difference), QT interval, QRS complex, RR interval etc.

Software and systems are designed to perform digitization of Paper-ECG. Mostly MATLAB software is used, but some other systems are also used such as image system, MS-DOS turbo C, Windows, GUI, PC system, self developed system etc. The result of numerous papers is summarized in Table1. Number of samples indicates the number of ECG signals which is used to perform work of papers. ECG signals are used for digitization of Paper-ECG and to validate the existing algorithm.

CHALLENGES

There are various challenges arises during the digitization of Paper-ECG. It discusses various challenges of papers with their possible solution. The initial challenge is involved in the process of Paper-ECG digitization is gridline removal. In this process information of ECG signal is also removed, none of the existing methods provide flawless gridline removal. Zang et al. (1987) presents a pre-processing to eliminate noise. These noises include shadows created by uneven illumination, gridlines are inherent to Paper-ECG record and gray values of images are randomly changes. Ideal thresholding techniques can be implemented on digitized images to reduce unwanted noise. Widman et al. (1991) finds limitation of the study that it is cannot be expanded easily to high frequency signals. It provides triangle approximation method to overcome that limitation. Bhullar et al. (1993) realizes noise during gridline removal, which is removed by using color-filter during scanning. It reduces high frequency noise by using low-pass filtering and smoothing function. Lawson et al. (1995) proposes a thresholding operation to remove surface noise such as gridlines. Wang et al. realizes noise during gridline removal, which is removed on the basis of line-width dissimilarity between grid and signal. Chia et al. (1996) presents a pre-processing to eliminate random noise and fill minor gaps. Components of high frequency noise which is introduced by thinning, scanning and extraction process is removed by filtering process. Kao et al. (2001) uses color filter, area filter and morphological operation to remove the background of Paper-ECG. Noisy spots and characters are removed by using a background noise filter. Lobodzinski et al. (2003) fails to show robust conversion of long Paper-ECG record. This problem can be overcome by testing on a limited number of samples. Mitra et al. (2004) realizes noise during gridline removal, which is removed by using Background separation. ECG signal in time domain is corrupted by various noises such as Electromyography (EMG) signal, 50 Hz power line interference, abrupt baseline shift, etc., which is reduced by analyzing the ECG signal in the frequency domain. Conversion of ECG signals from time domain into the frequency domain by Fast Fourier Transform (FFT). Badilini et al. (2005) uses various grid modes to identify the background of ECG such as gridlines. It requires user-interaction which means system is not fully automatic, due to which relative timing between leads is not attained correctly.

Badilini et al. (2005) and Karsikas et al. (2007) both have alignment issues which can be eliminated by using multi-lead functionality. Chebil et al. (2008) proposes a gray level thresholding operation to remove surrounding noise such as gridlines. Ravichandran et al. (2013) introduces interpolation and median filter to eliminate salt and pepper noise introduces during gridline removal. The Paper-ECG used in hospitals differ in shape, size, sampling rate, formats, so main challenge in the process of digital conversion of Paper-ECG is to achieve a worldwide ECG format.

Previously this chapter discusses the challenges for Paper-ECG conversion. Now this chapter discusses the challenges for ECG as biometric, which are mentioned below:

Challenges for ECG as Biometric

- Efficient and reliable methods have to be designed to assess the effects of heart diseases, mental and physical stress, exercise and other influences including diet, medications, and drugs.
- Heart rate variability due to physical, mental or other stressful activities can have the effect of changing the morphology of the ECG.
- Database size in an important issue, to form a temple ECG of an individual a larger database compared to other methods of authentication is required.
- QRS-complex is most distinctive wave in the ECG, but it is uncertain whether it's solely can support ECG biometric recognition (Odinaka et al., 2012, Singh et al., 2012).

CONCLUSION

In order to diagnose the possible cardiac disorder, the ECG (electrocardiogram) signals are usually recorded on standard grid papers in hospitals. Many efforts have been made to advance the technology in order to improve the diagnosis and management of cardiovascular disease. This chapter discusses the various challenges and a systematic study of different methods that have been used so far to convert Paper-ECG records into digitized form so that they can be retrieved efficiently. It presents a literature review on existing algorithms for digital conversion of Paper-ECG. The various algorithms for conversion of Paper-ECG into digital form have been proposed in the most efficient means to store ECG records and analyze heart activity. Digitization of Paper-ECG shows a major role in the examination of various cardiovascular diseases, this can also be further extended to understand the cardiac electrophysiology, arrhythmia mechanisms and the relation to clinical diseases.

REFERENCES

Badilini, F., Erdem, T., Zareba, W., & Moss, A. J. (2005). ECGScan: A method for conversion of paper electrocardiographic printouts to digital electrocardiographic files. *Journal of Electrocardiology*, *38*(4), 310–318. doi:10.1016/j.jelectrocard.2005.04.003 PMID:16216602

Barold, S. S. (2003). Willem Einthoven and the birth of clinical electrocardiography a hundred years ago. *Cardiac Electrophysiology Review*, *7*(1), 99–104. doi:10.1023/A:1023667812925 PMID:12766530

Bhullar, H. K., Fothergill, J. C., Goddard, W. P., & de Bono, D. P. (1993). Automated measurement of QT interval dispersion from hard-copy ECGs. *Journal of Electrocardiology*, *26*(4), 321–331. doi:10.1016/0022-0736(93)90053-G PMID:8228721

Chebil, J., & Maitah, J. (2008). A Novel Method for Digitizing Standard ECG Papers. *IEEE International Conference on Computer and Communication Engineering*. 10.1109/ICCCE.2008.4580816

Chia, P., Ong, K., & Bhullar, H., K. (1996). An Accurate and Automatic System for Extracting Features from ECG Paper Recordings: Its Use in Clinical Studies and Telemedicine. *IEEE Computers in Cardiology*.

Garg, D., K., Thakur, D., Sharma, S., & Bhardwaj, S. (2012). ECG Paper Records Digitization through Image Processing Techniques. *International Journal of Computer Applications, 48*(13).

Gutta, S., & Cheng, Q. (2016). Joint feature extraction and classifier design for ECG-based biometric recognition. *IEEE Journal of Biomedical and Health Informatics*, *20*(2), 460–468. doi:10.1109/JBHI.2015.2402199 PMID:25680220

Guyton, A. C., & Hall, J. E. (2006). *Textbook of medical physiology*. Saunders/Elsevier.

Hampton, J., R. (2013). *The ECG made easy*. Toronto: Churchill Livingstone Elsevier.

Hejazi, M., Al-Haddad, S. A., Singh, Y. P., Hashim, S. J., & Aziz, A. F. (2016). ECG biometric authentication based on non-fiducial approach using kernel methods. *Digital Signal Processing*, *1*(52), 72–86. doi:10.1016/j.dsp.2016.02.008

Iaizzo, P. A. (2005). *Handbook of cardiac anatomy, physiology and devices*. Humana Press. doi:10.1007/978-1-59259-835-9

Islam, M. S., Ammour, N., Alajlan, N., & Abdullah-al-wadud, M. (2017). Selection of Heart-Biometric Templates for Fusion. *IEEE Access: Practical Innovations, Open Solutions*, *5*, 2169–3536. doi:10.1109/ACCESS.2017.2667224

Kao, T., Hwang, L. J., Lin, Y. H., Lin, T. H., & Hsiao, C. H. (2001). Computer analysis of the electrocardiograms from ECG paper recordings. *Proceedings of the 23rd Annual EMBS International Conference*.

Karsikas, M., Huikuri, H., Juha, S., Lehtola, L., & Tapio, S. (2007). Influence of Paper Electrocardiogram Digitizing on T Wave and QRS Complex Morphology Parameters. *Original Articles, 12*(4).

Khandpur, R. S. (2003). *Biomedical Instrumentation: Technology and Applications*. New Delhi: Tata McGraw-Hill Publishing Company Limited.

Lawson, W., T., Wagner, G., S., Selvester, R., S., & Ybarra, G., A. (1995). New Method for Digitization and Computerized Analysis of Paper Recordings of Standard 12-Lead Electrocardiograms. *IEEE Computers in Cardiology.*

Leung, S. W., & Zhang, Y. T. (1998). Digitization of electrocardiogram (ECG) signals using deltasigma modulation. *Proceedings of the 20th Annual International Conference of the ZEEE Engineering in Medicine and Biology Society*, *20*(4), 0-7803-5164-9.

Lobodzinski, S., M., Teppner, U., & Laks, M. (2003). State of the art techniques for preservation and reuse of hard copy electrocardiograms. *Journal of Electrocardiology, 36.*

Lumini, A., & Nanni, L. (2017). Overview of the combination of biometric matchers. *Information Fusion*, *1*(33), 71–85. doi:10.1016/j.inffus.2016.05.003

Merone, M., Soda, P., Sansone, M., & Sansone, C. (2017). ECG databases for biometric systems: A systematic review. *Expert Systems with Applications*, *1*(67), 189–202. doi:10.1016/j.eswa.2016.09.030

Mitra, S., & Mitra, M. (2003). An automated data extraction system from 12 lead ECG images. *Computer Methods and Programs in Biomedicine*, *71*(1), 33–38. doi:10.1016/S0169-2607(02)00032-9 PMID:12725963

Mitra, S., Mitra, M., & Chaudhuri, B. B. (2004). Generation of digital time database from paper ECG records and Fourier transform-based analysis for disease identification. *Computers in Biology and Medicine*, *34*(7), 551–560. doi:10.1016/j.compbiomed.2003.08.001 PMID:15369707

Morales, E., Sevilla, D., Pierluissi, J. H., & Nazeran, H. (2005). Digitization and Synchronization Method for Electrocardiogram Printouts. *IEEE Engineering in Medicine and Biology 27th Annual Conference.*

Najarian, K., & Splinter, R. (2012). *Biomedical Signal and Image Processing*. New York: Taylor and Francis group.

O'keefe, J. H., Hammill, S. C., Freed, M. S., & Pogwizd, S. M. (2009). *The Complete guide to ECGs: A comprehensive study guide to improve ECG interpretation skills*. Jones and Bartlett publishers.

Odinaka, I., Lai, P. H., Kaplan, A. D., O'Sullivan, J. A., Sirevaag, E. J., & Rohrbaugh, J. W. (2012). ECG biometric recognition: A comparative analysis. *IEEE Transactions on Information Forensics and Security*, *7*(6), 1812–1824. doi:10.1109/TIFS.2012.2215324

Rangayyan, R. M. (2002). *Biomedical Signal Analysis: A Case-Study Approach*. New York: John Wiley & sons.

Ravichandran, L., Harless, C., Shah, A. J., Wick, C. A., Mcclellan, J. H., & Tridandapani, S. (2013). Novel tool for complete digitization of paper electrocardiography data. *IEEE Journal of Translational Engineering in Health and Medicine*, *1*, 2168–2372. doi:10.1109/JTEHM.2013.2262024 PMID:26594601

Sajjan, M. (2013). *Learn ECG in a day: A systematic approach*. New Delhi: Jaypee Brothers Medical Publishers.

Sharma, L. D., & Sunkaria, R. K. (2017). P and T-Wave Characterization in the Presence of U-Wave in Electrocardiogram. *Current Trends in Signal Processing. STM Journals*, *7*(1), 29–34.

Sidek, K. A., Khalil, I., & Jelinek, H. F. (2014). ECG biometric with abnormal cardiac conditions in remote monitoring system. *IEEE Transactions on Systems, Man, and Cybernetics. Systems*, *44*(11), 1498–1509. doi:10.1109/TSMC.2014.2336842

Singh, Y. N., Singh, S. K., & Ray, A. K. (2012). Bioelectrical signals as emerging biometrics: Issues and challenges. *ISRN Signal Processing*, *26*.

Thaler, M. S. (2015). *The only EKG book you'll ever need.* Tokyo: Wolters Kluwer Health.

Waits, G., S., & Soliman, E., Z. (2016). Digitizing paper electrocardiogram: Status and challenges. *Journal of Electrocardiology*.

Wang, J. T., & Mital, D. P. (1996). A microcomputer-based prototype for ECG paper record conversion. *Journal of Network and Computer Applications*, *19*(3), 295–307. doi:10.1006/jnca.1996.0020

Widman, L. E., & Hines, L. S. (1991). Digitization of electrocardiograms by desktop optical scanner. *Journal of Electrocardiology*, *24*(4), 325–338. doi:10.1016/0022-0736(91)90015-E PMID:1744545

Williams & Wilkins. (2011). *ECG interpretation made incredibly easy*. Tokyo: Wolters Kluwer Health.

Zhang, Z., Zhang, H., & Zhuang, T. (1987). One-dimensional signal extraction of paper-written ECG imagen and its archiving. *SPIE Visual Communications and image processing II, 845*, 419–22.

KEY TERMS AND DEFINITIONS

Depolarization: An excited cell which shows action potential is known as depolarized and the process is called depolarization.

Diastole: Relaxing or filling phase of a cardiac chamber.

Electrocardiogram: Electrical activity of the heart is recorded in the form of a waveform.

Histogram: Histogram of an image plots the number of pixels for each intensity level.

Repolarization: After getting into the depolarized state for a definite interval of time, the cells are again polarized and go back into resting potential state through repolarization process.

Systole: Contracting or pumping phase of a cardiac chamber.

Vectorcardiography: It is the technique to examine the electrical activity of heart by acquiring ECG along three axes.

Compilation of References

Abate, A. F., Nappi, M., Riccio, D., & Sabatino, G. (2007). 2D and 3D face recognition: A survey. *Pattern Recognition Letters*, *28*(14), 1885–1906. doi:10.1016/j.patrec.2006.12.018

Abdel-Hamid, O., Mohamed, A-R., & Jiang, H. (2014). Convolutional neural networks for speech recognition. *IEEE/ACM Trans Audio Speech Lang Process*, *22*, 1533–1545.

Adey, P. (2009). Facing airport security: Affect, biopolitics, and the preemptive securitization of the mobile body. *Proceedings of Environment and Planning: Society and Space*, *27*(2), 274–295.

Ahmed, F., Paul, P. P., & Gavrilova, M. (2016). Joint-triplet motion image and local binary pattern for 3d action recognition using Kinect. In *Proceedings of 29th Int. Conference on Computer Animation and Social Agents* (pp. 111-119). Academic Press.

Ahmed, F., Paul, P. P., & Gavrilova, M. (2015). DTW-based Kernel and Rank Level Fusion for 3D Gait Recognition using Kinect. *The Visual Computer*, *31*(6-8), 915–924. doi:10.100700371-015-1092-0

Ahmed, F., Paul, P. P., & Gavrilova, M. (2016). Adaptive Pooling of the Most Relevant Spatio-Temporal Features for Action Recognition. In *Proceedings of 2016 IEEE International Symposium on Multimedia (ISM 2016)* (pp. 177-180). IEEE. 10.1109/ISM.2016.0042

Ahmed, F., Sieu, B., & Gavrilova, M. (2018). Score and Rank-Level Fusion for Emotion Recognition Using Genetic Algorithm. In *Proceedings of 17th IEEE International Conference on Cognitive Informatics & Cognitive Computing* (pp. 46-53). IEEE.

Aishwarya, D., Gowri, M., & Saranya, R. K. (2016). Palm print recognition using liveness detection technique. In *Science Technology Engineering and Management (ICONSTEM), Second International Conference on* (pp. 109–114). IEEE.

Alan, G., & Decker, C. (2012). Cash at your fingertips: Biometric technology for transfers in developing countries. *The Review of Policy Research*, *29*(1), 91–117. doi:10.1111/j.1541-1338.2011.00539.x

Aldridge, K., Boyadjiev, S. A., Capone, G. T., DeLeon, V. B., & Richtsmeier, J. T. (2005). Precision and error of three-dimensional phenotypic measures acquired from 3dMD photogrammetric images. *American Journal of Medical Genetics. Part A*, *138*(3), 247–253. doi:10.1002/ajmg.a.30959 PMID:16158436

Aldridge, K., George, I. D., Cole, K. K., Austin, J. R., Takahashi, T. N., Duan, Y., & Miles, J. H. (2011). Facial phenotypes in subgroups of prepubertal boys with autism spectrum disorders are correlated with clinical phenotypes. *Molecular Autism*, *2*(1), 15. doi:10.1186/2040-2392-2-15 PMID:21999758

Alegre, F., Amehraye, A., & Evans, N. (2013). Spoofing countermeasures to protect automatic speaker verification from voice conversion. *Proceedings of IEEE ICASSP*, *13*, 3068–3072.

Aley-Raz, A., Krause, M., & Salmon, I. (2013). *Device, system, and method of liveness detection using voice biometrics*. United States Patent, No. USOO8442824B2.

Alford, A., Hansen, C., Dozier, G., Bryant, K., Kelly, J., Abegaz, T., . . . Woodard, D. L. (2011, June). GEC-based multi-biometric fusion. In *Evolutionary Computation (CEC), 2011 IEEE Congress on* (pp. 2071-2074). IEEE. 10.1109/CEC.2011.5949870

Alhussein, M. (2016). Automatic facial emotion recognition using weber local descriptor for e-Healthcare system. *Cluster Computing*, *19*(1), 99–108. doi:10.100710586-016-0535-3

Ali, H. S., & Abdalla, M. I. (2011). Score-level fusion for efficient multimodal person identification using face and speech. *International Journal of Computer Science and Information Security*, *9*(4), 48.

Alipanahi, B., Delong, A., Weirauch, M. T., & Frey, B. J. (2015). Predicting the sequence specificities of DNA-and RNA-binding proteins by deep learning. *Nature Biotechnology*, *33*(8), 831–838. doi:10.1038/nbt.3300 PMID:26213851

Al-Kaff, A., Moreno, F. M., Escalera, A. D. L., & Armingol, J. M. (2017). Intelligent vehicle for search, rescue and transportation purposes. In *Proceedings of IEEE International Symposium on Safety, Security and Rescue Robotics (SSRR)* (pp. 110-115). IEEE. 10.1109/SSRR.2017.8088148

Ambinder, E. P. (2005). Electronic health records. *Journal of Oncology Practice / American Society of Clinical Oncology*, *1*(2), 57–63. doi:10.1200/jop.2005.1.2.57 PMID:20871681

Amir, S., Waqar, A., Siddiqui, M. A., Majeed, M., Haseeb, H., & Mehmood, S. H. (2016). Kinect Controlled UGV. *Proceedings of Wireless Personal Communications*, *95*(2), 631–640. doi:10.100711277-016-3915-3

Anand, V., & Kanhangad, V. (2017). Pore based indexing for High-Resolution Fingerprints. In *IEEE International Conference on Identity, Security and Behavior Analysis (ISBA)* (pp. 1-6). IEEE.

Angermueller, C., Parnamaa, T., Parts, L., & Stegle, O. (2016). Deep learning for computational biology. *Molecular Systems Biology*, *12*(7), 878. doi:10.15252/msb.20156651 PMID:27474269

Angkustsiri, K., Krakowiak, P., Moghaddam, B., Wardinsky, T., Gardner, J., Kalamkarian, N., ... Hansen, R. L. (2011). Minor physical anomalies in children with autism spectrum disorders. *Autism*, *15*(6), 746–760. doi:10.1177/1362361310397620 PMID:21610186

Ansari, S., & Gupta, P. (2007). Localization of Ear using Outher Helix Curve of the Ear. *IEEE Proceedings of the International Conference on Computing: Theory and Applications (ICCTA'07)*, 1–5.

Anwar, A. S., Ghany, K. K. A., & Elmahdy, H. (2015). Human Ear Recognition Using Geometrical Features Extraction. *Procedia Computer Science, 65*, 529–537. doi:10.1016/j.procs.2015.09.126

Aravinth, J., & Valarmathy, S. (2016). Multi classifier-based score level fusion of multi-modal biometric recognition and its application to remote biometrics authentication. *Imaging Science Journal*, *64*(1), 1–14. doi:10.1080/13682199.2015.1104067

Aristidou, A., Charalambous, P., & Chrysanthou, Y. (2015). Emotion analysis and classification: Understanding the performers' emotions using the LMA entities. *Proceedings of Computer Graphics Forum*, *34*(6), 262–276. doi:10.1111/cgf.12598

Aronson, S., & Rehm, H. (2015). Building the foundation for genomics in precision medicine. *Nature*, *526*(7573), 336–342. doi:10.1038/nature15816 PMID:26469044

Arrieta, M. I., Martinez, B., Criado, B., Simon, A., Salazar, L., & Lostao, C. M. (1990). Dermatoglyphic analysis of autistic Basque children. *American Journal of Medical Genetics. Part A*, *35*(1), 1–9. doi:10.1002/ajmg.1320350102 PMID:2301457

Askar, H. A., Radi, N., & MacDermott, A. (2016). Recurrent Neural Networks in Medical Data Analysis and Classifications. Emerging Topics in Computer Science and Applied Computing, 147-165.

Azam, S., & Gavrilova, M. (2017a). Person Identification Using Discriminative Visual Aesthetic, 30th Canadian Conference on Artificial Intelligence. In *Proceedings of AI'17 Springer, LNAI 10233* (pp. 15-26). Springer.

Azam, S., & Gavrilova, M. (2017b). Biometric Pattern Recognition from Social Media Aesthetics. *International Journal of Cognitive Informatics and Natural Intelligence, IGI Global*, *11*(3), 1–16. doi:10.4018/IJCINI.2017070101

Badilini, F., Erdem, T., Zareba, W., & Moss, A. J. (2005). ECGScan: A method for conversion of paper electrocardiographic printouts to digital electrocardiographic files. *Journal of Electrocardiology*, *38*(4), 310–318. doi:10.1016/j.jelectrocard.2005.04.003 PMID:16216602

Badrinath, G. S., Gupta, P., & Mehrotra, H. (2013). Score level fusion of voting strategy of geometric hashing and SURF for an efficient palmprint-based identification. *Journal of Real-Time Image Processing*, *8*(3), 265–284. doi:10.100711554-011-0229-2

Bagar, M., Ghani, A., Aftab, A., Arbab, S., & Yasin, S. (2016). Deep belief networks for iris recognition based on contour detection. *International Conference on Open Source Systems & Technologies (ICOSST)*. 10.1109/ICOSST.2016.7838580

Bai, C., Wang, W., Zhao, T., & Li, M. (2018). Fast exact fingerprint indexing based on Compact Binary Minutia Cylinder Codes. *Neurocomputing*, *275*, 1711–1724. doi:10.1016/j.neucom.2017.10.027

Baltrušaitis, T., McDuff, D., Banda, N., Mahmoud, M., El Kaliouby, R., Robinson, P., & Picard, R. (2011). Real-time inference of mental states from facial expressions and upper body gestures. In *Proceedings of Automatic Face & Gesture Recognition (FG 2011), 2011 IEEE International Conference on, IEEE* (pp. 909–914). IEEE.

Banerjee, R., Sinha, A., & Chakravarty, K. (2013). Gait based people identification system using multiple switching Kinects. In *Proceedings of 13th International Conference on Intelligent Systems Design and Applications* (pp. 182-187). Academic Press.

Bansal, R., Sehgal, P., & Bedi, P. (2011). *Minutiae extraction from fingerprint images-a review.* arXiv preprint arXiv:1201.1422.

Bänziger, T., Grandjean, D., & Scherer, K. (2009). Emotion recognition from expressions in face, voice, and body. In *Proceedings of the Multimodal Emotion Recognition Test (MERT) Emotion (Washington, D.C.)*, *9*(5), 691–698. doi:10.1037/a0017088 PMID:19803591

Barold, S. S. (2003). Willem Einthoven and the birth of clinical electrocardiography a hundred years ago. *Cardiac Electrophysiology Review*, *7*(1), 99–104. doi:10.1023/A:1023667812925 PMID:12766530

Bartlett, R. (2017). *Enlitic uses deep learning to make doctors faster and more accurate.* Retrieved from http://www.enlitic.com/index.html

Baveye, Y., Dellandrea, E., Chamaret, C., & Chen, L. (2015). Liris-accede: A video database for affective content analysis. *IEEE Transactions on Affective Computing*, *6*(1), 43–55. doi:10.1109/TAFFC.2015.2396531

Bebis, G., Deaconu, T., & Georgiopoulos, M. (1999). Fingerprint identification using delaunay triangulation. International Conference on Information Intelligence and Systems. *IEEE Proceedings*, 452–459.

Bellazzi, R., & Zupan, B. (2008). Predictive data mining in clinical medicine: Current issues and guidelines. *International Journal of Medical Informatics*, *77*(2), 81–97. doi:10.1016/j.ijmedinf.2006.11.006 PMID:17188928

Bengio, Y. (2009). Learning deep architectures for AI. *Foundations and Trends® in Machine Learning*, *2*, 1–127.

Bengio, Y. (2012). Practical recommendations for gradient-based training of deep architectures. *Neural Networks*, *2*, 437–478.

Bengio, Y., Courville, A., & Vincent, P. (2013). Representation learning: A review and new perspectives. *IEEE Transactions on Pattern Analysis and Machine Intelligence*, *35*(8), 1798–1828. doi:10.1109/TPAMI.2013.50 PMID:23787338

Bengio, Y., Lamblin, P., & Popovici, D. (2007). Greedy layer-wise training of deep networks. *Advances in Neural Information Processing Systems*, *19*, 153–160.

Beyl, T., Nicolai, P., Raczkowsky, J., Worn, H., Comparetti, M. D., & De Momi, E. (2013). Multi Kinect people detection for intuitive and safe human robot cooperation in the operating room. In *Proceedings of 16th International Conference on Advanced Robotics* (pp.1–6). Academic Press.

Bhanu, B., & Tan, X. (2003). Fingerprint indexing based on novel features of minutiae triplets. *IEEE Transactions on Pattern Analysis and Machine Intelligence*, *25*(5), 616–622. doi:10.1109/TPAMI.2003.1195995

Bhullar, H. K., Fothergill, J. C., Goddard, W. P., & de Bono, D. P. (1993). Automated measurement of QT interval dispersion from hard-copy ECGs. *Journal of Electrocardiology, 26*(4), 321–331. doi:10.1016/0022-0736(93)90053-G PMID:8228721

Biggio, B., fumera, g.,, Russu, P., Didaci, L., & Roli, F. (2015). Adversarial Biometric Recognition: A review on biometric system security from the adversarial machine-learning perspective. *IEEE Signal Processing Magazine, 32*(5), 31–41. doi:10.1109/MSP.2015.2426728

Bigy, A., Banitsas, K., Badii, A., & Cosmas, J. (2015). Recognition of postures and freezing of gait in Parkinson's disease patients using Microsoft Kinect sensor. In *Proceedings of 7th Annual International IEEE EMBS Conference on Neural Engineering* (pp. 731–734).IEEE.

Bishop, C. M. (2006). Pattern Recognition and Machine Learning (Information Science and Statistics) (2nd ed.). Springer.

Boom in healthcare biometric market. (2018). *Biometric Technology Today, 2018*(4), 3.

Bowyer, K. W., Hollingsworth, K., & Flynn, P. J. (2008). Image understanding for iris biometrics: A survey. *Computer Vision and Image Understanding, 110*(2), 281–307. doi:10.1016/j.cviu.2007.08.005

Brazelle, M., & Brazelle, S. (2018). Hand Determination of Whorl Patterns Using Axis Slant. *Journal of Forensic Identification, 68*(1), 77.

Brindha, V. (2012). Multi-Modal Biometric Template Security: Fingerprint and Palmprint Based Fuzzy Vault. *Journal of Biometrics & Biostatistics, 03*(06). doi:10.4172/2155-6180.1000150

Broman, V., Marshall, V., Bellamkonda, S., Levya, M., & Hanson, C. (2015, August 18). *Speaker recognition and denial of a transaction based on matching a known voice print.* Google Patents.

Brosch, T., & Tam, R. (2013). *Manifold learning of brain MRIs by deep learning* (Vol. 16). Medical Image Computing and Computer- Assisted Intervention.

Burge, M., & Burger, W. (2000). Ear biometrics in computer vision. *Proceedings 15th International Conference on Pattern Recognition. ICPR-2000*, 2, 822–826. 10.1109/ICPR.2000.906202

Burget, L., Matějka, P., Schwarz, P., Glembek, O., & Cernocky, J. H. (2007). Analysis of feature extraction and channel compensation in a GMM speaker recognition system. *IEEE Transactions on Audio, Speech, and Language Processing, 15*(7), 1979–1986. doi:10.1109/TASL.2007.902499

Campbell, J. P. (1997). Speaker recognition: A tutorial. *Proceedings of the IEEE, 85*(9), 1437–1462. doi:10.1109/5.628714

Campbell, W. M., Sturim, D. E., & Reynolds, D. A. (2006). Support vector machines using GMM supervectors for speaker verification. *IEEE Signal Processing Letters, 13*(5), 308–311. doi:10.1109/LSP.2006.870086

Cao, K., & Jain, A. K. (2017). Fingerprint indexing and matching: An integrated approach. In *IEEE International Joint Conference on Biometrics (IJCB)* (pp. 437-445). IEEE. 10.1109/BTAS.2017.8272728

Cappelli, R. (2011). Fast and accurate fingerprint indexing based on ridge orientation and frequency. *IEEE Transactions on Systems, Man, and Cybernetics. Part B, Cybernetics*, *41*(6), 1511–1521. doi:10.1109/TSMCB.2011.2155648 PMID:21708504

Cappelli, R., Ferrara, M., & Maltoni, D. (2011). Fingerprint indexing based on minutia cylinder-code. *IEEE Transactions on Pattern Analysis and Machine Intelligence*, *33*(5), 1051–1057. doi:10.1109/TPAMI.2010.228 PMID:21173446

Castellano, G., Villalba, S. D., & Camurri, A. (2007). Recognising human emotions from body movement and gesture dynamics. In *Proceedings of International Conference on Affective Computing and Intelligent Interaction* (pp. 71–82). Springer. 10.1007/978-3-540-74889-2_7

Chandra, A., Durand, R., & Weaver, S. (2008). The uses and potential of biometrics in health care: Are consumers and providers ready for it? *International Journal of Pharmaceutical and Healthcare Marketing*, *2*(1), 22–34. doi:10.1108/17506120810865406

Chang, Y. J., Chen, S. F., & Huang, J. D. (2011). *A Kinect-based system for physical rehabilitation: A pilot study for young adults with motor disabilities*. Academic Press.

Chatzis, V., Bors, A. G., & Pitas, I. (1999). Multimodal decision-level fusion for person authentication. *IEEE Transactions on Systems, Man, and Cybernetics-Part A: Systems and Humans, 29*(6), 674-680.

Chebil, J., & Maitah, J. (2008). A Novel Method for Digitizing Standard ECG Papers. *IEEE International Conference on Computer and Communication Engineering*. 10.1109/ICCCE.2008.4580816

Chellappa, R., Wilson, C. L., & Sirohey, S. (1995). Human and machine recognition of faces: A survey. *Proceedings of the IEEE*, *83*(5), 705–741. doi:10.1109/5.381842

Chen, C. K., Lin, C. L., & Chiu, Y. M. (2011). Individual identification based on chaotic electrocardiogram signals. *6th IEEE Conference on Industrial Electronics and Applications (ICIEA)*, 1771-1776.

Chen, C., Jafari, R., & Kehtarnavaz, N. (2015). A multimodal dataset for human action recognition utilizing a depth camera and a wearable inertial sensor. In *Proceedings of IEEE International Conference on Image Processing (ICIP)* (pp. 168-172). IEEE.

Cheng, J.-Z., Ni, D., Chou, Y.-H., Qin, J., Tiu, C. M., Chang, Y. C., ... Chen, C. M. (2016). Computer-aided diagnosis with deep learning architecture: Applications to breast lesions in US images and pulmonary nodules in CT scans. *Scientific Reports*, *6*(1), 24454. doi:10.1038rep24454 PMID:27079888

Cheng, Y., Wang, F., Zhang, P., & Hu, J. (2015). Risk prediction with electronic health records: a deep learning approach. *ACM International Conference on Knowledge Discovery and Data Mining*, 705–14.

Chen, S., Ren, K., & Piao, S. (2017). You can hear but you cannot steal: Defending against voice impersonation attacks on smartphones. In *Proceedings of IEEE 37th International Conference on Distributed Computing Systems* (pp. 183-195). Atlanta, GA: IEEE.

Chen, Y., Li, L., Zhang, G. Q., & Xu, R. (2015). Phenome-driven disease genetics prediction toward drug discovery. *Bioinformatics (Oxford, England), 31*(12), i276–i283. doi:10.1093/bioinformatics/btv245 PMID:26072493

Chia, P., Ong, K., & Bhullar, H., K. (1996). An Accurate and Automatic System for Extracting Features from ECG Paper Recordings: Its Use in Clinical Studies and Telemedicine. *IEEE Computers in Cardiology*.

Chingovska, I., Anjos, A., & Marcel, S. (2014). Biometrics evaluation under spoofing attacks. *IEEE Transactions on Information Forensics and Security*, *9*(12), 2264–2276. doi:10.1109/TIFS.2014.2349158

Cho, K., Courville, A., & Bengio, Y. (2015). *Describing multimedia content using attention-based encoder-decoder networks.* arXiv.:1507.01053v1.

Cho, K., van Merrienboer, B., Bahdanau, D., & Bengio, Y. (2014). *On the properties of neural machine translation: encoder-decoder approaches.* arXiv preprint arXiv:1409.1259.

Choi, E., Bahadori, M. T., Schuetz, A., Stewart, W. F., & Sun, J. (2015). *Doctor AI: predicting clinical events via recurrent neural networks.* arXiv: 1511.05942

Choi, H., Choi, K., & Kim, J. (2010). Mosaicing touchless and mirror-reflected fingerprint images. *IEEE Transactions on Information Forensics and Security*, *5*(1), 52–61. doi:10.1109/TIFS.2009.2038758

Choi, Y., Ms, C. Y.-I. C., & Sontag, D. (2016). Learning low-dimensional representations of medical concepts. *Proceedings of the AMIA Summit on Clinical Research Informatics.*

Choras, M. (2007, June). Image feature extraction methods for ear biometrics--a survey. In Null (pp. 261-265). IEEE.

Chowdhury, M., Islam, R., & Gao, J. (2017). Robust ear biometric recognition using neural network. In *Industrial Electronics and Applications (ICIEA), 2017 12th IEEE Conference on* (pp. 1855–1859). IEEE.

Chung, W. (2016). Social media analytics: Security and privacy issues. *Journal Of Information Privacy And Security*, *12*(3), 105–106. doi:10.1080/15536548.2016.1213994

Cidon, D. (2018). Making IT better: how biometrics can cure healthcare. *Biometric Technology Today, 2018*(7), 5–8.

Clark, J. (2016). *Google's Deep Mind forms health unit to build medical software*. Retrieved from https://www.bloomberg.com/news/articles/2016-02-24/google-s-deepmind-forms-health-unit-to-buildmedical-software

Collins, F. S., & Varmus, H. (2015). A new initiative on precision medicine. *The New England Journal of Medicine*, *372*, 793–795.

Czyzewski, A., Kostek, B., Bratoszewski, P., Kotus, J., & Szykulski, M. (2017). An audio-visual corpus for multimodal automatic speech recognition. *Journal of Intelligent Information Systems*, *49*(2), 167–192. doi:10.100710844-016-0438-z

Dalal, N., Triggs, B., & Schmid, C. (2006). Human detection using oriented histograms of flow and appearance. In *Proceedings of European Conference on Computer Vision* (pp. 428–441). Springer.

De Bruin, E. I., Graham, J. H., Louwerse, A., & Huizink, A. C. (2014). Mild dermatoglyphic deviations in adolescents with autism spectrum disorders and average intellectual abilities as compared to typically developing boys. *Autism Research and Treatment*. PMID:25478224

De Gelder, B. (2009). Why bodies? Twelve reasons for including bodily expressions in affective neuroscience. *Philosophical Transactions of the Royal Society of London. Series B, Biological Sciences*, *364*(1535), 3475–3484. doi:10.1098/rstb.2009.0190 PMID:19884142

De Leon, P., Pucher, M., Yamagishi, J., Hernaez, I., & Saratxaga, I. (2012). Evaluation of speaker verification security and detection of HMM-Based synthetic speech. *IEEE Transactions on Audio, Speech, and Language Processing*, *20*(8), 2280–2290. doi:10.1109/TASL.2012.2201472

Deep Learning. (2016). *Advances in Medicine | NVIDIA*. Retrieved from http://www.nvidia.com/object/deep-learning-in-medicine.html

Dehak, N., Kenny, P. J., Dehak, R., Dumouchel, P., & Ouellet, P. (2011). Front-end factor analysis for speaker verification. *IEEE Transactions on Audio, Speech, and Language Processing*, *19*(4), 788–798. doi:10.1109/TASL.2010.2064307

DeMyer, W., Zeman, W., & Palmer, C. G. (1964). The face predicts the brain: Diagnostic significance of median facial anomalies for holoprosencephaly (arhinencephaly). *Pediatrics*, *34*(2), 256–263. PMID:14211086

Deng, L., & Li, X. (2013). Machine learning paradigms for speech recognition: An overview. *IEEE Transactions on Audio, Speech, and Language Processing*, *21*(5), 1060–1089. doi:10.1109/TASL.2013.2244083

Dernoncourt, F., Lee, J. Y., Uzuner, O., & Szolovits, P. (2016). De-identification of patient notes with recurrent neural networks. *Journal of the American Medical Informatics Association*, *24*(3), 596–606. PMID:28040687

Dindaroğlu, F., Kutlu, P., Duran, G. S., Görgülü, S., & Aslan, E. (2016). Accuracy and reliability of 3D stereophotogrammetry: A comparison to direct anthropometry and 2D photogrammetry. *The Angle Orthodontist*, *86*(3), 487–494. doi:10.2319/041415-244.1 PMID:26267357

Diwakar, M., & Maharshi, M. (2013). An extraction and recognition of tongue-print images for biometrics authentication system. *International Journal of Computers and Applications*, *61*(3).

Duda, M., Ma, R., Haber, N., & Wall, D. P. (2017). Use of machine learning for behavioral distinction of autism and ADHD. *Translational Psychiatry*, *6*(2), e732. doi:10.1038/tp.2015.221 PMID:26859815

El-Latif, A. A. A., Hossain, M. S., & Wang, N. (2017). Score level multibiometrics fusion approach for healthcare. *Cluster Computing*, 1–12.

Elmouhtadi, M., Aboutajdine, D., & El Fkihi, S. (2015). Fingerprint indexing based barycenter triangulation. In *Third World Conference on Complex Systems (WCCS)* (pp. 1-6). IEEE.

Emersic, Z., Gabriel, L. L., Struc, V., & Peer, P. (2017). *Pixel-wise Ear Detection with Convolutional Encoder-Decoder Networks*. Retrieved from http://arxiv.org/abs/1702.00307

Esteva, A., Kuprel, B., Novoa, R. A., Ko, J., Swetter, S. M., Blau, H. M., & Thrun, S. (2017). Dermatologist-level classification of skin cancer with deep neural networks. *Nature*, *542*(7639), 115–118. doi:10.1038/nature21056 PMID:28117445

Fakoor, R., Ladhak, F., & Nazi, A. (2013). Using deep learning to enhance cancer diagnosis and classification. *International Conference on Machine Learning*.

Fananas, L., Moral, P., & Bertranpetit, J. (1990). Quantitative dermatoglyphics in schizophrenia: Study of family history subgroups. *Human Biology*, 421–427. PMID:2373511

Fan, Y., Lu, X., Li, D., & Liu, Y. (2016). Video-based emotion recognition using CNN-RNN and C3D hybrid networks. In *Proceedings of the 18th ACM International Conference on Multimodal Interaction* (pp. 445-450). ACM.

Farhan, W., Wang, Z., Huang, Y., Wang, S., Wang, F., & Jiang, X. (2016). A predictive model for medical events based on contextual embedding of temporal sequences. *Journal of Medical Internet Research*, *4*, e39. PMID:27888170

Farkas, L. (1994). Anthropometry of the head and face in clinical practice. *Anthropometry of the Head and Face*, *2*, 71–77.

Fasel, B., & Luettin, J. (2003). Automatic facial expression analysis: A survey. *Pattern Recognition*, *36*(1), 259–275. doi:10.1016/S0031-3203(02)00052-3

Fatjó-Vilas, M., Gourion, D., Campanera, S., Mouaffak, F., Levy-Rueff, M., Navarro, M. E., & Fañanás, L. (2008). New evidences of gene and environment interactions affecting prenatal neurodevelopment in schizophrenia-spectrum disorders: A family dermatoglyphic study. *Schizophrenia Research*, *103*(1), 209–217. doi:10.1016/j.schres.2008.04.015 PMID:18583099

Faundez-Zanuy, M. (2005). Signature recognition state-of-the-art. *IEEE Aerospace and Electronic Systems Magazine*, *20*(7), 28–32. doi:10.1109/MAES.2005.1499249

Faundez-Zanuy, M., Hussain, A., Mekyska, J., Sesa-Nogueras, E., Monte-Moreno, E., Esposito, E., & Lopez-de-Ipiña, A. K. (2013). Biometric applications related to human beings, there is life beyond security. *Cognitive Computation*, *5*(1), 136–151. doi:10.100712559-012-9169-9

Fearon, P., Lane, A., Airie, M., Scannell, J., McGowan, A., Byrne, M., & Waddington, J. (2001). Is reduced dermatoglyphic a–b ridge count a reliable marker of developmental impairment in schizophrenia? *Schizophrenia Research*, *50*(3), 151–157. doi:10.1016/S0920-9964(00)00089-X PMID:11439235

Felzenszwalb, P., McAllester, D., & Ramanan, D. (2008). A discriminatively trained, multiscale, deformable part model. In *Proceedings of CVPR* (pp. 1–8). IEEE.

Fernandez-Baena, A., Susín, A., & Lligadas, X. (2012). Biomechanical Validation of Upper-Body and Lower-Body Joint Movements of Kinect Motion Capture Data for Rehabilitation Treatments. In *Proceedings of 4th International Conference on Intelligent Networking and Collaborative Systems* (pp. 656-661). Academic Press.

Fierrez-Aguilar, J., Ortega-Garcia, J., Gonzalez-Rodriguez, J., & Bigun, J. (2004). Kernel-based multimodal biometric verification using quality signals. *Proceedings of the Society for Photo-Instrumentation Engineers*, *5404*, 544–554. doi:10.1117/12.542800

Fomichev, M., Álvarez, F., Steinmetzer, D., Gardner-Stephen, P., & Hollick, M. (2018). Survey and Systematization of Secure Device Pairing. *IEEE Communications Surveys and Tutorials*, *20*(1), 517–550. doi:10.1109/COMST.2017.2748278

Font, R., Espín, J. M., & Cano, M. J. (2017). Experimental analysis of features for replay attack detection-results on the ASVspoof 2017 Challenge. *Proceedings of INTERSPEECH*, *17*, 7–11. doi:10.21437/Interspeech.2017-450

Forbes, A. P. (1964). Fingerprints and palm prints (dermatoglyphics) and palmar-flexion creases in gonadal dysgenesis, pseudohypoparathyroidism and Klinefelter's syndrome. *The New England Journal of Medicine*, *270*(24), 1268–1277. doi:10.1056/NEJM196406112702402 PMID:14133664

Fothergill, S., Mentis, H., Kohli, P., & Nowozin, S. (2012). Instructing people for training gestural interactive systems. In *Proceedings of the SIGCHI Conference on Human Factors in Computing Systems* (pp. 1737–1746). ACM.

Fragopanagos, N., & Taylor, J. G. (2005). Emotion recognition in human–computer interaction. *Neural Networks*, *18*(4), 389–405. doi:10.1016/j.neunet.2005.03.006 PMID:15921887

Froba, B., Rothe, C., & Kublbeck, C. (2000). Evaluation of sensor calibration in a biometric person recognition framework based on sensor fusion. In *Automatic Face and Gesture Recognition, 2000. Proceedings. Fourth IEEE International Conference on* (pp. 512-517). IEEE.

Gao, X., Chen, X., Cao, J., Deng, Z., Liu, C., & Feng, J. (2010, September). A novel method of fingerprint minutiae extraction based on Gabor phase. In *Image Processing (ICIP), 2010 17th IEEE International Conference on* (pp. 3077-3080). IEEE.

Garg, D., K., Thakur, D., Sharma, S., & Bhardwaj, S. (2012). ECG Paper Records Digitization through Image Processing Techniques. *International Journal of Computer Applications, 48*(13).

Garofolo, J. S., Lamel, L. F., & Fisher, W. M. (1993). *TIMIT acoustic-phonetic continuous speech corpus*. Philadelphia: Linguistic Data Consortium. doi:10.6028/NIST.IR.4930

Gavrilescu, M. (2015). Recognizing emotions from videos by studying facial expressions, body postures and hand gestures. In *Proceedings of Telecommunications Forum (TELFOR)* (pp. 720-723). IEEE.

Gavrilova, M., Ahmed, F., Azam, S., Paul, P., Sultana, M., Rahman, W., & Zohra, F. T. (2016). Emerging Trends in Security System Design using the Concept of Social Behavioral Biometrics. Information Fusion for Cyber-Security Analytics, 691, 229-252.

Gavrilova, M., & Monwar, M. (2012). *Multimodal Biometrics and Intelligent Image Processing for Security Systems*. IGI Global.

Gavrilova, M., Wang, Y., Ahmed, F., & Paul, P. P. (2017). Kinect Sensor Gesture and Activity Recognition. *IEEE Consumer Electronics Magazine*, *7*(1), 88–94. doi:10.1109/MCE.2017.2755498

Gawande, U., Zaveri, M., & Kapur, A. (2011). Multi-algorithmic approaches to iris recognition. *Biometric Technology Today, 2011*(4), 8-10. doi:10.10160969-4765(11)70075-3

Gawande, U., & Golhar, Y. (2018). Biometric security system: A rigorous review of unimodal and multimodal biometrics techniques. *International Journal of Biometrics*, *10*(2), 142. doi:10.1504/IJBM.2018.091629

Gehring, W. J., Yan, Q. Q., Billeter, M., Furukubo-Tokunaga, K., Schier, A. F., Resendez-Perez, D., . . . Wüthrich, K. (1995). Homeodomain-DNA recognition. In Nmr In Structural Biology: A Collection of Papers by Kurt Wüthrich (pp. 493-505). Academic Press. doi:10.1142/9789812795830_0042

Germain, R. S., Califano, A., & Colville, S. (1997). Fingerprint matching using transformation parameter clustering. *IEEE Computational Science & Engineering*, *4*(4), 42–49. doi:10.1109/99.641608

Gerstung, M., Papaemmanuil, E., Martincorena, I., Bullinger, L., Gaidzik, V. I., Paschka, P., ... Campbell, P. J. (2017). Precision oncology for acute myeloid leukemia using a knowledge bank approach. *Nature Genetics*, *49*(3), 332–340. doi:10.1038/ng.3756 PMID:28092685

Ghouti, L., & Bahjat, A. A. (2009, December). Iris fusion for multibiometric systems. In *Signal Processing and Information Technology (ISSPIT), 2009 IEEE International Symposium on* (pp. 248-253). IEEE. 10.1109/ISSPIT.2009.5407577

Gilani, S. Z., Tan, D. W., Russell-Smith, S. N., Maybery, M. T., Mian, A., Eastwood, P. R., ... Whitehouse, A. J. (2015). Sexually dimorphic facial features vary according to level of autistic-like traits in the general population. *Journal of Neurodevelopmental Disorders*, *7*(1), 14. doi:10.118611689-015-9109-6 PMID:25901187

Girshick, R., Donahue, J., Darrell, T., & Malik, J. (2014). Rich feature hierarchies for accurate object detection and semantic segmentation. In *Proceedings of the IEEE conference on computer vision and pattern recognition* (pp. 580–587). IEEE.

Gold, S. (2012). Biometrics fit to secure the London Olympics. *Biometric Technology Today, 2012*(3), 5-8. doi:10.10160969-4765(12)70072-3

Gold, S. (2013). Healthcare biometrics: solving the staff and patient security governance challenge. *Biometric Technology Today, 2013*(8), 5–9.

Gotsis, M., Tasse, A., Sweden, M., Lympouridis, V., Poulos, I., Thin, A., ... Jordan-Marsh, M. (2012). Mixed reality game prototypes for upper body exercise and rehabilitation. In *Proceedings of IEEE Virtual Reality Workshops (VRW)* (pp. 181-182). Costa Mesa, CA: IEEE.

Gottlieb, A., Stein, G. Y., Ruppin, E., Altman, R. B., & Sharan, R. (2013). A method for inferring medical diagnoses from patient similarities. *BMC Medicine*, *11*(1), 194. doi:10.1186/1741-7015-11-194 PMID:24004670

Graham, J. H., Raz, S., Hel-Or, H., & Nevo, E. (2010). Fluctuating asymmetry: Methods, theory, and applications. *Symmetry*, *2*(2), 466–540. doi:10.3390ym2020466

Gulshan, V., Peng, L., Coram, M., Stumpe, M. C., Wu, D., Narayanaswamy, A., ... Webster, D. R. (2016). Development and validation of a deep learning algorithm for detection of diabetic retinopathy in retinal fundus photographs. *Journal of the American Medical Association*, *316*(22), 2402–2410. doi:10.1001/jama.2016.17216 PMID:27898976

Gunes, H., & Piccardi, M. (2005). Affect recognition from face and body: early fusion vs. late fusion. In *Proceedings of 2005 IEEE International Conference on Systems, Man and Cybernetics* (vol. 4, pp. 3437–3443). IEEE. 10.1109/ICSMC.2005.1571679

Guo, Y., Liu, Y., Oerlemans, A., Lao, S., Wu, S., & Lew, M. S. (2016). Deep learning for visual understanding: A review. *Neurocomputing*, *187*, 27–48. doi:10.1016/j.neucom.2015.09.116

Gupta, R. K., & Gupta, A. K. (2013). New, easy and effective method to take dermatoglyphic prints. *National Journal of Medical Research*, *3*(1), 45–47.

Gutta, S., & Cheng, Q. (2016). Joint feature extraction and classifier design for ECG-based biometric recognition. *IEEE Journal of Biomedical and Health Informatics*, *20*(2), 460–468. doi:10.1109/JBHI.2015.2402199 PMID:25680220

Guyton, A. C., & Hall, J. E. (2006). *Textbook of medical physiology*. Saunders/Elsevier.

Haitsma, J., & Kalker, T. (2002). A highly robust audio fingerprinting system. *Proceedings of the 3rd International Conference on Music Information Retrieval.*

Halawani, A. (2016). *Human Ear Localization: A Template-Based Approach*. Academic Press. doi:10.18178/ijsps.4.3.258-262

Hammerla, N. Y., Halloran, S., & Ploetz, T. (2016). *Deep, convolutional, and recurrent models for human activity recognition using wearables.* arXiv preprint arXiv:1604.08880.

Hammond, P., Forster-Gibson, C., Chudley, A. E., Allanson, J. E., Hutton, T. J., Farrell, S. A., ... Lewis, M. E. S. (2008). Face–brain asymmetry in autism spectrum disorders. *Molecular Psychiatry*, *13*(6), 614–623. doi:10.1038/mp.2008.18 PMID:18317467

Hammond, P., Hutton, T. J., Allanson, J. E., Campbell, L. E., Hennekam, R., Holden, S., ... Murphy, K. C. (2004). 3D analysis of facial morphology. *American Journal of Medical Genetics. Part A*, *126*(4), 339–348. doi:10.1002/ajmg.a.20665 PMID:15098232

Hampton, J., R. (2013). *The ECG made easy*. Toronto: Churchill Livingstone Elsevier.

Hanmandlu, M., Grover, J., Gureja, A., & Gupta, H. M. (2011). Score level fusion of multimodal biometrics using triangular norms. *Pattern Recognition Letters*, *32*(14), 1843–1850. doi:10.1016/j.patrec.2011.06.029

Hannun, A., Case, C., Casper, J., Catanzaro, B., Diamos, G., Elsen, E., . . . Ng, A. Y. (2014). *Deep speech: scaling up end-to-end speech recognition.* arXiv:1412.5567.

Han, W. Y., & Lee, J. C. (2012). Palm vein recognition using adaptive Gabor filter. *Expert Systems with Applications, 39*(18), 13225–13234. doi:10.1016/j.eswa.2012.05.079

Harb, A., Abbas, M., Cherry, A., Jaber, H., & Ayache, M. (2015, September). Palm print recognition. In *Advances in Biomedical Engineering (ICABME), 2015 International Conference on* (pp. 13-16). IEEE. 10.1109/ICABME.2015.7323239

Hariri, M. (2011). Possibility of spoof attack against robustness of multibiometric authentication systems. *Optical Engineering (Redondo Beach, Calif.), 50*(7), 079001. doi:10.1117/1.3599874

Hariri, M., & Shokouhi, S. (2011). Robustness of Multi Biometric Authentication Systems against Spoofing. *Computer And Information Science, 5*(1). doi:10.5539/cis.v5n1p77

Hartigan, J. A., & Wong, M. A. (1979). Algorithm AS 136: A k-means clustering algorithm. *Journal of the Royal Statistical Society. Series C, Applied Statistics, 28*(1), 100–108.

Hartin, P. J., & Barry, R. J. (1979). A comparative dermatoglyphic study of autistic, retarded, and normal children. *Journal of Autism and Developmental Disorders, 9*(3), 233–246. doi:10.1007/BF01531738 PMID:489510

Hashemian, M., & Pourghassem, H. (2017). Decision-Level Fusion-Based Structure of Autism Diagnosis Using Interpretation of EEG Signals Related to Facial Expression Modes. *Neurophysiology, 49*(1), 59–71. doi:10.100711062-017-9630-1

Hassan, B. (2003). Examining data accuracy and authenticity with leading digit frequency analysis. *Industrial Management & Data Systems, 103*(2), 121–125. doi:10.1108/02635570310463447

Hatch, A. O., Kajarekar, S., & Stolcke, A. (2006). Within-class covariance normalization for SVM-based speaker recognition. *Proceedings of INTERSPEECH, 06*, 1471–1474.

Hayat, M., Khan, S. H., & Bennamoun, M. (2017). Empowering simple binary classifiers for image set based face recognition. *International Journal of Computer Vision, 123*(3), 479–498. doi:10.100711263-017-1000-3

Healthcare Biometrics Market by Technology (Single-factor (Fingerprint, Face, Iris, Palm, Behavioral (Signature, Voice)), Multifactor, Multimodal), Application (Workforce Management), End User (Hospitals, Healthcare Institutions) - Global Forecast to 2021. (2016). Retrieved from marketsandmarkets.com

He, H., She, Y., Xiahou, J., Yao, J., Li, J., Hong, Q., & Ji, Y. (2018). Real-Time Eye-Gaze Based Interaction for Human Intention Prediction and Emotion Analysis. *Proceedings of Computer Graphics International, 2018*, 185–194.

Hei, X., Du, X., Lin, S., & Lee, I. (2013). PIPAC: Patient Infusion Pattern based Access Control Scheme for Wireless Insulin Pump System. In Proceedings of IEEE INFOCOM'13 (pp. 3030-3038). Turin, Italy: IEEE.

Hei, X., & Du, X. (2011). Biometric-based two-level secure access control for implantable medical devices during emergencies. *Proceedings - IEEE INFOCOM*, *11*, 346–350.

Hei, X., Du, X., Wu, J., & Hu, F. (2010). Defending Resource Depletion Attacks on Implantable Medical Devices. *Proceedings of IEEE Globecom*, *10*, 1–5.

Hejazi, M., Al-Haddad, S. A., Singh, Y. P., Hashim, S. J., & Aziz, A. F. (2016). ECG biometric authentication based on non-fiducial approach using kernel methods. *Digital Signal Processing*, *1*(52), 72–86. doi:10.1016/j.dsp.2016.02.008

Hinton, G. E., Osindero, S., & Teh, Y.-W. (2006). A fast learning algorithm for deep belief nets. *Neural Computation*, *18*(7), 1527–1554. doi:10.1162/neco.2006.18.7.1527 PMID:16764513

Hinton, G. E., & Salakhutdinov, R. R. (2006). Reducing the dimensionality of data with neural networks. *Science*, *313*(5786), 504–507. doi:10.1126cience.1127647 PMID:16873662

Hochreiter, S., & Schmidhuber, J. (1997). Long short-term memory. *Neural Computation*, *9*(8), 1735–1780. doi:10.1162/neco.1997.9.8.1735 PMID:9377276

Hoerbst, A., & Ammenwerth, E. (2010). Electronic health records. *Methods of Information in Medicine*, *49*(04), 320–336. doi:10.3414/ME10-01-0038 PMID:20603687

Holland, S. M. (2008). *Principal components analysis (PCA)*. Athens, GA: Department of Geology, University of Georgia.

Horng, S. J., Chen, Y. H., Run, R. S., Chen, R. J., Lai, J. L., & Sentosal, K. O. (2009, December). An improved score level fusion in multimodal biometric systems. In *Parallel and Distributed Computing, Applications and Technologies, 2009 International Conference on* (pp. 239-246). IEEE.

Hossain, E., & Chetty, G. (2011, October). Multimodal face-gait fusion for biometric person authentication. In *Embedded and Ubiquitous Computing (EUC), 2011 IFIP 9th International Conference on* (pp. 332-337). IEEE.

Hripcsak, G., & Albers, D. J. (2013). Next-generation phenotyping of electronic health records. *Journal of the American Medical Informatics Association*, *20*(1), 117–121. doi:10.1136/amiajnl-2012-001145 PMID:22955496

Huang, G. B., Ramesh, M., Berg, T., & Learned-Miller, E. (2007). *Labeled faces in the wild: A database for studying face recognition in unconstrained environments.* Technical Report 07-49. University of Massachusetts, Amherst.

Huang, M. C., Xu, W., Su, Y., Lange, B., Chang, C. Y., & Sarrafzadeh, M. (2012). Smartglove for upper extremities rehabilitative gaming assessment. In *Proceedings of 5th International Conference on Pervasive Technologies Related to Assistive Environments, PETRA '12* (pp. 20:1–20:4). ACM.

Hubel, D. H., & Wiesel, T. N. (1968). Receptive fields and functional architecture of monkey striate cortex. *The Journal of Physiology*, *195*(1), 215–243. doi:10.1113/jphysiol.1968.sp008455 PMID:4966457

Hunt, A. J., & Black, A. W. (1996). Unit selection in a concatenative speech synthesis system using a large speech database. *Proceedings of IEEE ICASSP*, *96*, 373–376.

Hurley, D. J., Arbab-Zavar, B., & Nixon, M. S. (2008). The ear as a biometric. In *Handbook of biometrics* (pp. 131-150). Springer.

Hus, V., Gotham, K., & Lord, C. (2014). Standardizing ADOS domain scores: Separating severity of social affect and restricted and repetitive behaviors. *Journal of Autism and Developmental Disorders*, *44*(10), 2400–2412. doi:10.100710803-012-1719-1 PMID:23143131

Iaizzo, P. A. (2005). *Handbook of cardiac anatomy, physiology and devices*. Humana Press. doi:10.1007/978-1-59259-835-9

Iannarelli, A. V. (1989). *Ear Identification*. Paramont Publishing Company.

IBM News Room. (2018). *IBM Watson Health Showcases Health Technology Advancements at American Diabetes Association's 78th Scientific Sessions, June 26, 2018. ICD-9 Code*. Retrieved from https://www.cms.gov/Medicare/Coding/ICD9ProviderDiagnosticCodes/index.html(02 June 2014, date last accessed)

Inchingolo, P., Bergamasco, S., & Bon, M. (2001). Medical data protection with a new generation of hardware authentication tokens. *Proceedings of Mediterranean Conf. on Medical and Biological Engineering and Computing*.

Ishii, M., Imaoka, H., & Sato, A. (2017). Fast k-Nearest Neighbor Search for Face Identification Using Bounds of Residual Score. In *Automatic Face & Gesture Recognition (FG 2017), 2017 12th IEEE International Conference on* (pp. 194-199). IEEE.

Islam, M. S., Ammour, N., Alajlan, N., & Abdullah-al-wadud, M. (2017). Selection of Heart-Biometric Templates for Fusion. *IEEE Access: Practical Innovations, Open Solutions*, *5*, 2169–3536. doi:10.1109/ACCESS.2017.2667224

Islam, S. M. S., Davies, R., Bennamoun, M., & Mian, A. S. (2011). Efficient Detection and Recognition of 3D Ears. *International Journal of Computer Vision*, *95*(1), 52–73. doi:10.100711263-011-0436-0

Isobe, Y., Ohki, T., & Komatsu, N. (2013). Security performance evaluation for biometric template protection techniques. *International Journal of Biometrics*, *5*(1), 53. doi:10.1504/IJBM.2013.050733

Jafri, R., & Arabnia, H. R. (2009). A survey of face recognition techniques. *JIPS*, *5*(2), 41–68.

Jahan, S., Chowdhury, M., & Islam, R. (2017, December). Robust fingerprint verification for enhancing security in healthcare system. In *2017 International Conference on Image and Vision Computing New Zealand (IVCNZ)* (pp. 1-5). IEEE.

Jain, A. K., Arora, S. S., Cao, K., Best-Rowden, L., & Bhatnagar, A. (2017). Fingerprint recognition of young children. *IEEE Transactions on Information Forensics and Security*, *12*(7), 1501–1514. doi:10.1109/TIFS.2016.2639346

Jain, A. K., Bolle, R., & Pankanti, S. (2006). *Biometrics: personal identification in networked society* (Vol. 479). Springer Science & Business Media.

Jain, A. K., Flynn, P., & Ross, A. A. (Eds.). (2007). *Handbook of biometrics*. Springer Science & Business Media.

Jain, A. K., & Ross, A. (2004). Multibiometrics systems. *Communications of the ACM*, *47*(1), 34–40. doi:10.1145/962081.962102

Jain, A. K., Ross, A., & Prabhakar, S. (2004). An introduction to biometrics recognition. *IEEE Transactions on Circuits and Systems for Video Technology*, *14*(1), 4–20. doi:10.1109/TCSVT.2003.818349

Jain, A., Flynn, P., & Ross, A. (2008). Handbook of Biometrics: Eds. doi:10.1007/978-0-387-71041-9

Jain, A., Nandakumar, K., & Ross, A. (2005). Score normalization in multimodal biometric systems. *Pattern Recognition*, *38*(12), 2270–2285. doi:10.1016/j.patcog.2005.01.012

Jaswal, G., Nath, R., & Kaul, A. (2015). Texture based palm Print recognition using 2-D Gabor filter and sub space approaches. In *Signal Processing, Computing and Control (ISPCC), 2015 International Conference on* (pp. 344–349). IEEE.

Jaswal, G., Nigam, A., & Nath, R. (2017). DeepKnuckle: Revealing the human identity. *Multimedia Tools and Applications*, *76*(18), 18955–18984. doi:10.100711042-017-4475-6

Jayaraman, U., Gupta, A. K., & Gupta, P. (2014). An efficient minutiae based geometric hashing for fingerprint database. *Neurocomputing*, *137*, 115–126. doi:10.1016/j.neucom.2013.02.059

Jayaraman, U., Prakash, S., & Gupta, P. (2010). An iris retrieval technique based on color and texture. In *Proceedings of the Seventh Indian Conference on Computer Vision, Graphics and Image Processing* (pp. 93-100). ACM.

Jensen, P. B., Jensen, L. J., & Brunak, S. (2012). Mining electronic health records: Towards better research applications and clinical care. *Nature Reviews. Genetics*, *13*(6), 395–405. doi:10.1038/nrg3208 PMID:22549152

Jia, W., Cai, H. Y., Gui, J., Hu, R. X., Lei, Y. K., & Wang, X. F. (2012). Newborn footprint recognition using orientation feature. *Neural Computing & Applications*, *21*(8), 1855–1863. doi:10.100700521-011-0530-9

Jindal, V., Birjandtalab, J., Pouyan, M. B., & Nourani, M. (2016). An adaptive deep learning approach for PPG-based identification. *38th Annual International Conference of the IEEE Engineering in Medicine and Biology Society (EMBC)*, 6401–6404.

Jin, Z., Teoh, A. B. J., Ong, T. S., & Tee, C. (2010). A Revocable Fingerprint Template for Security and Privacy Preserving. *Transactions on Internet and Information Systems (Seoul)*, *4*(6).

Jordan, M. I., & Mitchell, T. M. (2015). Machine learning: Trends, perspectives, and prospects. *Science*, *349*(6245), 255–260. doi:10.1126cience.aaa8415 PMID:26185243

Kao, T., Hwang, L. J., Lin, Y. H., Lin, T. H., & Hsiao, C. H. (2001). Computer analysis of the electrocardiograms from ECG paper recordings. *Proceedings of the 23rd Annual EMBS International Conference.*

Karsikas, M., Huikuri, H., Juha, S., Lehtola, L., & Tapio, S. (2007). Influence of Paper Electrocardiogram Digitizing on T Wave and QRS Complex Morphology Parameters. *Original Articles, 12*(4).

Karu, K., & Jain, A. K. (1996). Fingerprint classification. *Pattern Recognition, 29*(3), 389–404. doi:10.1016/0031-3203(95)00106-9

Kashyap, R., & Piersson, A. (2018a). Big Data Challenges and Solutions in the Medical Industries. In Handbook of Research on Pattern Engineering System Development for Big Data Analytics. IGI Global.

Kashyap, R., Gautam, P., & Tiwari, V. (2018). Management and Monitoring Patterns and Future Scope. In Handbook of Research on Pattern Engineering System Development for Big Data Analytics. IGI Global.

Kashyap, R. (2018). Object boundary detection through robust active contour based method with global information. *International Journal of Image Mining, 3*(1), 22. doi:10.1504/IJIM.2018.093008

Kashyap, R., & Gautam, P. (2016). Fast level set method for segmentation of medical images. In *Proceedings of the International Conference on Informatics and Analytics (ICIA-16).* ACM. 10.1145/2980258.2980302

Kashyap, R., & Gautam, P. (2017a). 'Fast Medical Image Segmentation Using Energy-Based Method', Biometrics. *Concepts, Methodologies, Tools, and Applications, 3*(1), 1017–1042. doi:10.4018/978-1-5225-0983-7.ch040

Kashyap, R., & Gautam, P. (2017b). *Fast Medical Image Segmentation Using Energy-Based Method.* In *Pattern and Data Analysis in Healthcare Settings.* Medical Information Science Reference.

Kashyap, R., & Piersson, A. (2018b). *Impact of Big Data on Security. In Handbook of Research on Network Forensics and Analysis Techniques* (pp. 283–299). IGI Global. doi:10.4018/978-1-5225-4100-4.ch015

Kashyap, R., & Tiwari, V. (2017). Energy-based active contour method for image segmentation. *International Journal of Electronic Healthcare, 9*(2–3), 210–225. doi:10.1504/IJEH.2017.083165

Kashyap, R., & Tiwari, V. (2018). Active contours using global models for medical image segmentation. *International Journal of Computational Systems Engineering, 4*(2/3), 195. doi:10.1504/IJCSYSE.2018.091404

Kaur, J., & Batra, A. P. S. (2013). Role of dermatoglyphics in medical disorders. *Indian Journal of Fundamental and Applied Life Sciences, 3*(3), 536.

Kaushik, V. D., Umarani, J., Gupta, A. K., Gupta, A. K., & Gupta, P. (2013). An efficient indexing scheme for face database using modified geometric hashing. *Neurocomputing, 116*, 208–221. doi:10.1016/j.neucom.2011.12.056

Kavati, I., Prasad, M. V., & Bhagvati, C. (2013, August). Vein pattern indexing using texture and hierarchical decomposition of delaunay triangulation. In *International Symposium on Security in Computing and Communication* (pp. 213-222). Springer. 10.1007/978-3-642-40576-1_21

Kavati, I., Prasad, M. V., & Bhagvati, C. (2014). A new indexing method for biometric databases using match scores and decision level fusion. In *Advanced Computing, Networking and Informatics-Volume 1* (pp. 493–500). Cham: Springer. doi:10.1007/978-3-319-07353-8_57

Kazemi, M., Fayyazi-Bordbar, M. R., & Mahdavi-Shahri, N. (2017). Comparative dermatoglyphic study between autistic patients and normal people in Iran. *Iranian Journal of Medical Sciences*, *42*(4), 392. PMID:28761206

Kelley, D. R., Snoek, J., & Rinn, J. L. (2016). Basset: Learning the regulatory code of the accessible genome with deep convolutional neural networks. *Genome Research*, *26*(7), 990–999. doi:10.1101/gr.200535.115 PMID:27197224

Kenny, P., Ouellet, P., Dehak, N., Gupta, V., & Dumouchel, P. (2008). A study of interspeaker variability in speaker verification. *IEEE Transactions on Audio, Speech, and Language Processing*, *16*(5), 980–988. doi:10.1109/TASL.2008.925147

Khalaf, E., & Sulaiman, N. (2015). A New Secure Storing System for Biometric Templates Based Encryption and Concealment. *Journal of Applied Sciences (Faisalabad)*, *15*(5), 773–782. doi:10.3923/jas.2015.773.782

Khandpur, R. S. (2003). *Biomedical Instrumentation: Technology and Applications*. New Delhi: Tata McGraw-Hill Publishing Company Limited.

Khodadoust, J., & Khodadoust, A. M. (2017). Fingerprint indexing based on expanded Delaunay triangulation. *Expert Systems with Applications*, *81*, 251–267. doi:10.1016/j.eswa.2017.03.048

Kim, J. M. S., Kang, U., & Lee, Y. (2017). Statistics and Deep Belief Network-Based Cardiovascular Risk Prediction. *Healthcare Informatics Research*, *23*(3), 169–175. doi:10.4258/hir.2017.23.3.169 PMID:28875051

King, R. (2014). *India to use Aadhaar to provide universal healthcare*. Retrieved from http://www.biometricupdate.com/201410/india-to-use-aadhaar-to-provide-universal-healthcare

Kinnunen, T., & Li, H. (2010). An overview of text-independent speaker recognition: From features to supervectors. *Speech Communication*, *52*(1), 12–40. doi:10.1016/j.specom.2009.08.009

Kinnunen, T., Sahidullah, M., Delgado, H., Todisco, M., Evans, N., Yamagishi, J., & Lee, K. A. (2017a). The ASVspoof 2017 Challenge: Assessing the limits of replay spoofing attack detection. *Proceedings of INTERSPEECH*, *17*, 2–6. doi:10.21437/Interspeech.2017-1111

Kinnunen, T., Sahidullah, M., & Falcone, M. (2017b). RedDots replayed: A new replay spoofing attack corpus for text-dependent speaker verification research. *Proceedings of IEEE ICASSP*, *17*, 5395–5399.

Kisku, D. R., Gupta, P., & Sing, J. K. (2013). *Advances in biometrics for secure human authentication and recognition*. CRC Press. doi:10.1201/b16247

Klich, A. (2017). Electronic health records (ehr)- selected legal issues. *Ekonomiczne Problemy Usług*, *126*, 351–360. doi:10.18276/epu.2017.126/2-35

Koh, P. W., Pierson, E., & Kundaje, A. (2016). Denoising genome-wide histone ChIP-seq with convolutional neural networks. *bioRxiv*. doi:10.1101/052118

Koolagudi, S., & Rao, K. (2012). Emotion recognition from speech: A review. *Proceedings of International Journal of Speech Technology*, *15*(2), 99–117. doi:10.100710772-011-9125-1

Kosti, R., Alvarez, J. M., Recasens, A., & Lapedriza, A. (2017). Emotion recognition in context. *The IEEE Conference on Computer Vision and Pattern Recognition (CVPR)*. doi: 10.1109/CVPR.2017.212

Kostyuk, N., Rajnarayanan, R. V., Isokpehi, R. D., & Cohly, H. H. (2010). Autism from a biometric perspective. *International Journal of Environmental Research and Public Health*, *7*(5), 1984–1995. doi:10.3390/ijerph7051984 PMID:20623006

Krizhevsky, A., Sutskever, I., & Hinton, G. E. (2012). Imagenet classification with deep convolutional neural networks. In Advances in neural information processing systems (pp. 1097–1105). Academic Press.

Kryszczuk, K., Richiardi, J., Prodanov, P., & Drygajlo, A. (2005). Error handling in multimodal biometric systems using reliability measures. *Proc. 13th Eur. Signal Process. Conf.*, 1-4.

Kukula, E., Sutton, M., & Elliott, S. (2010). The Human–Biometric-Sensor Interaction Evaluation Method: Biometric Performance and Usability Measurements. *IEEE Transactions on Instrumentation and Measurement*, *59*(4), 784–791. doi:10.1109/TIM.2009.2037878

Kumar, A., Hanmandlu, M., Sanghvi, H., & Gupta, H. M. (2010, September). Decision level biometric fusion using Ant Colony Optimization. In *Image Processing (ICIP), 2010 17th IEEE International Conference on* (pp. 3105-3108). IEEE.

Kumar, T., Braeken, A., Liyanage, M., & Ylianttila, M. (2017, May). Identity privacy preserving biometrics based authentication scheme for Naked healthcare environment. In *Communications (ICC), 2017 IEEE International Conference on* (pp. 1-7). IEEE.

Kumar, V. A., & Ramakrishnan, M. (2010). Footprint recognition using modified sequential haar energy transform (MSHET). *International Journal of Computer Science Issues*, *7*(3).

Kumar, A., Hanmandlu, M., Sharma, V., & Gupta, H. M. (2011, December). Rank based hybrid multimodal fusion using PSO. In *International Conference on Swarm, Evolutionary, and Memetic Computing* (pp. 217-224). Springer. 10.1007/978-3-642-27172-4_27

Kumar, A., & Zhang, D. (2006). Personal recognition using hand shape and texture. *IEEE Transactions on Image Processing*, *15*(8), 2454–2461. doi:10.1109/TIP.2006.875214 PMID:16900698

Laban, R., & Ullmann, L. (2011). *The Mastery of Movement* (4th ed.). Binsted, UK: Dance Books Ltd.

Lajus, D. L., Graham, J. H., & Kozhara, A. V. (2003). Developmental instability and the stochastic component of total phenotypic variance. *Developmental instability: causes and consequences*, 343-363.

Lane, N. D., Bhattacharya, S., Georgiev, P., Forlivesi, C., Jiao, L., & Kawsar, F. (2016). DeepX: a software accelerator for low-power deep learning inference on mobile devices. *ACM/IEEE International Conference on Information Processing in Sensor Networks*, 1–12.

Lane, N. D., & Georgiev, P. (2015). Can deep learning revolutionize mobile sensing? *International Workshop on Mobile Computing Systems and Applications*, 117–22.

Lange, B., Chang, C. Y., Suma, E., Newman, B., Rizzo, A. S., & Bolas, M. (2011). Development and evaluation of low-cost game-based balance rehabilitation tool using the Microsoft Kinect sensor. In *Proceedings of Engineering in Medicine and Biology Society, EMBC, 2011 Annual International Conference of the IEEE* (pp. 1831-1834). IEEE.

Larcher, A., Lee, K., & Martinez, P. L. S. (2014). Extended RSR2015 for text-dependent speaker verification over VHF channel. *Proceedings of INTERSPEECH*, *14*, 1322–1326.

Lasko, T. A., Denny, J. C., & Levy, M. A. (2013). Computational phenotype discovery using unsupervised feature learning over noisy, sparse, and irregular clinical data. *PLoS One*, *8*(6), e66341. doi:10.1371/journal.pone.0066341 PMID:23826094

Lawson, W., T., Wagner, G., S., Selvester, R., S., & Ybarra, G., A. (1995). New Method for Digitization and Computerized Analysis of Paper Recordings of Standard 12-Lead Electrocardiograms. *IEEE Computers in Cardiology.*

LeCun, Y., Bengio, Y., & Hinton, G. (2015). Deep learning. *Nature*, *521*(7553), 436–444. doi:10.1038/nature14539 PMID:26017442

Lecun, Y., Bottou, L., Bengio, Y., & Haffner, P. (1998). Gradient-based learning applied to document recognition. *Proceedings of the IEEE*, *86*(11), 2278–2324. doi:10.1109/5.726791

LeCun, Y., Jackel, L. D., Bottou, L., Cortes, C., Denker, J. S., Drucker, H., ... Simard, P. (1995). Learning algorithms for classification: A comparison on handwritten digit recognition. *Neural Networks: The Statistical Mechanics Perspective*, *261*, 276.

Lee, D. T., & Schachter, B. J. (1980). Two algorithms for constructing a Delaunay triangulation. *International Journal of Computer & Information Sciences*, *9*(3), 219–242. doi:10.1007/BF00977785

Lee, K., Larcher, A., & Wang, G. (2015). The reddots data collection for speaker recognition. *Proceedings of INTERSPEECH*, *15*, 2996–3000.

Lee, S. O., Kim, Y. G., & Park, G. T. (2005). A feature map consisting of orientation and inter-ridge spacing for fingerprint retrieval. In *International Conference on Audio-and Video-Based Biometric Person Authentication* (pp. 184-190). Springer. 10.1007/11527923_19

Lee, W.-H., & Lee, R. B. (2017). Implicit Smartphone User Authentication with Sensors and Contextual Machine Learning. *47th Annual IEEE/IFIP International Conference on Dependable Systems and Networks (DSN).*

Lee, Y. S., & Chung, W. Y. (2012). Visual sensor based abnormal event detection with moving shadow removal in home healthcare applications. *Sensors (Basel)*, *12*(1), 573–584. doi:10.3390120100573 PMID:22368486

Lei, Y., Scheffer, N., & Ferrer, L. (2014). A novel scheme for speaker recognition using a phonetically-aware deep neural network. *Proceedings of IEEE ICASSP*, *14*, 1695–1699.

Leonard, D. C., Pons, A. P., & Asfour, S. S. (2009). Realization of a universal patient identifier for electronic medical records through biometrics technology. *IEEE Transactions on Information Technology in Biomedicine*, *13*(4), 494–500. doi:10.1109/TITB.2008.926438 PMID:19273015

Leung, S. W., & Zhang, Y. T. (1998). Digitization of electrocardiogram (ECG) signals using deltasigma modulation. *Proceedings of the 20th Annual International Conference of the ZEEE Engineering in Medicine and Biology Society*, *20*(4), 0-7803-5164-9.

Leung, M. K. K., Delong, A., Alipanahi, B., & Frey, J. (2016). Machine learning in genomic medicine: A review of computational problems and data sets. *Proceedings of the IEEE*, *104*(1), 176–197. doi:10.1109/JPROC.2015.2494198

Li, G., Yang, B., & Busch, C. (2015). A novel fingerprint indexing approach focusing on minutiae location and direction. In *Identity, IEEE International Conference on Security and Behavior Analysis (ISBA), 2015* (pp. 1-6). IEEE.

Li, R., Nguyen, M., & Yan, W. Q. (2017). Computer Input of Morse Codes Using Finger Gesture Recognition. In *Proceedings of International Conference on Digital Image Computing: Techniques and Applications (DICTA)* (pp. 1-8). Academic Press.

Liang, Z., Zhang, G., Huang, J. X., & Hu, Q. V. (2014). Deep learning for healthcare decision making with EMRs. *IEEE International Conference on Bioinformatics and Biomedicine*, 556–9.

Libbrecht, M. W., & Noble, W. S. (2015). Machine learning applications in genetics and genomics. *Nature Reviews. Genetics*, *16*(6), 321–332. doi:10.1038/nrg3920 PMID:25948244

Li, C., Raghunathan, A., & Jha, N. K. (2011). Hijacking an insulin pump: Security attacks and defenses for a diabetes therapy system. In *Proceedings of the 13th IEEE Intl. Conf. on e-Health NAS* (pp. 150-156). IEEE.

Li, L., Cheng, W.-Y., Glicksberg, B. S., Gottesman, O., Tamler, R., Chen, R., ... Dudley, J. T. (2015). Identification of type 2 diabetes subgroups through topological analysis of patient similarity. *Science Translational Medicine*, *7*(311), 311ra174. doi:10.1126citranslmed.aaa9364 PMID:26511511

Lim, E., Jiang, X., & Yau, W. (2002). Fingerprint quality and validity analysis. In *Image Processing. 2002. Proceedings. 2002 International Conference on* (Vol. 1, pp. I-I). IEEE.

Li, P., Fu, Y., & Mohammed, U. (2012). Probabilistic models for Inference about Identity. *IEEE Transactions on Pattern Analysis and Machine Intelligence*, *34*(1), 144–157. doi:10.1109/TPAMI.2011.104 PMID:21576751

Lipton, Z. C., Kale, D. C., & Elkan, C. (2015). Learning to diagnose with LSTM recurrent neural networks. *International Conference on Learning Representations*, 1–18.

Li, S. Z., & Jain, A. K. (2009). *Encyclopedia of Biometrics: I-Z* (Vol. 1). Springer Science & Business Media. doi:10.1007/978-0-387-73003-5

Li, S., & Chan, A. B. (2014). 3d human pose estimation from monocular images with deep convolutional neural network. In *Proceedings of Asian Conference on Computer Vision* (pp. 332–347). Springer.

Liu, S., Cai, W., Pujol, S., Kikinis, R., & Feng, D. D. (2014). Early diagnosis of Alzheimer's disease with deep learning. *International Symposium on Biomedical Imaging*, 1015–18.

Liu, H., Zhang, Y., Li, Y., Zhou, Q., Tadoh, R., Uemura, T., ... Serikawa, S. (2017). Depth Map Reconstruction for Underwater Kinect Camera Using Inpainting and Local Image Mode Filtering. *IEEE Access: Practical Innovations, Open Solutions*, *5*, 7115–7122. doi:10.1109/ACCESS.2017.2690455

Liu, Y., Zhang, B., & Zhang, D. (2015). Ear-parotic face angle: A unique feature for 3D ear recognition. *Pattern Recognition Letters*, *53*, 9–15. doi:10.1016/j.patrec.2014.10.014

Li, W., You, J., & Zhang, D. (2005). Texture-based palmprint retrieval using a layered search scheme for personal identification. *IEEE Transactions on Multimedia*, *7*(5), 891–898. doi:10.1109/TMM.2005.854380

Lobodzinski, S., M., Teppner, U., & Laks, M. (2003). State of the art techniques for preservation and reuse of hard copy electrocardiograms. *Journal of Electrocardiology, 36.*

Lo, N., Wu, C., & Chuang, Y. (2017). An authentication and authorization mechanism for long-term electronic health records management. *Procedia Computer Science*, *111*, 145–153. doi:10.1016/j.procs.2017.06.021

Long, T. B., & Hanh, T. (2012, September). Multimodal biometric person authentication using fingerprint, face features. In *Pacific Rim International Conference on Artificial Intelligence* (pp. 613-624). Springer.

Lumini, A., & Nanni, L. (2017). Overview of the combination of biometric matchers. *Information Fusion*, *1*(33), 71–85. doi:10.1016/j.inffus.2016.05.003

Luo, J., Wu, M., Gopukumar, D., & Zhao, Y. (2016). Big data application in biomedical research and healthcare: A literature review. *Biomedical Informatics Insights*, *8*, 1–10. doi:10.4137/BII.S31559 PMID:26843812

Lyman, G. H., & Moses, H. L. (2016). Biomarker tests for molecularly targeted therapies - the key to unlocking precision medicine. *The New England Journal of Medicine*, *375*, 4–6.

Lyons, J., Dehzangi, A., Heffernan, R., Sharma, A., Paliwal, K., Sattar, A., ... Yang, Y. (2014). Predicting backbone Ca angles and dihedrals from protein sequences by stacked sparse auto-encoder deep neural network. *Journal of Computational Chemistry*, *35*(28), 2040–2046. doi:10.1002/jcc.23718 PMID:25212657

Ma, L., Wang, Y., & Tan, T. (2002). Iris recognition based on multichannel Gabor filtering. In *Proc. Fifth Asian Conf. Computer Vision* (*Vol. 1*, pp. 279-283). Academic Press.

Mahnken, S. (2014). Today's authentication options: the need for adaptive multifactor authentication. *Biometric Technology Today, 2014*(7), 8-10. doi:10.10160969-4765(14)70126-2

Mai, G., Lim, M., & Yuen, P. (2017). Binary feature fusion for discriminative and secure multi-biometric cryptosystems. *Image and Vision Computing*, *58*, 254–265. doi:10.1016/j.imavis.2016.11.011

Maio, D., Maltoni, D., Cappelli, R., Wayman, J. L., & Jain, A. K. (2004). FVC2004: Third fingerprint verification competition. In *Biometric Authentication* (pp. 1–7). Berlin: Springer. doi:10.1007/978-3-540-25948-0_1

Maity, S., & Abdel-Mottaleb, M. (2015). 3D ear segmentation and classification through indexing. *IEEE Transactions on Information Forensics and Security*, *10*(2), 423–435. doi:10.1109/TIFS.2014.2379437

Ma, J., Sheridan, R. P., Liaw, A., Dahl, G. E., & Svetnik, V. (2015). Deep neural nets as a method for quantitative structure – activity relationships. *Journal of Chemical Information and Modeling*, *55*(2), 263–274. doi:10.1021/ci500747n PMID:25635324

Mali, K., & Bhattacharya, S. (2013). Comparative study of different biometric features. *International Journal of Advanced Research in Computer and Communication Engineering*, *2*(7), 2776–2784.

Maltoni, D., Maio, D., Jain, A. K., & Prabhakar, S. (2009). *Handbook of fingerprint recognition*. Springer Science & Business Media. doi:10.1007/978-1-84882-254-2

Mamoshina, P., Vieira, A., Putin, E., & Zhavoronkov, A. (2016). Applications of deep learning in biomedicine. *Molecular Pharmaceutics*, *13*(5), 1445–1454. doi:10.1021/acs.molpharmaceut.5b00982 PMID:27007977

Manning, C. D., Raghavan, P., & Schutze, H. (2008). *Introduction to information retrieval* (Vol. 3). Cambridge, UK: Cambridge University Press. doi:10.1017/CBO9780511809071

Manouilenko, I., Eriksson, J. M., Humble, M. B., & Bejerot, S. (2014). Minor physical anomalies in adults with autism spectrum disorder and healthy controls. *Autism Research and Treatment*. PMID:24782925

Mansfield-Devine, S. (2011). Biometrics for mobile devices struggle to go mainstream. *Biometric Technology Today*, *2011*(9), 10-11. doi:10.10160969-4765(11)70172-2

Marera, D. O., Oyieko, W., & Agumba, G. (2015). Variation in Dermatoglyphic patterns among diabetics in Western Uganda population. *African Journal of Science and Research*, *7*(3), 20–25.

Maret, Y., Oberson, D., & Gavrilova, M. (2018a). Identifying an Emotional State from Body Movements Using Genetic-based Algorithms. In *Proceedings of 17th International Conference on Artificial Intelligence and Soft Computing* (vol. 10841, pp. 474-485). Springer. 10.1007/978-3-319-91253-0_44

Maret, Y., Oberson, D., & Gavrilova, M. (2018b). Real-Time Embedded System for Gesture Recognition. *Proceedings of IEEE Conference on Systems, Man and Cybernatics.*

Marfella, L., Marasco, E., & Sansone, C. (2012, September). Liveness-based fusion approaches in multibiometrics. In *Biometric Measurements and Systems for Security and Medical Applications (BIOMS), 2012 IEEE Workshop on* (pp. 1-7). IEEE. 10.1109/BIOMS.2012.6345779

Marin, E., Singelée, D., Yang, B., Verbauwhede, I., & Preneel, B. (2016). On the Feasibility of Cryptography for a Wireless Insulin Pump System. In *Proceedings of ACM CODASPY '16* (pp. 113-120). ACM. 10.1145/2857705.2857746

Marohn, D. (2006). Biometrics in healthcare. *Biometric Technology Today*, *14*(9), 9–11. doi:10.1016/S0969-4765(06)70592-6

Mayhew, S. (2018). *HID Global reports strong biometrics adoption in Latin America.* Retrieved from https://www.biometricupdate.com/201804/hid-global-reports-strong-biometrics-adoption-in-latin-america

Meraoumia, A., Chitroub, S., & Bouridane, A. (2012, June). Multimodal biometric person recognition system based on fingerprint & finger-knuckle-print using correlation filter classifier. In *Communications (ICC), 2012 IEEE International Conference on* (pp. 820-824). IEEE.

Merone, M., Soda, P., Sansone, M., & Sansone, C. (2017). ECG databases for biometric systems: A systematic review. *Expert Systems with Applications*, *1*(67), 189–202. doi:10.1016/j.eswa.2016.09.030

Miotto, R., Li, L., & Dudley, J. T. (2016). Deep learning to predict patient future diseases from the electronic health records. *European Conference in Information Retrieval*, 768–774.

Miotto, R., Li, L., Kidd, B. A., & Dudley, J. T. (2016). Deep patient: An unsupervised representation to predict the future of patients from the electronic health records. *Scientific Reports*, *6*(1), 26094. doi:10.1038rep26094 PMID:27185194

Miotto, R., & Weng, C. (2015). Case-based reasoning using electronic health records efficiently identifies eligible patients for clinical trials. *Journal of the American Medical Informatics Association*, *22*(e1), e141–e150. doi:10.1093/jamia/ocu050 PMID:25769682

Mirmohamadsadeghi, L., & Drygajlo, A. (2011, October). Palm vein recognition with local binary patterns and local derivative patterns. In *Biometrics (IJCB), 2011 International Joint Conference on* (pp. 1-6). IEEE.

Mitra, S., & Mitra, M. (2003). An automated data extraction system from 12 lead ECG images. *Computer Methods and Programs in Biomedicine*, *71*(1), 33–38. doi:10.1016/S0169-2607(02)00032-9 PMID:12725963

Mitra, S., Mitra, M., & Chaudhuri, B. B. (2004). Generation of digital time database from paper ECG records and Fourier transform-based analysis for disease identification. *Computers in Biology and Medicine*, *34*(7), 551–560. doi:10.1016/j.compbiomed.2003.08.001 PMID:15369707

Mogli, G. (2012). Role of Biometrics in healthcare privacy and security management system. *Sri Lanka Journal of Bio-Medical Informatics*, *2*(4), 156. doi:10.4038ljbmi.v2i4.2245

Mohammedi, M., Omar, M., Aitabdelmalek, W., Mansouri, A., & Bouabdallah, A. (2018, April). Secure and lightweight biometrics-based remote patient authentication scheme for home healthcare systems. *In Programming and Systems (ISPS), 2018 International Symposium on* (pp. 1-6). IEEE.

Mohan, A., Blough, D. M., Kurc, T., Post, A., & Saltz, J. (2011). Detection of conflicts and inconsistencies in taxonomy-based authorization policies. *IEEE International Conference on Bioinformatics and Biomedicine*, 590–594.

Molchanov, P., Gupta, S., Kim, K., & Kautz, J. (2015). Hand gesture recognition with 3d convolutional neural networks. In *Proceedings of the IEEE Conference on Computer Vision and Pattern Recognition Workshops* (pp. 1–7). IEEE.

Morales, E., Sevilla, D., Pierluissi, J. H., & Nazeran, H. (2005). Digitization and Synchronization Method for Electrocardiogram Printouts. *IEEE Engineering in Medicine and Biology 27th Annual Conference.*

Morocho, D., Morales, A., Fierrez, J., & Vera-Rodriguez, R. (2016). Towards human-assisted signature recognition: improving biometric systems through attribute-based recognition. In *Identity, Security and Behavior Analysis (ISBA), 2016 IEEE International Conference on* (pp. 1–6). IEEE.

Mukherjee, R., & Ross, A. (2008). Indexing iris images. In *Pattern Recognition, 2008. ICPR 2008. 19th International Conference on* (pp. 1-4). Academic Press. 10.1109/ICPR.2008.4761880

Munier, S., Arora, R., Hesling, C., Li, J., Francis, J., Shelton, C., ... Berges, M. (2017). Real-time fine-grained occupancy estimation using depth sensors on ARM embedded platforms. *Proceedings of the 23rd IEEE Real Time and Embedded Technology and Application Symposium.*

Muñoz-Briseño, A., Hernández-Palancar, J., & Gago-Alonso, A. (2015). Minutiae based palmprint indexing. In *Computational Forensics* (pp. 10–19). Cham: Springer. doi:10.1007/978-3-319-20125-2_2

Muramatsu, D., Shiraishi, A., Makihara, Y., Uddin, M. Z., & Yagi, Y. (2015). Gait-based person recognition using arbitrary view transformation model. *IEEE Transactions on Image Processing*, *24*(1), 140–154. doi:10.1109/TIP.2014.2371335 PMID:25423652

Murphy, K. P. (2012). *Machine learning: A probabilistic perspective.* MIT Press.

Mwaura, G. W., Mwangi, W., & Otieno, C. (2017). Multimodal Biometric System:-Fusion Of Face And Fingerprint Biometrics At Match Score Fusion Level. *International Journal of Scientific & Technology Research*, *6*(4), 41–49.

Najarian, K., & Splinter, R. (2012). *Biomedical Signal and Image Processing.* New York: Taylor and Francis group.

Nakada, M., Wang, H., & Terzopoulos, D. (2017). AcFR: active face recognition using convolutional neural networks. In *Computer Vision and Pattern Recognition Workshops (CVPRW), 2017 IEEE Conference on* (pp. 35–40). IEEE.

Nandakumar, K., Chen, Y., Dass, S. C., & Jain, A. K. (2008). Likelihood ratio-based biometric score fusion. *IEEE Transactions on Pattern Analysis and Machine Intelligence*, *30*(2), 342–347. doi:10.1109/TPAMI.2007.70796 PMID:18084063

Neverova, N., Wolf, C., Paci, G., Sommavilla, G., Taylor, G., & Nebout, F. (2013). A multi-scale approach to gesture detection and recognition. In *Proceedings of the IEEE International Conference on Computer Vision Workshops* (pp. 484–491). IEEE.

Neves, J., Moreno, J., & Proença, H. (2017). *QUIS-CAMPI: an annotated multi-biometrics data feed from surveillance scenarios.* IET Biometrics; doi:10.1049/iet-bmt.2016.0178

Newell, A., Yang, K., & Deng, J. (2016). Stacked hourglass networks for human pose estimation. In *Proceedings of European Conference on Computer Vision* (pp. 483–499). Springer.

Ng, R. Y. F., Tay, Y. H., & Mok, K. M. (2008, August). A review of iris recognition algorithms. In *Information Technology, 2008. ITSim 2008. International Symposium on* (Vol. 2, pp. 1-7). IEEE. 10.1109/ITSIM.2008.4631656

Nguyen, P., Tran, T., Wickramasinghe, N., & Venkatesh, S. (2017). Deepr: A Convolutional Net for Medical Records. *IEEE Journal of Biomedical and Health Informatics, 21*(1), 22–30. doi:10.1109/JBHI.2016.2633963 PMID:27913366

Noroozi, F., Corneanu, C., Kamińska, D., Sapiński, T., Escalera, S., & Anbarjafari, G. (2018). *Survey on emotional body gesture recognition. IEEE Transactions on Affective Computing.*

Nurse, E., Mashford, B. S., Yepes, A. J., Kornek, I. K., Harrer, S., & Freestone, D. R. (2016). Decoding EEG and LFP signals using deep learning: heading TrueNorth. *ACM International Conference on Computing Frontiers*, 259–266. 10.1145/2903150.2903159

NVIDIA news. (2015). *NVIDIA Announces Financial Results for Second Quarter Fiscal 2016.* Retrieved from https://nvidianews.nvidia.com/news/nvidia-announces-financial-results-for-second-quarter-fiscal-2016

Nwosu, K. C., & Igbonagwam, O. (2014, October). BEPOC: A biometrics-enabled point-of-care patients and electronic medical records management system for medical missions. In *Healthcare Innovation Conference (HIC), 2014 IEEE* (pp. 223-226). IEEE.

O'keefe, J. H., Hammill, S. C., Freed, M. S., & Pogwizd, S. M. (2009). *The Complete guide to ECGs: A comprehensive study guide to improve ECG interpretation skills.* Jones and Bartlett publishers.

Obafemi-Ajayi, T., Miles, J. H., Takahashi, T. N., Qi, W., Aldridge, K., Zhang, M., ... Duan, Y. (2015). Facial structure analysis separates autism spectrum disorders into meaningful clinical subgroups. *Journal of Autism and Developmental Disorders, 45*(5), 1302–1317. doi:10.100710803-014-2290-8 PMID:25351828

Odinaka, I., Lai, P. H., Kaplan, A. D., O'Sullivan, J. A., Sirevaag, E. J., & Rohrbaugh, J. W. (2012). ECG biometric recognition: A comparative analysis. *IEEE Transactions on Information Forensics and Security, 7*(6), 1812–1824. doi:10.1109/TIFS.2012.2215324

Offit, K. (2011). Personalized medicine: New genomics, old lessons. *Human Genetics, 130*(1), 3–14. doi:10.100700439-011-1028-3 PMID:21706342

Ogundipe, O. (2011). Electronic Medical Record Keeping: Eleven Years Experience at Life Support Eye Clinic, Lagos, Nigeria. *Nigerian Journal of Ophthalmology, 19*(1). doi:10.4314/njo.v19i1.70771

Ojansivu, V., & Heikkilä, J. (2008, July). Blur insensitive texture classification using local phase quantization. In *International conference on image and signal processing* (pp. 236-243). Springer. 10.1007/978-3-540-69905-7_27

Okoh Peter, D., Oghenemavwe Loveday, E., & Yorkum Leyira, K. (2013). *Dermatoglyphic Patterns of Autistic Children in Nigeria*. Academic Press.

Omara, I., Li, F., Zhang, H., & Zuo, W. (2016). A novel geometric feature extraction method for ear recognition. *Expert Systems with Applications*, *65*, 127–135. doi:10.1016/j.eswa.2016.08.035

Osman, H. I., Hashim, F. H., Zaki, W. M. D. W., & Huddin, A. B. (2017). Entryway detection algorithm using Kinect depth camera for UAV application. In *Proceedings of IEEE 8th Control and System Graduate Research Colloquium (ICSGRC)* (pp. 77-80). IEEE.

Ozgen, H., Hellemann, G. S., De Jonge, M. V., Beemer, F. A., & van Engeland, H. (2013). Predictive value of morphological features in patients with autism versus normal controls. *Journal of Autism and Developmental Disorders*, *43*(1), 147–155. doi:10.100710803-012-1554-4 PMID:22669539

Paliwal, A., Jayaraman, U., & Gupta, P. (2010, September). A score based indexing scheme for palmprint databases. In *Image Processing (ICIP), 2010 17th IEEE International Conference on* (pp. 2377-2380). IEEE.

Papagiannopoulou, E. A., Chitty, K. M., Hermens, D. F., Hickie, I. B., & Lagopoulos, J. (2014). A systematic review and meta-analysis of eye-tracking studies in children with autism spectrum disorders. *Social Neuroscience*, *9*(6), 610–632. PMID:24988218

Paper, C., Tiwari, S., Shankaracharya, S., Campus, T., Kumar, S., Naya, I., … Cad, R. (2015). *Ear Recognition for Newborns*. Academic Press.

Park, Y., & Kellis, M. (2015). Deep learning for regulatory genomics. *Nature Biotechnology*, *33*(8), 825–826. doi:10.1038/nbt.3313 PMID:26252139

Parmar, P. A., & Degadwala, S. D. (2016). A feature level fusion fingerprint indexing approach based on MV and MCC using SVM classifier. In *International Conference on Communication and Signal Processing (ICCSP), 2016* (pp. 1024-1028). IEEE.

Pease, B., & Pease, A. (2004). *The definitive book of body language*. Bantam.

Pflug, A., Busch, C., & Ross, A. (2014). 2D ear classification based on unsupervised clustering. In *Biometrics (IJCB), 2014 IEEE International Joint Conference on* (pp. 1–8). IEEE.

Pham, T., Tran, T., Phung, D., & Venkatesh, S. (2016). *DeepCare: a deep dynamic memory model for predictive medicine*. arXiv:1602.00357.

Phillips, P. J. (2004). *The facial recognition technology (FERET) database*. Academic Press.

Phillips, P. J., Wechsler, H., Huang, J., & Rauss, P. J. (1998). The FERET database and evaluation procedure for face-recognition algorithms. *Image and Vision Computing*, *16*(5), 295–306. doi:10.1016/S0262-8856(97)00070-X

Piwek, L., Ellis, D. A., Andrews, S., & Jainson, A. (2016). The rise of consumer health wearables: Promises and barriers. *PLoS Medicine*, *13*(2), e1001953. doi:10.1371/journal.pmed.1001953 PMID:26836780

Pöhlmann, S., Harkness, E., Taylor, C., & Astley, S. M. (2016). Evaluation of Kinect 3D Sensor for Healthcare Imaging. *Journal of Medical and Biological Engineering*, *36*(6), 857–870. doi:10.100740846-016-0184-2 PMID:28111534

Pointner, P. (2017). Hybrid tech: the future of biometric ID verification? *Biometric Technology Today*, *2017*(8), 8-10. doi:10.10160969-4765(17)30183-2

Ponemon Institute Report. (2016). *National Patient Misidentification Report*. Ponemon Institute LLC.

Prabhu, N., Issrani, R., Mathur, S., Mishra, G., & Sinha, S. (2014). Dermatoglyphics in health and oral diseases-A review. *JSM Dent*, *2*(4), 1044.

Prasoon, A., Petersen, K., Igel, C., Lauze, F., Dam, E., & Nielsen, M. (2013). *Deep feature learning for knee cartilage segmentation using a triplanar convolutional neural network* (Vol. 16). Medical Image Computing and Computer- Assisted Intervention.

Proenca, H. (2013). Iris biometrics: Indexing and retrieving heavily degraded data. *IEEE Transactions on Information Forensics and Security*, *8*(12), 1975–1985. doi:10.1109/TIFS.2013.2283458

Proença, H., & Alexandre, L. A. (2005, September). UBIRIS: A noisy iris image database. In *International Conference on Image Analysis and Processing* (pp. 970-977). Springer.

Puhan, N. B., & Sudha, N. (2011). Coarse indexing of iris database based on iris colour. *International Journal of Biometrics*, *3*(4), 353–375. doi:10.1504/IJBM.2011.042817

R, K., & G, R. (2011). *Biometrics-A Vision for the Future*. Retrieved from www.hcl.com

Rabuzin, K., Bača, M., & Maleković, M. (2007). A Multimodal Biometric System Implemented within an Active Database Management System. *Journal of Software*, *2*(4). doi:10.4304/jsw.2.4.24-31

Ranganathan, H., Chakraborty, S., & Panchanathan, S. (2016). Multimodal emotion recognition using deep learning architectures. In *Proceedings of 2016 IEEE Winter Conference on Applications of Computer Vision (WACV)* (pp. 1–9). IEEE.

Rangayyan, R. M. (2002). *Biomedical Signal Analysis: A Case-Study Approach*. New York: John Wiley & sons.

Rani, K., & Dahiya, S. (2017). Comparative analysis of fingerprint recognition using neural network and genetic algorithm. *Indian Journal of Scientific Research*, 84–90.

Raposo, R., Hoyle, E., Peixinho, A., & Proença, H. (2011). UBEAR: A dataset of ear images captured on-the-move in uncontrolled conditions. In *Computational Intelligence in Biometrics and Identity Management (CIBIM), 2011 IEEE Workshop on* (pp. 84–90). IEEE.

Ratha, N. K., Connell, J. H., & Bolle, R. M. (2001). Enhancing security and privacy in biometrics-based authentication systems. *IBM Systems Journal*, *40*(3), 614–634. doi:10.1147j.403.0614

Rathgeb, C., Breitinger, F., Baier, H., & Busch, C. (2015). Towards bloom filter-based indexing of iris biometric data. In *International Conference on Biometrics (ICB), 2015* (pp. 422-429). IEEE.

Ravichandran, L., Harless, C., Shah, A. J., Wick, C. A., Mcclellan, J. H., & Tridandapani, S. (2013). Novel tool for complete digitization of paper electrocardiography data. *IEEE Journal of Translational Engineering in Health and Medicine*, *1*, 2168–2372. doi:10.1109/JTEHM.2013.2262024 PMID:26594601

Ravi, D., Wong, C., Lo, B., & Yang, G. Z. (2017). A deep learning approach to on-node sensor data analytics for mobile or wearable devices. *IEEE Journal of Biomedical and Health Informatics*, *21*(1), 56–64. doi:10.1109/JBHI.2016.2633287 PMID:28026792

Razavian, N., Marcus, J., & Sontag, D. (2016). Multi-task prediction of disease onsets from longitudinal laboratory tests. *Proceedings of the 1st Machine Learning for Healthcare Conference*, 73–100.

Reed, T., Reichmann, A., & Palmer, C. G. (1977). Dermatoglyphic differences between 45, X and other chromosomal abnormalities of Turner syndrome. *Human Genetics*, *36*(1), 13–23. doi:10.1007/BF00390431 PMID:858621

Ren, S., He, K., Girshick, R., & Sun, J. (2017). Faster R-CNN: Towards Real-Time Object Detection with Region Proposal Networks. *IEEE Transactions on Pattern Analysis and Machine Intelligence*, *39*(6), 1137–1149. doi:10.1109/TPAMI.2016.2577031 PMID:27295650

Reynolds, D. A., Quatieri, T. F., & Dunn, R. B. (2000). Speaker verification using adapted Gaussian Mixture Models. *Digital Signal Processing*, *10*(1-3), 19–41. doi:10.1006/dspr.1999.0361

Reynolds, D. A., & Rose, R. C. (1995). Robust text-independent speaker identification using Gaussian Mixture Speaker Models. *IEEE Transactions on Speech and Audio Processing*, *3*(1), 72–83. doi:10.1109/89.365379

Riera, A., Soria-Frisch, A., Caparrini, M., Grau, C., & Ruffini, G. (2008). Unobtrusive biometrics system based on electroencephalogram analysis. *EURASIP Journal on Advances in Signal Processing*, *2008*, 18.

Roeschlin, M., Martinovic, I., & Rasmussen, K. B. (2018). Device pairing at the touch of an electrode. *Proceedings of NDSS'18*.

Ross, A. A., & Govindarajan, R. (2005, March). Feature level fusion of hand and face biometrics. In *Biometric Technology for Human Identification II* (Vol. 5779, pp. 196–205). International Society for Optics and Photonics. doi:10.1117/12.606093

Ross, A. A., Nandakumar, K., & Jain, A. K. (2006). *Handbook of multibiometrics* (Vol. 6). Springer Science & Business Media.

Ross, A., & Jain, A. (2003). Information fusion in biometrics. *Pattern Recognition Letters, 24*(13), 2115–2125. doi:10.1016/S0167-8655(03)00079-5

Ross, A., & Jain, A. K. (2004). Multimodal biometrics: An overview. *Proceedings of the 12th European Signal Processing Conference, 1221*-1224.

Roy, A. K., Soni, Y., & Dubey, S. (2013). Enhancing effectiveness of motor rehabilitation using Kinect motion sensing technology. In *Proceedings of 2013 IEEE Global Humanitarian Technology Conference: South Asia Satellite (GHTC-SAS)* (pp. 298-304). IEEE.

Rumelhart, D. E., Hinton, G. E., & Williams, R. J. (1986). Learning representations by back-propagating errors. *Nature, 323*(6088), 533–536. doi:10.1038/323533a0

Safavi, S., Gan, H., Mporas, I., & Sotudeh, R. (2016). Fraud detection in voice-based identity authentication applications and services. In *Data Mining Workshops (ICDMW), 2016 IEEE 16th International Conference on* (pp. 1074–1081). IEEE.

Safie, S. I., Soraghan, J. J., & Petropoulakis, L. (2011). Electrocardiogram (ECG) Biometric Authentication Using Pulse Active Ratio (PAR). *IEEE Transactions on Information Forensics and Security, 6*(4), 1315–1322. doi:10.1109/TIFS.2011.2162408

Sagar, V. K., Ngo, D. B. L., & Foo, K. C. K. (1995, October). Fuzzy feature selection for fingerprint identification. In *Security Technology, 1995. Proceedings. Institute of Electrical and Electronics Engineers 29th Annual 1995 International Carnahan Conference on* (pp. 85-90). IEEE. 10.1109/CCST.1995.524738

Saha, S., Datta, S., Konar, A., & Janarthanan, R. (2014). A study on emotion recognition from body gestures using kinect sensor. In *Proceedings of Communications and Signal Processing (ICCSP), 2014 International Conference on, IEEE* (pp. 056–060). IEEE.

Sahidullah, M., Delgado, H., Todisco, M., Yu, H., Kinnunen, T., Evans, N., & Tan, Z.-H. (2016). Integrated spoofing countermeasures and automatic speaker verification: An evaluation on ASVspoof 2015. *Proceedings of INTERSPEECH, 16*, 1700–1704. doi:10.21437/Interspeech.2016-1280

Sahidullah, Md., Kinnunen, T., & Hanilci, C. (2015). A comparison of features for synthetic speech detection. *Proceedings of INTERSPEECH, 15*, 2087–2091.

Sajjan, M. (2013). *Learn ECG in a day: A systematic approach.* New Delhi: Jaypee Brothers Medical Publishers.

Salah, A. A. (2010). Machine Learning for Biometrics. In Handbook of Research on Machine Learning Applications and Trends: Algorithms, Methods, and Techniques. Centre for Mathematics and Computer Science (CWI). Doi:10.4018/978-1-60566-766-9.ch026

Salakhutdinov, R., Mnih, A., & Hinton, G. (2007). Restricted Boltzmann machines for collaborative filtering. *Proceedings of the 24th International Conference on Machine Learning*, 791–798.

Sanchez-Reillo, R., Sanchez-Avila, C., & Gonzalez-Marcos, A. (2000). Biometrics identification through hand geometry measurements. *IEEE Transactions on Pattern Analysis and Machine Intelligence*, *22*(10), 1168–1171. doi:10.1109/34.879796

Sarangi, P. P., Panda, M., Mishra, B. S. P., & Dehuri, S. (2017). An automated ear localization technique based on modified Hausdorff distance. In *Proceedings of International Conference on Computer Vision and Image Processing* (pp. 229–240). Springer. 10.1007/978-981-10-2107-7_21

Sathyanarayana, A., Joty, S., Fernandez-Luque, L., Ofli, F., Srivastava, J., Elmagarmid, A., ... Taheri, S. (2016). Correction of: Sleep quality prediction from wearable data using deep learning. *JMIR Mhealth Uhealth*, *4*(4), e130. doi:10.2196/mhealth.6953 PMID:27885989

Schaumann, B., & Alter, M. (1976). Embryogenesis and genetics of epidermal ridges. In *Dermatoglyphics in Medical Disorders* (pp. 1–11). Berlin: Springer. doi:10.1007/978-3-642-51620-7_1

Schaumann, B., & Alter, M. (2012). *Dermatoglyphics in medical disorders.* Springer Science & Business Media.

Scheirer, W. J., Rocha, A., Micheals, R. J., & Boult, T. E. (2011). Meta-recognition: The theory and practice of recognition score analysis. *IEEE Transactions on Pattern Analysis and Machine Intelligence*, *33*(8), 1689–1695. doi:10.1109/TPAMI.2011.54 PMID:21422483

Schmidhuber, J. (2015). Deep learning in neural networks: An overview. *Neural Networks*, *61*, 85–117. doi:10.1016/j.neunet.2014.09.003 PMID:25462637

Schürmann, D., & Sigg, S. (2013). Secure communication based on ambient audio. *IEEE Transactions on Mobile Computing*, *12*(2), 358–370. doi:10.1109/TMC.2011.271

Segun, O. F., & Olawale, F. B. (2017). *Healthcare data breaches: Biometric technology to the rescue.* Academic Press.

Senecal, S., Cuel, L., Aristidou, A., & Magnenat-Thalmann, N. (2016). Continuous body emotion recognition system during theater performances. *Proceedings of Computer Animation and Virtual Worlds*, *27*(3-4), 311–320. doi:10.1002/cav.1714

Shakil, K. A., Zareen, F. J., Alam, M., & Jabin, S. (2017). BAMHealthCloud: A Biometric Authentication and Data Management System for Healthcare Data in Cloud. *Journal of King Saud University - Computer and Information Sciences*, 1-9.

Shameer, K., Badgeley, M. A., Miotto, R., Glicksberg, B. S., Morgan, J. W., & Dudley, J. T. (2017). Translational bioinformatics in the era of real-time biomedical, healthcare and wellness data streams. *Briefings in Bioinformatics*, *18*(1), 1105–1124. doi:10.1093/bib/bbv118 PMID:26876889

Sharma, L. D., & Sunkaria, R. K. (2017). P and T-Wave Characterization in the Presence of U-Wave in Electrocardiogram. *Current Trends in Signal Processing. STM Journals*, *7*(1), 29–34.

Shawl, D. (2013). *Biometrics–Implementing into the Healthcare Industry Increases the Security For The Doctors, Nurses, and Patients* (Master's thesis). Davenport University.

Shen, T. W. (2005). *Biometrics identity verification based on electrocardiogram (ECG)* (PhD dissertation). University of Wisconsin.

Siddiqui, M., & Medioni, G. (2010). *Human pose estimation from a single view point, real-time range sensor. In CVPRW* (pp. 1–8). IEEE.

Sidek, K. A., Khalil, I., & Jelinek, H. F. (2014). ECG biometric with abnormal cardiac conditions in remote monitoring system. *IEEE Transactions on Systems, Man, and Cybernetics. Systems*, *44*(11), 1498–1509. doi:10.1109/TSMC.2014.2336842

Sidlauskas, D. P., & Tamer, S. (2008). Hand geometry recognition. In *Handbook of Biometrics* (pp. 91–107). Boston, MA: Springer. doi:10.1007/978-0-387-71041-9_5

Simonyan, K., & Zisserman, A. (2014). *Very deep convolutional networks for large-scale image recognition.* ArXiv Preprint ArXiv:1409.1556.

Singh, Y. N., Singh, S. K., & Ray, A. K. (2012). Bioelectrical signals as emerging biometrics: Issues and challenges. *ISRN Signal Processing*, *26*.

Sizov, A., Khoury, E., & Kinnunen, T. (2015). Joint speaker verification and anti-spoofing in the *i*-vector space. *IEEE Transactions on Information Forensics and Security*, *10*(4), 821–832. doi:10.1109/TIFS.2015.2407362

Slaney, M., & Casey, M. (2008). Locality-sensitive hashing for finding nearest neighbors. *IEEE Signal Processing Magazine*, *25*(2), 128–131. doi:10.1109/MSP.2007.914237

Smari, K., & Bouhlel, M. (2016). Gesture recognition system and finger tracking with kinect Steps. In *Proceedings of 7th International Conference on Sciences of Electronics, Technologies of Information and Telecommunications (SETIT)* (pp. 544-548). Academic Press.

Smolensky, P. (1986). *Information processing in dynamical systems: Foundations of harmony theory (No. CU-CS-321-86).* Colorado Univ at Boulder Dept of Computer Science.

SNOMED CT. (2016). *U.S. National Library of Medicine.* Retrieved from https://www.nlm.nih.gov/healthit/snomedct/

Stone, E., & Skubic, M. (2012). Passive, in-home gait measurement using an inexpensive depth camera. In *Proceedings of Initial results: International Conference on Pervasive Computing Technologies for Healthcare* (pp. 183-186). IEEE.

Storisteanu, D. M., Norman, T. L., Grigore, A., & Norman, T. L. (2015). Biometric fingerprint system to enable rapid and accurate identification of beneficiaries. *Global Health, Science and Practice*, *3*(1), 135–137. doi:10.9745/GHSP-D-15-00010 PMID:25745127

Stošljević, M., & Adamović, M. (2013). Dermatoglyphic characteristics of digito-palmar complex in autistic boys in Serbia. *Vojnosanitetski Pregled. Military-Medical and Pharmaceutical Review*, *70*(4), 386–390. doi:10.2298/VSP1304386S PMID:23700943

Sumathi, S., & Malini, R. R. (2010, April). Face recognition system to enhance E Health. In *E-Health Networking, Digital Ecosystems and Technologies (EDT), 2010 International Conference on* (Vol. 1, pp. 195-198). IEEE.

Su, Y., Feng, J., & Zhou, J. (2016). Fingerprint indexing with pose constraint. *Pattern Recognition, 54*, 1–13. doi:10.1016/j.patcog.2016.01.006

Tabatabaee, H., Milani-Fard, A., & Jafariani, H. (2006, April). A novel human identifier system using retina image and fuzzy clustering approach. In *Proceedings of the 2nd IEEE International Conference on Information and Communication Technologies* (pp. 1031-1036). IEEE.

Taha, K. E., & Norrozila, S. (2015). A survey of multi-biometrics and fusion levels. *Indian Journal of Science and Technology*, *8*(32). doi:10.17485/ijst/2015/v8i32/91488

Tan, D. W., Russell-Smith, S. N., Simons, J. M., Maybery, M. T., Leung, D., Ng, H. L., & Whitehouse, A. J. (2015). Perceived gender ratings for high and low scorers on the Autism-Spectrum Quotient consistent with the extreme male brain account of autism. *PLoS One*, *10*(7), e0131780. doi:10.1371/journal.pone.0131780 PMID:26186689

Tang, D., Huang, B., Li, R., & Li, W. (2010, August). A person retrieval solution using finger vein patterns. In *Pattern Recognition (ICPR), 2010 20th International Conference on* (pp. 1306-1309). IEEE.

Tan, R., & Perkowski, M. (2017). Toward Improving Electrocardiogram (ECG) Biometric Verification using Mobile Sensors: A Two-Stage Classifier Approach. *Sensors (Basel)*, *17*(2), 410. doi:10.339017020410 PMID:28230745

Tanu & Kakkar, D. (2018). Diagnostic Assessment Techniques and Non-invasive Biomarkers for Autism Spectrum Disorder. *International Journal of E-Health and Medical Communications*.

Tatonetti, N. P., Ye, P. P., Daneshjou, R., & Altman, R. B. (2012). Data-driven prediction of drug effects and interactions. *Science Translational Medicine*, *4*(125), 125ra31. doi:10.1126citranslmed.3003377 PMID:22422992

Thaler, M. S. (2015). *The only EKG book you'll ever need.* Tokyo: Wolters Kluwer Health.

Thanh, L., & Hải, N. (2012). Developping Kerberos-role authentication protocol for resource management system. *Journal Of Computer Science And Cybernetics*, *20*(4). doi:10.15625/1813-9663/20/4/1899

Thomas, C. D. L., Claes, P. D., Shaweesh, A. I., & Clement, J. G. (2011). Three-dimensional facial shape archetypes for identification and diagnosis. In Biological Shape Analysis (pp. 131-144). Academic Press.

Tian, L., & Mu, Z. (2017). Ear recognition based on deep convolutional network. In *Proceedings - 2016 9th International Congress on Image and Signal Processing, BioMedical Engineering and Informatics, CISP-BMEI 2016* (pp. 437–441). 10.1109/CISP-BMEI.2016.7852751

Tiquet, A. (2016). *Horus Among Seven Prize-Winning Startups at Emerging Companies Summit.* Retrieved from https://blogs.nvidia.com/blog/2016/04/07/horus-help-blind-people-see/

Tiwari, K., & Gupta, P. (2015). Indexing fingerprint database with minutiae based coaxial Gaussian track code and quantized lookup table. In *IEEE International Conference on Image Processing (ICIP)* (pp. 4773-4777). IEEE.

Toh, K.-A., Yau, W.-Y., Lim, E., Chen, L., & Ng, C.-H. (2004). *Fusion of auxiliary information for multi-modal biometrics authentication in Bio-metric Authentication*. Berlin, Germany: Springer. doi:10.1007/978-3-540-25948-0_92

Toshev, A., & Szegedy, C. (2014). Deeppose: Human pose estimation via deep neural networks. In *Proceedings of the IEEE Conference on Computer Vision and Pattern Recognition* (pp. 1653–1660). IEEE.

Trokielewicz, M., Czajka, A., & Maciejewicz, P. (2017). Implications of ocular pathologies for iris recognition reliability. *Image and Vision Computing*, *58*, 158–167. doi:10.1016/j.imavis.2016.08.001

Turek, S. (1990). Dermatoglyphics and schizophrenia: Analysis of quantitative traits. *Collegium Antropologicum*, *14*(1), 137–150.

UMLS (Unified Medical Language System). (n.d.). Retrieved from https://www.nlm. nih.gov/research/umls/

USTB. (2004). *University of Science & Technology Beijing (USTB)*. Retrieved from http://www1.ustb.edu.cn/resb/

Vakanski, A., Jun, H., Paul, D., & Baker, R. R. (2018). *A Data Set of Human Body Movements for Physical Rehabilitation Exercises*. Academic Press. doi:10.3390/data3010002

Valenzuela, R., Schwartz, W., & Pedrini, H. (2014). Linear dimensionality reduction applied to scale invariant feature transformation and speeded up robust feature descriptors. *Journal of Electronic Imaging*, *23*(3), 033017. doi:10.1117/1.JEI.23.3.033017

Varol, S., Orhan, F., Tuncer, S., & Akyuz, S. (2016). Biometric systems in the context of information security in health institutions. *Health Care Academician Journal*, *3*(4), 155. doi:10.5455ad.13-1483706096

Vincent, P., Larochelle, H., Lajoie, I., Lajoie, I., Bengio, Y., & Mazagol, P. A. (2010). Stacked denoising autoencoders: Learning useful representations in a deep network with a local denoising criterion. *Journal of Machine Learning Research*, *11*, 3371–3408.

Vogt, R., & Kajarekar, S. (2008). Paper. *Proceedings of Odyssey*, 21–24.

Waits, G., S., & Soliman, E., Z. (2016). Digitizing paper electrocardiogram: Status and challenges. *Journal of Electrocardiology*.

Wang, H., Zhang, C., Song, Y., & Pang, B. (2017). Information-fusion based robot simultaneous localization and mapping adapted to search and rescue cluttered environment. In *Proceedings of 18th International Conference on Advanced Robotics (ICAR)* (pp. 511-517). Academic Press. 10.1109/ICAR.2017.8023658

Wang, B., Mezlini, A. M., Demir, F., Fiume, M., Tu, Z., Brudno, M., ... Goldenberg, A. (2014). Similarity network fusion for aggregating data types on a genomic scale. *Nature Methods*, *11*(3), 333–337. doi:10.1038/nmeth.2810 PMID:24464287

Wang, F., Zhang, P., Wang, X., & Hu, J. (2014). Clinical risk prediction by exploring high-order feature correlations. *AMIA Annual Symposium*, 1170–1179.

Wang, J. T., & Mital, D. P. (1996). A microcomputer-based prototype for ECG paper record conversion. *Journal of Network and Computer Applications*, *19*(3), 295–307. doi:10.1006/jnca.1996.0020

Wang, L., & Leedham, G. (2006). Near- and Far- Infrared Imaging for Vein Pattern Biometrics. *2006 IEEE International Conference on Video and Signal Based Surveillance*, 52-52. 10.1109/AVSS.2006.80

Waoo, N., Kashyap, R., & Jaiswal, A. (2010). DNA nano array analysis using hierarchical quality threshold clustering. In *2010 2nd IEEE International Conference on Information Management and Engineering*. IEEE.

Webster, D., & Celik, O. (2014). Systematic review of Kinect applications in elderly care and stroke rehabilitation. *Journal of Neuroengineering and Rehabilitation*, *11*(1), 1–24. doi:10.1186/1743-0003-11-108 PMID:24996956

Wechsler, H. (2012). Biometric Security and Privacy Using Smart Identity Management and Interoperability: Validation and Vulnerabilities of Various Techniques. *The Review of Policy Research*, *29*(1), 63–89. doi:10.1111/j.1541-1338.2011.00538.x

Weinberg, S. M., Scott, N. M., Neiswanger, K., Brandon, C. A., & Marazita, M. L. (2004). Digital three-dimensional photogrammetry: Evaluation of anthropometric precision and accuracy using a Genex 3D camera system. *The Cleft Palate-Craniofacial Journal*, *41*(5), 507–518. doi:10.1597/03-066.1 PMID:15352857

Whitehouse, A. J., Gilani, S. Z., Shafait, F., Mian, A., Tan, D. W., Maybery, M. T., ... Eastwood, P. (2015). Prenatal testosterone exposure is related to sexually dimorphic facial morphology in adulthood. *Proc. R. Soc. B*, *282*(1816), 20151351.

Widman, L. E., & Hines, L. S. (1991). Digitization of electrocardiograms by desktop optical scanner. *Journal of Electrocardiology*, *24*(4), 325–338. doi:10.1016/0022-0736(91)90015-E PMID:1744545

Williams & Wilkins. (2011). *ECG interpretation made incredibly easy*. Tokyo: Wolters Kluwer Health.

Williams, R. J., & Zipser, D. (1989). A learning algorithm for continually running fully recurrent neural networks. *Neural Computation*, *1*(2), 270–280. doi:10.1162/neco.1989.1.2.270

Wilson, C. L., Candela, G. T., & Watson, C. I. (1994). Neural network fingerprint classification. *Journal of Artificial Neural Networks*, *1*(2), 203–228.

Wójtowicz, W., & Ogiela, M. (2014). Biometric watermarks based on face recognition methods for authentication of digital images. *Security and Communication Networks*, *8*(9), 1672–1687. doi:10.1002ec.1114

Wolman, S. R., Campbell, M., Marchi, M. L., Deutsch, S. I., & Gershon, T. D. (1990). Dermatoglyphic study in autistic children and controls. *Journal of the American Academy of Child and Adolescent Psychiatry*, *29*(6), 878–884. doi:10.1097/00004583-199011000-00006 PMID:2273014

Wong, C., Yang, E., Yan, X. T., & Gu, D. (2017). An overview of robotics and autonomous systems for harsh environments. In *Proceedings of 23rd International Conference on Automation and Computing (ICAC)* (pp. 1-8). Academic Press.

World Bank Group. (2017). *ID4D Identification for Development 2017 Annual Report*. World Bank.

Wu, Z., Khodabakhsh, A., & Demiroglu, C. (2015b). Spoofing and Anti-Spoofing (SAS) corpus v1.0 [dataset]. University of Edinburgh, The Centre for Speech Technology Research (CSTR) doi:. 10.7488/ds/252

Wu, D., José, J. V., Nurnberger, J. I., & Torres, E. B. (2018). A biomarker characterizing neurodevelopment with applications in autism. *Scientific Reports*, *8*(1), 614. doi:10.103841598-017-18902-w PMID:29330487

Wu, G., Wang, J., Zhang, Y., & Jiang, S. (2018). A Continuous Identity Authentication Scheme Based on Physiological and Behavioral Characteristics. *Sensors (Basel)*, *18*(1). doi:10.339018010179 PMID:29320463

Wu, Z., Evans, N., Kinnunen, T., Yamagishi, J., Alegre, F., & Li, H. (2015a). Spoofing and countermeasures for speaker verification: A survey. *Speech Communication*, *66*, 130–153. doi:10.1016/j.specom.2014.10.005

Wu, Z., Ke, Q., Sun, J., & Shum, H. Y. (2011). Scalable face image retrieval with identity-based quantization and multireference reranking. *IEEE Transactions on Pattern Analysis and Machine Intelligence*, *33*(10), 1991–2001. doi:10.1109/TPAMI.2011.111 PMID:21646678

Wu, Z., Yamagishi, J., Kinnunen, T., Hanilci, C., Sahidullah, M., Sizov, A., ... Delgado, H. (2016). ASVspoof: The automatic speaker verification spoofing and countermeasures challenge. *IEEE Journal of Selected Topics in Signal Processing*, *11*(4), 588–604. doi:10.1109/JSTSP.2017.2671435

Xia, L., Chen, C., & Aggarwal, J. (2012). View invariant human action recognition using histograms of 3d join. In *Proceedings of IEEE Conference CVPR* (pp. 20–27). IEEE.

Xu, R., Li, L., & Wang, Q. (2014). dRiskKB: A large-scale disease-disease risk relationship knowledge base constructed from biomedical text. *BMC Bioinformatics*, *15*(1), 105. doi:10.1186/1471-2105-15-105 PMID:24725842

Yager, N., & Amin, A. (2004). Fingerprint classification: A review. *Pattern Analysis & Applications*, *7*(1), 77–93. doi:10.100710044-004-0204-7

Yang, X., & Tian, Y. (2014). Effective 3d action recognition using eigenjoints. *Journal of Visual Communication and Image Representation*, *25*(1), 2–11. doi:10.1016/j.jvcir.2013.03.001

Yoo, Y., Brosch, T., Traboulsee, A., Li, D. K. B., & Tam, R. (2014). Deep learning of image features from unlabeled data for multiple sclerosis lesion segmentation. *International Workshop on Machine Learning in Medical Imaging*, 117–124.

You, J., Kong, W. K., Zhang, D., & Cheung, K. H. (2004). On hierarchical palmprint coding with multiple features for personal identification in large databases. *IEEE Transactions on Circuits and Systems for Video Technology*, *14*(2), 234–243. doi:10.1109/TCSVT.2003.821978

Yum, D. H., Kim, J. S., Hong, S. J., & Lee, P. J. (2011). Distance Bounding Protocol with Adjustable False Acceptance Rate. *IEEE Communications Letters*, *15*(4), 434–436. doi:10.1109/LCOMM.2011.030911.110066

Zen, H., Tokuda, K., & Black, A. W. (2009). Statistical parametric speech synthesis. *Speech Communication*, *51*(11), 1039–1064. doi:10.1016/j.specom.2009.04.004

Zhai, Y., Cao, H., Cao, L., Ma, H., Gan, J., & Zeng, J., … Zhi, Y. (2018). A Novel Finger-Knuckle-Print Recognition Based on Batch-Normalized CNN. In *Chinese Conference on Biometric Recognition* (pp. 11–21). Springer. 10.1007/978-3-319-97909-0_2

Zhang, C., Tian, Y., & Capezuti, E. (2012). Article. In Lecture Notes in Computer Science: Vol. 7382. Privacy preserving automatic fall detection for elderly using RGBD cameras. Springer-Verlag.

Zhang, G. H., Poon, C. C., & Zhang, Y. T. (2009, November). A biometrics based security solution for encryption and authentication in tele-healthcare systems. In *Applied Sciences in Biomedical and Communication Technologies, 2009. ISABEL 2009. 2nd International Symposium on* (pp. 1-4). IEEE.

Zhang, Z., Zhang, H., & Zhuang, T. (1987). One-dimensional signal extraction of paper-written ECG imagen and its archiving. *SPIE Visual Communications and image processing II, 845*, 419–22.

Zhang, D., Liu, Z., & Yan, J. Q. (2010). Dynamic tongueprint: A novel biometrics identifier. *Pattern Recognition*, *43*(3), 1071–1082. doi:10.1016/j.patcog.2009.09.002

Zhang, D., Liu, Z., Yan, J. Q., & Shi, P. F. (2007, August). Tongue-print: a novel biometrics pattern. In *International Conference on Biometrics* (pp. 1174-1183). Springer.

Zhang, D., Lu, G., & Zhang, L. (2018). Local Features for Finger-Knuckle-Print Recognition. In *Advanced Biometrics* (pp. 111–130). Springer. doi:10.1007/978-3-319-61545-5_6

Zhang, L., Tan, S., & Yang, J. (2017). Hearing your voice is not enough: An articulatory gesture based liveness detection for voice authentication. *Proceedings of CCS*, *17*, 55–71.

Zhang, L., Tan, S., Yang, J., & Chen, Y. (2016). VoiceLive: A phoneme localization based liveness detection for voice authentication on smartphones. *Proceedings of CCS*, *16*, 1080–1091. doi:10.1145/2976749.2978296

Zhang, Q., Xiao, Y., Dai, W., Suo, J., Wang, C., Shi, J., & Zheng, H. (2016). Deep learning based classification of breast tumors with shear-wave elastography. *Ultrasonics*, *72*, 150–157. doi:10.1016/j.ultras.2016.08.004 PMID:27529139

Zhang, Y., & Mu, Z. (2017). Ear Detection under Uncontrolled Conditions with Multiple Scale Faster Region - Based Convolutional. *Neural Networks*. doi:10.3390ym9040053

Zhang, Y., Mu, Z., Yuan, L., Zeng, H., & Chen, L. (2017). 3D Ear Normalization and Recognition Based on Local Surface Variation. *Applied Sciences*, *7*(1), 104. doi:10.3390/app7010104

Zhang, Z., Liu, A., Gao, Z., & Liu, C. (2013). A Kinect-based 3D sensing and human action recognition solution for urban search and rescue environments. In *Proceedings of IEEE RO-MAN* (pp. 344-345). IEEE. 10.1109/ROMAN.2013.6628488

Zhao, W., Feng, H., Lun, R., Espy, D. D., & Reinthal, M. A. (2014). A Kinect-based rehabilitation exercise monitoring and guidance system. In *Proceedings of 2014 IEEE 5th International Conference on Software Engineering and Service Science* (pp. 762-765). IEEE.

Zhou, Y., & Zaferiou, S. (2017). Deformable Models of Ears in-the-Wild for Alignment and Recognition. *Proceedings - 12th IEEE International Conference on Automatic Face and Gesture Recognition, FG 2017 - 1st International Workshop on Adaptive Shot Learning for Gesture Understanding and Production, ASL4GUP 2017, Biometrics in the Wild, Bwild 2017, Heteroge*, 626–633. 10.1109/FG.2017.79

Zhou, J., Cadavid, S., & Abdel-Mottaleb, M. (2012). An efficient 3-D ear recognition system employing local and holistic features. *IEEE Transactions on Information Forensics and Security*, *7*(3), 978–991. doi:10.1109/TIFS.2012.2189005

Zhou, J., & Troyanskaya, O. G. (2015). Predicting effects of noncoding variants with deep learning-based sequence model. *Nature Methods*, *12*(10), 931–934. doi:10.1038/nmeth.3547 PMID:26301843

Zhou, W., Hu, J., Wang, S., Petersen, I., & Bennamoun, M. (2016). Partial fingerprint indexing: A combination of local and reconstructed global features. *Concurrency and Computation*, *28*(10), 2940–2957. doi:10.1002/cpe.3600

Zhou, Y., Liu, Y., Feng, Q., Yang, F., Huang, J., & Nie, Y. (2014). Palm-vein classification based on principal orientation features. *PLoS One*, *9*(11), e112429. doi:10.1371/journal.pone.0112429 PMID:25383715

Zhu, J., Pande, A., Mohapatra, P., & Han, J. J. (2015). Using deep learning for energy expenditure estimation with wearable sensors. *17th International Conference on E-health Networking, Application Services (HealthCom),* 501–6.

Zhu, Q., & Mu, Z. (2018, August). An Efficient 3D Ear Recognition System Based on Indexing. In *Chinese Conference on Biometric Recognition* (pp. 507-516). Springer. 10.1007/978-3-319-97909-0_54

Related References

To continue our tradition of advancing information science and technology research, we have compiled a list of recommended IGI Global readings. These references will provide additional information and guidance to further enrich your knowledge and assist you with your own research and future publications.

Abu Seman, S. A., & Ramayah, T. (2017). Are We Ready to App?: A Study on mHealth Apps, Its Future, and Trends in Malaysia Context. In J. Pelet (Ed.), *Mobile Platforms, Design, and Apps for Social Commerce* (pp. 69–83). Hershey, PA: IGI Global. doi:10.4018/978-1-5225-2469-4.ch005

Aggarwal, S., & Azad, V. (2017). A Hybrid System Based on FMM and MLP to Diagnose Heart Disease. In S. Bhattacharyya, S. De, I. Pan, & P. Dutta (Eds.), *Intelligent Multidimensional Data Clustering and Analysis* (pp. 293–325). Hershey, PA: IGI Global. doi:10.4018/978-1-5225-1776-4.ch011

Akaichi, J., & Mhadhbi, L. (2016). A Clinical Decision Support System: Ontology-Driven Approach for Effective Emergency Management. In J. Moon & M. Galea (Eds.), *Improving Health Management through Clinical Decision Support Systems* (pp. 270–294). Hershey, PA: IGI Global. doi:10.4018/978-1-4666-9432-3.ch013

Akerkar, R. (2016). Towards an Intelligent Integrated Approach for Clinical Decision Support. In A. Aggarwal (Ed.), *Managing Big Data Integration in the Public Sector* (pp. 187–205). Hershey, PA: IGI Global. doi:10.4018/978-1-4666-9649-5.ch011

Al-Busaidi, S. S. (2018). Interdisciplinary Relationships Between Medicine and Social Sciences. In M. Al-Suqri, A. Al-Kindi, S. AlKindi, & N. Saleem (Eds.), *Promoting Interdisciplinarity in Knowledge Generation and Problem Solving* (pp. 124–137). Hershey, PA: IGI Global. doi:10.4018/978-1-5225-3878-3.ch009

Al Kareh, T., & Thoumy, M. (2018). The Impact of Health Information Digitization on the Physiotherapist-Patient Relationship: A Pilot Study of the Lebanese Community. *International Journal of Healthcare Information Systems and Informatics, 13*(2), 29–53. doi:10.4018/IJHISI.2018040103

Alenzuela, R. (2017). Research, Leadership, and Resource-Sharing Initiatives: The Role of Local Library Consortia in Access to Medical Information. In S. Ram (Ed.), *Library and Information Services for Bioinformatics Education and Research* (pp. 199–211). Hershey, PA: IGI Global. doi:10.4018/978-1-5225-1871-6.ch012

Ameri, H., Alizadeh, S., & Noughabi, E. A. (2017). Application of Data Mining Techniques in Clinical Decision Making: A Literature Review and Classification. In E. Noughabi, B. Raahemi, A. Albadvi, & B. Far (Eds.), *Handbook of Research on Data Science for Effective Healthcare Practice and Administration* (pp. 257–295). Hershey, PA: IGI Global. doi:10.4018/978-1-5225-2515-8.ch012

Anand, S. (2017). Medical Image Enhancement Using Edge Information-Based Methods. In B. Singh (Ed.), *Computational Tools and Techniques for Biomedical Signal Processing* (pp. 123–148). Hershey, PA: IGI Global. doi:10.4018/978-1-5225-0660-7.ch006

Angjellari-Dajci, F., Sapienza, C., Lawless, W. F., & Kavanagh, K. (2016). Human Capital Accumulation in Medical Simulated Learning Environments: A Framework for Economic Evaluation. In M. Russ (Ed.), *Quantitative Multidisciplinary Approaches in Human Capital and Asset Management* (pp. 65–88). Hershey, PA: IGI Global. doi:10.4018/978-1-4666-9652-5.ch004

Anil, M., Ayyildiz-Tamis, D., Tasdemir, S., Sendemir-Urkmez, A., & Gulce-Iz, S. (2016). Bioinspired Materials and Biocompatibility. In M. Bououdina (Ed.), *Emerging Research on Bioinspired Materials Engineering* (pp. 294–322). Hershey, PA: IGI Global. doi:10.4018/978-1-4666-9811-6.ch011

Anton, J. L., Soriano, J. V., Martinez, M. I., & Garcia, F. B. (2018). Comprehensive E-Learning Appraisal System. In M. Khosrow-Pour, D.B.A. (Ed.), Encyclopedia of Information Science and Technology, Fourth Edition (pp. 5787-5799). Hershey, PA: IGI Global. doi:10.4018/978-1-5225-2255-3.ch503

Arabi, M. (2016). Redesigning the Organizational Structure to Reach Efficiency: The Case of Ministry of Health and Medical Education of Iran. In R. Gholipour & K. Rouzbehani (Eds.), *Social, Economic, and Political Perspectives on Public Health Policy-Making* (pp. 257–294). Hershey, PA: IGI Global. doi:10.4018/978-1-4666-9944-1.ch012

Assis-Hassid, S., Heart, T., Reychav, I., & Pliskin, J. S. (2016). Modelling Factors Affecting Patient-Doctor-Computer Communication in Primary Care. *International Journal of Reliable and Quality E-Healthcare*, *5*(1), 1–17. doi:10.4018/IJRQEH.2016010101

Atzori, B., Hoffman, H. G., Vagnoli, L., Messeri, A., & Grotto, R. L. (2018). Virtual Reality as Distraction Technique for Pain Management in Children and Adolescents. In M. Khosrow-Pour, D.B.A. (Ed.), Encyclopedia of Information Science and Technology, Fourth Edition (pp. 5955-5965). Hershey, PA: IGI Global. doi:10.4018/978-1-5225-2255-3.ch518

Audibert, M., Mathonnat, J., Pélissier, A., & Huang, X. X. (2018). The Impact of the New Rural Cooperative Medical Scheme on Township Hospitals' Utilization and Income Structure in Weifang Prefecture, China. In I. Management Association (Ed.), Health Economics and Healthcare Reform: Breakthroughs in Research and Practice (pp. 109-121). Hershey, PA: IGI Global. doi:10.4018/978-1-5225-3168-5.ch007

Ayyachamy, S. (2016). Analysis and Comparison of Developed 2D Medical Image Database Design using Registration Scheme, Retrieval Scheme, and Bag-of-Visual-Words. In N. Dey & A. Ashour (Eds.), *Classification and Clustering in Biomedical Signal Processing* (pp. 149–168). Hershey, PA: IGI Global. doi:10.4018/978-1-5225-0140-4.ch007

Babalola, A. C. (2016). Literacy and Decision Making on Health Issues among Married Women in Southwest Nigeria. In V. Wang (Ed.), *Handbook of Research on Advancing Health Education through Technology* (pp. 191–212). Hershey, PA: IGI Global. doi:10.4018/978-1-4666-9494-1.ch009

Babu, A., & Ayyappan, S. (2017). A Methodological Evaluation of Crypto-Watermarking System for Medical Images. In C. Bhatt & S. Peddoju (Eds.), *Cloud Computing Systems and Applications in Healthcare* (pp. 189–217). Hershey, PA: IGI Global. doi:10.4018/978-1-5225-1002-4.ch010

Bakke, A. (2017). Ethos in E-Health: From Informational to Interactive Websites. In M. Folk & S. Apostel (Eds.), *Establishing and Evaluating Digital Ethos and Online Credibility* (pp. 85–103). Hershey, PA: IGI Global. doi:10.4018/978-1-5225-1072-7.ch005

Bali, S. (2018). Enhancing the Reach of Health Care Through Telemedicine: Status and New Possibilities in Developing Countries. In U. Pandey & V. Indrakanti (Eds.), *Open and Distance Learning Initiatives for Sustainable Development* (pp. 339–354). Hershey, PA: IGI Global. doi:10.4018/978-1-5225-2621-6.ch019

Barrett, T. E. (2018). Essentials for Education and Training for Tomorrow's Physicians. In C. Smith (Ed.), *Exploring the Pressures of Medical Education From a Mental Health and Wellness Perspective* (pp. 230–252). Hershey, PA: IGI Global. doi:10.4018/978-1-5225-2811-1.ch010

Barrett, T. J. (2016). Knowledge in Action: Fostering Health Education through Technology. In V. Wang (Ed.), *Handbook of Research on Advancing Health Education through Technology* (pp. 39–62). Hershey, PA: IGI Global. doi:10.4018/978-1-4666-9494-1.ch003

Berrahal, S., & Boudriga, N. (2017). The Risks of Wearable Technologies to Individuals and Organizations. In A. Marrington, D. Kerr, & J. Gammack (Eds.), *Managing Security Issues and the Hidden Dangers of Wearable Technologies* (pp. 18–46). Hershey, PA: IGI Global. doi:10.4018/978-1-5225-1016-1.ch002

Bhagat, A. P., & Atique, M. (2016). Medical Image Mining Using Fuzzy Connectedness Image Segmentation: Efficient Retrieval of Patients' Stored Images. In W. Karâa & N. Dey (Eds.), *Biomedical Image Analysis and Mining Techniques for Improved Health Outcomes* (pp. 184–209). Hershey, PA: IGI Global. doi:10.4018/978-1-4666-8811-7.ch009

Bhargava, P. (2018). Blended Learning: An Effective Application to Clinical Teaching. In S. Tang & C. Lim (Eds.), *Preparing the Next Generation of Teachers for 21st Century Education* (pp. 302–320). Hershey, PA: IGI Global. doi:10.4018/978-1-5225-4080-9.ch018

Binh, N. T., & Tuyet, V. T. (2016). The Combination of Adaptive Filters to Improve the Quality of Medical Images in New Wavelet Domain. In N. Dey & A. Ashour (Eds.), *Classification and Clustering in Biomedical Signal Processing* (pp. 46–76). Hershey, PA: IGI Global. doi:10.4018/978-1-5225-0140-4.ch003

Bird, J. L. (2017). Writing Healing Narratives. In V. Bryan & J. Bird (Eds.), *Healthcare Community Synergism between Patients, Practitioners, and Researchers* (pp. 1–28). Hershey, PA: IGI Global. doi:10.4018/978-1-5225-0640-9.ch001

Bottrighi, A., Leonardi, G., Piovesan, L., & Terenziani, P. (2016). Knowledge-Based Support to the Treatment of Exceptions in Computer Interpretable Clinical Guidelines. *International Journal of Knowledge-Based Organizations, 6*(3), 1–27. doi:10.4018/IJKBO.2016070101

Bougoulias, K., Kouris, I., Prasinos, M., Giokas, K., & Koutsouris, D. (2017). Ob/Gyn EMR Software: A Solution for Obstetricians and Gynecologists. In A. Moumtzoglou (Ed.), *Design, Development, and Integration of Reliable Electronic Healthcare Platforms* (pp. 101–111). Hershey, PA: IGI Global. doi:10.4018/978-1-5225-1724-5.ch006

Bouslimi, R., Ayadi, M. G., & Akaichi, J. (2018). Medical Image Retrieval in Healthcare Social Networks. *International Journal of Healthcare Information Systems and Informatics, 13*(2), 13–28. doi:10.4018/IJHISI.2018040102

Bouzaabia, O., Bouzaabia, R., & Mejri, K. (2017). Role of Internet in the Development of Medical Tourism Service in Tunisia. In A. Capatina & E. Rancati (Eds.), *Key Challenges and Opportunities in Web Entrepreneurship* (pp. 211–241). Hershey, PA: IGI Global. doi:10.4018/978-1-5225-2466-3.ch009

Briz-Ponce, L., Juanes-Méndez, J. A., & García-Peñalvo, F. J. (2016). The Role of Gender in Technology Acceptance for Medical Education. In M. Cruz-Cunha, I. Miranda, R. Martinho, & R. Rijo (Eds.), *Encyclopedia of E-Health and Telemedicine* (pp. 1013–1027). Hershey, PA: IGI Global. doi:10.4018/978-1-4666-9978-6.ch079

Briz-Ponce, L., Juanes-Méndez, J. A., & García-Peñalvo, F. J. (2016). A Research Contribution to the Analysis of Mobile Devices in Higher Education from Medical Students' Point of View. In L. Briz-Ponce, J. Juanes-Méndez, & F. García-Peñalvo (Eds.), *Handbook of Research on Mobile Devices and Applications in Higher Education Settings* (pp. 196–221). Hershey, PA: IGI Global. doi:10.4018/978-1-5225-0256-2.ch009

Brown-Jackson, K. L. (2018). Telemedicine and Telehealth: Academics Engaging the Community in a Call to Action. In S. Burton (Ed.), *Engaged Scholarship and Civic Responsibility in Higher Education* (pp. 166–193). Hershey, PA: IGI Global. doi:10.4018/978-1-5225-3649-9.ch008

Bwalya, K. J. (2017). Next Wave of Tele-Medicine: Virtual Presence of Medical Personnel. In K. Moahi, K. Bwalya, & P. Sebina (Eds.), *Health Information Systems and the Advancement of Medical Practice in Developing Countries* (pp. 168–180). Hershey, PA: IGI Global. doi:10.4018/978-1-5225-2262-1.ch010

Carlton, E. L., Holsinger, J. W. Jr, & Anunobi, N. (2016). Physician Engagement with Health Information Technology: Implications for Practice and Professionalism. *International Journal of Computers in Clinical Practice, 1*(2), 51–73. doi:10.4018/IJCCP.2016070103

Castillo-Page, L., Eliason, J., Conrad, S. S., & Nivet, M. A. (2016). Diversity in Undergraduate Medical Education: An Examination of Organizational Culture and Climate in Medical Schools. In C. Scott & J. Sims (Eds.), *Developing Workforce Diversity Programs, Curriculum, and Degrees in Higher Education* (pp. 304–326). Hershey, PA: IGI Global. doi:10.4018/978-1-5225-0209-8.ch016

Cebeci, H. I., & Hiziroglu, A. (2016). Review of Business Intelligence and Intelligent Systems in Healthcare Domain. In N. Celebi (Ed.), *Intelligent Techniques for Data Analysis in Diverse Settings* (pp. 192–206). Hershey, PA: IGI Global. doi:10.4018/978-1-5225-0075-9.ch009

Celik, G. (2018). Determining Headache Diseases With Genetic Algorithm. In U. Kose, G. Guraksin, & O. Deperlioglu (Eds.), *Nature-Inspired Intelligent Techniques for Solving Biomedical Engineering Problems* (pp. 249–262). Hershey, PA: IGI Global. doi:10.4018/978-1-5225-4769-3.ch012

Chakraborty, C., Gupta, B., & Ghosh, S. K. (2016). Mobile Telemedicine Systems for Remote Patient's Chronic Wound Monitoring. In A. Moumtzoglou (Ed.), *M-Health Innovations for Patient-Centered Care* (pp. 213–239). Hershey, PA: IGI Global. doi:10.4018/978-1-4666-9861-1.ch011

Chakraborty, S., Chatterjee, S., Ashour, A. S., Mali, K., & Dey, N. (2018). Intelligent Computing in Medical Imaging: A Study. In N. Dey (Ed.), *Advancements in Applied Metaheuristic Computing* (pp. 143–163). Hershey, PA: IGI Global. doi:10.4018/978-1-5225-4151-6.ch006

Chavez, A., & Kovarik, C. (2017). Open Source Technology for Medical Practice in Developing Countries. In K. Moahi, K. Bwalya, & P. Sebina (Eds.), *Health Information Systems and the Advancement of Medical Practice in Developing Countries* (pp. 33–59). Hershey, PA: IGI Global. doi:10.4018/978-1-5225-2262-1.ch003

Chen, E. T. (2016). Examining the Influence of Information Technology on Modern Health Care. In P. Manolitzas, E. Grigoroudis, N. Matsatsinis, & D. Yannacopoulos (Eds.), *Effective Methods for Modern Healthcare Service Quality and Evaluation* (pp. 110–136). Hershey, PA: IGI Global. doi:10.4018/978-1-4666-9961-8.ch006

Chin, S. (2016). Surviving Sandy: Recovering Collections after a Natural Disaster. In E. Decker & J. Townes (Eds.), *Handbook of Research on Disaster Management and Contingency Planning in Modern Libraries* (pp. 366–388). Hershey, PA: IGI Global. doi:10.4018/978-1-4666-8624-3.ch016

Chojnacki, M., & Wójcik, A. (2016). Security and the Role of New Technologies and Innovation in Medical Ethics. In A. Rosiek & K. Leksowski (Eds.), *Organizational Culture and Ethics in Modern Medicine* (pp. 52–77). Hershey, PA: IGI Global. doi:10.4018/978-1-4666-9658-7.ch003

Cılız, N., Yıldırım, H., & Temizel, Ş. (2016). Structure Development for Effective Medical Waste and Hazardous Waste Management System. In U. Akkucuk (Ed.), *Handbook of Research on Waste Management Techniques for Sustainability* (pp. 303–327). Hershey, PA: IGI Global. doi:10.4018/978-1-4666-9723-2.ch016

Ciufudean, C. (2018). Innovative Formalism for Biological Data Analysis. In M. Khosrow-Pour, D.B.A. (Ed.), Encyclopedia of Information Science and Technology, Fourth Edition (pp. 1814-1824). Hershey, PA: IGI Global. doi:10.4018/978-1-5225-2255-3.ch158

Ciufudean, C., & Ciufudean, O. (2016). Tele-Monitoring for Medical Diagnosis Availability. In M. Cruz-Cunha, I. Miranda, R. Martinho, & R. Rijo (Eds.), *Encyclopedia of E-Health and Telemedicine* (pp. 401–411). Hershey, PA: IGI Global. doi:10.4018/978-1-4666-9978-6.ch032

Colaguori, R., & Danesi, M. (2017). Medical Semiotics: A Revisitation and an Exhortation. *International Journal of Semiotics and Visual Rhetoric*, *1*(1), 11–18. doi:10.4018/IJSVR.2017010102

Cole, A. W., & Salek, T. A. (2017). Adopting a Parasocial Connection to Overcome Professional Kakoethos in Online Health Information. In M. Folk & S. Apostel (Eds.), *Establishing and Evaluating Digital Ethos and Online Credibility* (pp. 104–120). Hershey, PA: IGI Global. doi:10.4018/978-1-5225-1072-7.ch006

Contreras, E. C., & Puente, G. J. (2016). How to Identify Rheumatic Diseases by General Physicians. In T. Gasmelseid (Ed.), *Advancing Pharmaceutical Processes and Tools for Improved Health Outcomes* (pp. 136–166). Hershey, PA: IGI Global. doi:10.4018/978-1-5225-0248-7.ch006

D'Andrea, A., Ferri, F., & Grifoni, P. (2016). RFID Technologies in Healthcare Setting: Applications and Future Perspectives. *International Journal of Computers in Clinical Practice*, *1*(1), 15–27. doi:10.4018/IJCCP.2016010102

Damianakis, A., Kallonis, P., Loudos, G., Tsatsos, D., & Tsoukalis, A. (2016). Exploiting 3D Medical Equipment Simulations to Support Biomedical Engineering Academic Courses: Design Methodology and Implementation in a Small Scale National Project. In B. Khan (Ed.), *Revolutionizing Modern Education through Meaningful E-Learning Implementation* (pp. 277–295). Hershey, PA: IGI Global. doi:10.4018/978-1-5225-0466-5.ch015

Demiroz, E. (2016). Principles of Instructional Design for E-Learning and Online Learning Practices: Implications for Medical Education. In V. Wang (Ed.), *Handbook of Research on Advancing Health Education through Technology* (pp. 419–451). Hershey, PA: IGI Global. doi:10.4018/978-1-4666-9494-1.ch018

Deperlioglu, O. (2018). Intelligent Techniques Inspired by Nature and Used in Biomedical Engineering. In U. Kose, G. Guraksin, & O. Deperlioglu (Eds.), *Nature-Inspired Intelligent Techniques for Solving Biomedical Engineering Problems* (pp. 51–77). Hershey, PA: IGI Global. doi:10.4018/978-1-5225-4769-3.ch003

Dey, N., & Ashour, A. S. (2018). Meta-Heuristic Algorithms in Medical Image Segmentation: A Review. In N. Dey (Ed.), *Advancements in Applied Metaheuristic Computing* (pp. 185–203). Hershey, PA: IGI Global. doi:10.4018/978-1-5225-4151-6.ch008

Dey, N., Ashour, A. S., & Althoupety, A. S. (2017). Thermal Imaging in Medical Science. In V. Santhi (Ed.), *Recent Advances in Applied Thermal Imaging for Industrial Applications* (pp. 87–117). Hershey, PA: IGI Global. doi:10.4018/978-1-5225-2423-6.ch004

Di Virgilio, F., Camillo, A. A., & Camillo, I. C. (2017). The Impact of Social Network on Italian Users' Behavioural Intention for the Choice of a Medical Tourist Destination. *International Journal of Tourism and Hospitality Management in the Digital Age*, *1*(1), 36–49. doi:10.4018/IJTHMDA.2017010103

Dias, C. M., Ribeiro, A. G., & Furtado, S. F. (2016). An Overview about the Use of Healthcare Applications on Mobile Devices. In M. Cruz-Cunha, I. Miranda, R. Martinho, & R. Rijo (Eds.), *Encyclopedia of E-Health and Telemedicine* (pp. 285–298). Hershey, PA: IGI Global. doi:10.4018/978-1-4666-9978-6.ch024

Dogra, A. K., & Dogra, P. (2017). The Medical Tourism Industry in the BRIC Nations: An Indian Analysis. In M. Dhiman (Ed.), *Opportunities and Challenges for Tourism and Hospitality in the BRIC Nations* (pp. 320–336). Hershey, PA: IGI Global. doi:10.4018/978-1-5225-0708-6.ch020

Drowos, J. L., & Wood, S. K. (2017). Preparing Future Physicians to Adapt to the Changing Health Care System: Promoting Humanism through Curricular Design. In V. Bryan & J. Bird (Eds.), *Healthcare Community Synergism between Patients, Practitioners, and Researchers* (pp. 106–125). Hershey, PA: IGI Global. doi:10.4018/978-1-5225-0640-9.ch006

Dulam, K. (2017). Medical Patents and Impact on Availability and Affordability of Essential Medicines in India. In R. Aggarwal & R. Kaur (Eds.), *Patent Law and Intellectual Property in the Medical Field* (pp. 41–57). Hershey, PA: IGI Global. doi:10.4018/978-1-5225-2414-4.ch003

Dutta, P. (2017). Decision Making in Medical Diagnosis via Distance Measures on Interval Valued Fuzzy Sets. *International Journal of System Dynamics Applications*, *6*(4), 63–83. doi:10.4018/IJSDA.2017100104

Edoh, T. O., Pawar, P. A., Brügge, B., & Teege, G. (2016). A Multidisciplinary Remote Healthcare Delivery System to Increase Health Care Access, Pathology Screening, and Treatment in Developing Countries: The Case of Benin. *International Journal of Healthcare Information Systems and Informatics*, *11*(4), 1–31. doi:10.4018/IJHISI.2016100101

El Guemhioui, K., & Demurjian, S. A. (2017). Semantic Reconciliation of Electronic Health Records Using Semantic Web Technologies. *International Journal of Information Technology and Web Engineering*, *12*(2), 26–48. doi:10.4018/IJITWE.2017040102

Ellouze, N., Rekhis, S., & Boudriga, N. (2016). Forensic Investigation of Digital Crimes in Healthcare Applications. In O. Isafiade & A. Bagula (Eds.), *Data Mining Trends and Applications in Criminal Science and Investigations* (pp. 169–210). Hershey, PA: IGI Global. doi:10.4018/978-1-5225-0463-4.ch007

Eneanya, A. N. (2016). Health Policy Implementation and Its Barriers: The Case Study of US Health System. In R. Gholipour & K. Rouzbehani (Eds.), *Social, Economic, and Political Perspectives on Public Health Policy-Making* (pp. 42–63). Hershey, PA: IGI Global. doi:10.4018/978-1-4666-9944-1.ch003

Entico, G. J. (2016). Knowledge Management and the Medical Health Librarians: A Perception Study. In J. Yap, M. Perez, M. Ayson, & G. Entico (Eds.), *Special Library Administration, Standardization and Technological Integration* (pp. 52–77). Hershey, PA: IGI Global. doi:10.4018/978-1-4666-9542-9.ch003

Ferradji, M. A., & Zidani, A. (2016). Collaborative Environment for Remote Clinical Reasoning Learning. *International Journal of E-Health and Medical Communications*, *7*(4), 62–81. doi:10.4018/IJEHMC.2016100104

Fisher, J. (2018). Sociological Perspectives on Improving Medical Diagnosis Emphasizing CAD. In M. Khosrow-Pour, D.B.A. (Ed.), Encyclopedia of Information Science and Technology, Fourth Edition (pp. 1017-1024). Hershey, PA: IGI Global. doi:10.4018/978-1-5225-2255-3.ch088

Flores, C. D., Respício, A., Coelho, H., Bez, M. R., & Fonseca, J. M. (2016). Simulation for Medical Training. In M. Cruz-Cunha, I. Miranda, R. Martinho, & R. Rijo (Eds.), *Encyclopedia of E-Health and Telemedicine* (pp. 827–842). Hershey, PA: IGI Global. doi:10.4018/978-1-4666-9978-6.ch064

Frank, E. M. (2018). Healthcare Education: Integrating Simulation Technologies. In V. Bryan, A. Musgrove, & J. Powers (Eds.), *Handbook of Research on Human Development in the Digital Age* (pp. 163–182). Hershey, PA: IGI Global. doi:10.4018/978-1-5225-2838-8.ch008

Garner, G. (2018). Foundations for Yoga Practice in Rehabilitation. In S. Telles & N. Singh (Eds.), *Research-Based Perspectives on the Psychophysiology of Yoga* (pp. 263–307). Hershey, PA: IGI Global. doi:10.4018/978-1-5225-2788-6.ch015

Gasmelseid, T. M. (2016). Improving Pharmaceutical Care through the Use of Intelligent Pharmacoinformatics. In T. Gasmelseid (Ed.), *Advancing Pharmaceutical Processes and Tools for Improved Health Outcomes* (pp. 167–188). Hershey, PA: IGI Global. doi:10.4018/978-1-5225-0248-7.ch007

Gavurová, B., Kováč, V., & Šoltés, M. (2018). Medical Equipment and Economic Determinants of Its Structure and Regulation in the Slovak Republic. In M. Khosrow-Pour, D.B.A. (Ed.), Encyclopedia of Information Science and Technology, Fourth Edition (pp. 5841-5852). Hershey, PA: IGI Global. doi:10.4018/978-1-5225-2255-3.ch508

Ge, X., Wang, Q., Huang, K., Law, V., & Thomas, D. C. (2017). Designing Simulated Learning Environments and Facilitating Authentic Learning Experiences in Medical Education. In J. Stefaniak (Ed.), *Advancing Medical Education Through Strategic Instructional Design* (pp. 77–100). Hershey, PA: IGI Global. doi:10.4018/978-1-5225-2098-6.ch004

Gewald, H., & Gewald, C. (2018). Inhibitors of Physicians' Use of Mandatory Hospital Information Systems (HIS). *International Journal of Healthcare Information Systems and Informatics*, *13*(1), 29–44. doi:10.4018/IJHISI.2018010103

Ghosh, D., & Dinda, S. (2017). Health Infrastructure and Economic Development in India. In R. Das (Ed.), *Social, Health, and Environmental Infrastructures for Economic Growth* (pp. 99–119). Hershey, PA: IGI Global. doi:10.4018/978-1-5225-2364-2.ch006

Ghosh, K., & Sen, K. C. (2017). The Potential of Crowdsourcing in the Health Care Industry. In N. Wickramasinghe (Ed.), *Handbook of Research on Healthcare Administration and Management* (pp. 418–427). Hershey, PA: IGI Global. doi:10.4018/978-1-5225-0920-2.ch024

Ghrab, S., & Saad, I. (2016). Identifying Crucial Know-How and Knowing-That for Medical Decision Support. *International Journal of Decision Support System Technology*, *8*(4), 14–33. doi:10.4018/IJDSST.2016100102

Giokas, K., Tsirmpas, C., Anastasiou, A., Iliopoulou, D., Costarides, V., & Koutsouris, D. (2016). Contemporary Heart Failure Treatment Based on Improved Knowledge and Personalized Care of Co-morbidities. In D. Fotiadis (Ed.), *Handbook of Research on Trends in the Diagnosis and Treatment of Chronic Conditions* (pp. 301–314). Hershey, PA: IGI Global. doi:10.4018/978-1-4666-8828-5.ch014

Gopalan, V., Chan, E., & Ho, D. T. (2018). Deliberate Self-Harm and Suicide Ideology in Medical Students. In C. Smith (Ed.), *Exploring the Pressures of Medical Education From a Mental Health and Wellness Perspective* (pp. 122–143). Hershey, PA: IGI Global. doi:10.4018/978-1-5225-2811-1.ch005

Goswami, S., Dey, U., Roy, P., Ashour, A., & Dey, N. (2017). Medical Video Processing: Concept and Applications. In N. Dey, A. Ashour, & P. Patra (Eds.), *Feature Detectors and Motion Detection in Video Processing* (pp. 1–17). Hershey, PA: IGI Global. doi:10.4018/978-1-5225-1025-3.ch001

Goswami, S., Mahanta, K., Goswami, S., Jigdung, T., & Devi, T. P. (2018). Ageing and Cancer: The Epigenetic Basis, Alternative Treatment, and Care. In B. Prasad & S. Akbar (Eds.), *Handbook of Research on Geriatric Health, Treatment, and Care* (pp. 206–235). Hershey, PA: IGI Global. doi:10.4018/978-1-5225-3480-8.ch012

Gouva, M. I. (2017). The Psychological Impact of Medical Error on Patients, Family Members, and Health Professionals. In M. Riga (Ed.), *Impact of Medical Errors and Malpractice on Health Economics, Quality, and Patient Safety* (pp. 171–196). Hershey, PA: IGI Global. doi:10.4018/978-1-5225-2337-6.ch007

Gürcü, M., & Tengilimoğlu, D. (2017). Health Tourism-Based Destination Marketing. In A. Bayraktar & C. Uslay (Eds.), *Strategic Place Branding Methodologies and Theory for Tourist Attraction* (pp. 308–331). Hershey, PA: IGI Global. doi:10.4018/978-1-5225-0579-2.ch015

Gürsel, G. (2016). Mobility in Healthcare: M-Health. In A. Panagopoulos (Ed.), *Handbook of Research on Next Generation Mobile Communication Systems* (pp. 485–511). Hershey, PA: IGI Global. doi:10.4018/978-1-4666-8732-5.ch019

Gürsel, G. (2017). For Better Healthcare Mining Health Data. In S. Bhattacharyya, S. De, I. Pan, & P. Dutta (Eds.), *Intelligent Multidimensional Data Clustering and Analysis* (pp. 135–158). Hershey, PA: IGI Global. doi:10.4018/978-1-5225-1776-4.ch006

Guzman-Lugo, G., Lopez-Martinez, M., Lopez-Ramirez, B., & Avila-Vazquez, D. (2016). City-Driven Approach to Determine Health Services Based on Current User Location and Collaborative Information. *International Journal of Knowledge Society Research*, *7*(1), 53–62. doi:10.4018/IJKSR.2016010104

Gyaase, P. O., Darko-Lartey, R., William, H., & Borkloe, F. (2017). Towards an Integrated Electronic Medical Records System for Quality Healthcare in Ghana: An Exploratory Factor Analysis. *International Journal of Computers in Clinical Practice*, *2*(2), 38–55. doi:10.4018/IJCCP.2017070103

Handayani, P. W., Sandhyaduhita, P. I., Hidayanto, A. N., Pinem, A. A., Fajrina, H. R., Junus, K. M., ... Ayuningtyas, D. (2016). Integrated Hospital Information System Architecture Design in Indonesia. In T. Iyamu & A. Tatnall (Eds.), *Maximizing Healthcare Delivery and Management through Technology Integration* (pp. 207–236). Hershey, PA: IGI Global. doi:10.4018/978-1-4666-9446-0.ch013

Hanel, P. (2017). Is China Catching Up?: Health-Related Applications of Biotechnology. In T. Bas & J. Zhao (Eds.), *Comparative Approaches to Biotechnology Development and Use in Developed and Emerging Nations* (pp. 465–520). Hershey, PA: IGI Global. doi:10.4018/978-1-5225-1040-6.ch016

Harmsen, C. A., & Royle, R. N. (2016). St. Stephen's Hospital Hervey Bay: Study of Developing a Digital Hospital. In J. Moon & M. Galea (Eds.), *Improving Health Management through Clinical Decision Support Systems* (pp. 127–153). Hershey, PA: IGI Global. doi:10.4018/978-1-4666-9432-3.ch006

Harp, D., Shim, R. S., Johnson, J., Harp, J. A., Wilcox, W. C., & Wilcox, J. K. (2016). Race and Gender Inequalities in Medicine and Biomedical Research. In U. Thomas & J. Drake (Eds.), *Critical Research on Sexism and Racism in STEM Fields* (pp. 115–134). Hershey, PA: IGI Global. doi:10.4018/978-1-5225-0174-9.ch006

Hartman, A., & Brown, S. (2017). Synergism through Therapeutic Visual Arts. In V. Bryan & J. Bird (Eds.), *Healthcare Community Synergism between Patients, Practitioners, and Researchers* (pp. 29–48). Hershey, PA: IGI Global. doi:10.4018/978-1-5225-0640-9.ch002

Hazra, J., Chowdhury, A. R., & Dutta, P. (2016). Cluster Based Medical Image Registration Using Optimized Neural Network. In S. Bhattacharyya, P. Banerjee, D. Majumdar, & P. Dutta (Eds.), *Handbook of Research on Advanced Hybrid Intelligent Techniques and Applications* (pp. 551–581). Hershey, PA: IGI Global. doi:10.4018/978-1-4666-9474-3.ch018

Hung, S., Huang, W., Yen, D. C., Chang, S., & Lu, C. (2016). Effect of Information Service Competence and Contextual Factors on the Effectiveness of Strategic Information Systems Planning in Hospitals. *Journal of Global Information Management, 24*(1), 14–36. doi:10.4018/JGIM.2016010102

Iltchev, P., Śliwczyński, A., Szynkiewicz, P., & Marczak, M. (2016). Mobile Health Applications Assisting Patients with Chronic Diseases: Examples from Asthma Care. In A. Moumtzoglou (Ed.), *M-Health Innovations for Patient-Centered Care* (pp. 170–196). Hershey, PA: IGI Global. doi:10.4018/978-1-4666-9861-1.ch009

Iqbal, S., Ahmad, S., & Willis, I. (2017). Influencing Factors for Adopting Technology Enhanced Learning in the Medical Schools of Punjab, Pakistan. *International Journal of Information and Communication Technology Education, 13*(3), 27–39. doi:10.4018/IJICTE.2017070103

Jagan, J., Dalkiliç, Y., & Samui, P. (2016). Utilization of SVM, LSSVM and GP for Predicting the Medical Waste Generation. In G. Hua (Ed.), *Smart Cities as a Solution for Reducing Urban Waste and Pollution* (pp. 224–251). Hershey, PA: IGI Global. doi:10.4018/978-1-5225-0302-6.ch008

Jagiello, K., Sosnowska, A., Mikolajczyk, A., & Puzyn, T. (2017). Nanomaterials in Medical Devices: Regulations' Review and Future Perspectives. *Journal of Nanotoxicology and Nanomedicine*, *2*(2), 1–11. doi:10.4018/JNN.2017070101

Jena, T. K. (2018). Skill Training Process in Medicine Through Distance Mode. In U. Pandey & V. Indrakanti (Eds.), *Optimizing Open and Distance Learning in Higher Education Institutions* (pp. 228–243). Hershey, PA: IGI Global. doi:10.4018/978-1-5225-2624-7.ch010

Jesus, Â., & Gomes, M. J. (2016). Web 2.0 Tools in Biomedical and Pharmaceutical Education: Updated Review and Commentary. In T. Gasmelseid (Ed.), *Advancing Pharmaceutical Processes and Tools for Improved Health Outcomes* (pp. 52–78). Hershey, PA: IGI Global. doi:10.4018/978-1-5225-0248-7.ch003

Joseph, V., & Miller, J. M. (2018). Medical Students' Perceived Stigma in Seeking Care: A Cultural Perspective. In C. Smith (Ed.), *Exploring the Pressures of Medical Education From a Mental Health and Wellness Perspective* (pp. 44–67). Hershey, PA: IGI Global. doi:10.4018/978-1-5225-2811-1.ch002

Joyce, B. L., & Swanberg, S. M. (2017). Using Backward Design for Competency-Based Undergraduate Medical Education. In J. Stefaniak (Ed.), *Advancing Medical Education Through Strategic Instructional Design* (pp. 53–76). Hershey, PA: IGI Global. doi:10.4018/978-1-5225-2098-6.ch003

Kaljo, K., & Jacques, L. (2018). Flipping the Medical School Classroom. In M. Khosrow-Pour, D.B.A. (Ed.), Encyclopedia of Information Science and Technology, Fourth Edition (pp. 5800-5809). Hershey, PA: IGI Global. doi:10.4018/978-1-5225-2255-3.ch504

Karon, R. (2016). Utilisation of Health Information Systems for Service Delivery in the Namibian Environment. In T. Iyamu & A. Tatnall (Eds.), *Maximizing Healthcare Delivery and Management through Technology Integration* (pp. 169–183). Hershey, PA: IGI Global. doi:10.4018/978-1-4666-9446-0.ch011

Kasina, H., Bahubalendruni, M. V., & Botcha, R. (2017). Robots in Medicine: Past, Present and Future. *International Journal of Manufacturing, Materials, and Mechanical Engineering*, *7*(4), 44–64. doi:10.4018/IJMMME.2017100104

Katehakis, D. G. (2018). Electronic Medical Record Implementation Challenges for the National Health System in Greece. *International Journal of Reliable and Quality E-Healthcare*, *7*(1), 16–30. doi:10.4018/IJRQEH.2018010102

Katz, A., & Shtub, A. (2016). Design and Build a Wizard of Oz (WOZ) Telemedicine Simulator Platform. In M. Cruz-Cunha, I. Miranda, R. Martinho, & R. Rijo (Eds.), *Encyclopedia of E-Health and Telemedicine* (pp. 128–141). Hershey, PA: IGI Global. doi:10.4018/978-1-4666-9978-6.ch011

Kaur, P. D., & Sharma, P. (2017). Success Dimensions of ICTs in Healthcare. In B. Singh (Ed.), *Computational Tools and Techniques for Biomedical Signal Processing* (pp. 149–173). Hershey, PA: IGI Global. doi:10.4018/978-1-5225-0660-7.ch007

Kaushik, P. (2018). Comorbidity of Medical and Psychiatric Disorders in Geriatric Population: Treatment and Care. In B. Prasad & S. Akbar (Eds.), *Handbook of Research on Geriatric Health, Treatment, and Care* (pp. 448–474). Hershey, PA: IGI Global. doi:10.4018/978-1-5225-3480-8.ch025

Khachane, M. Y. (2017). Organ-Based Medical Image Classification Using Support Vector Machine. *International Journal of Synthetic Emotions*, *8*(1), 18–30. doi:10.4018/IJSE.2017010102

Kirci, P. (2018). Intelligent Techniques for Analysis of Big Data About Healthcare and Medical Records. In N. Shah & M. Mittal (Eds.), *Handbook of Research on Promoting Business Process Improvement Through Inventory Control Techniques* (pp. 559–582). Hershey, PA: IGI Global. doi:10.4018/978-1-5225-3232-3.ch029

Kldiashvili, E. (2016). Cloud Computing as the Useful Resource for Application of the Medical Information System for Quality Assurance Purposes. *International Journal of Computers in Clinical Practice*, *1*(2), 1–23. doi:10.4018/IJCCP.2016070101

Kldiashvili, E. (2018). Cloud Approach for the Medical Information System: MIS on Cloud. In M. Khosrow-Pour (Ed.), *Incorporating Nature-Inspired Paradigms in Computational Applications* (pp. 238–261). Hershey, PA: IGI Global. doi:10.4018/978-1-5225-5020-4.ch008

Kldiashvili, E., Burduli, A., Ghortlishvili, G., & Sheklashvili, I. (2016). Georgian Experience in Telecytology. In M. Cruz-Cunha, I. Miranda, R. Martinho, & R. Rijo (Eds.), *Encyclopedia of E-Health and Telemedicine* (pp. 62–71). Hershey, PA: IGI Global. doi:10.4018/978-1-4666-9978-6.ch006

Ko, H., Mesicek, L., Choi, J., Choi, J., & Hwang, S. (2018). A Study on Secure Contents Strategies for Applications With DRM on Cloud Computing. *International Journal of Cloud Applications and Computing*, *8*(1), 143–153. doi:10.4018/IJCAC.2018010107

Komendziński, T., Dreszer-Drogorób, J., Mikołajewska, E., Mikołajewski, D., & Bałaj, B. (2016). Interdisciplinary Education for Research and Everyday Clinical Practice: Lessons Learned from InteRDoCTor Project. In A. Rosiek & K. Leksowski (Eds.), *Organizational Culture and Ethics in Modern Medicine* (pp. 78–110). Hershey, PA: IGI Global. doi:10.4018/978-1-4666-9658-7.ch004

Komendziński, T., Mikołajewska, E., & Mikołajewski, D. (2018). Cross-Cultural Decision-Making in Healthcare: Theory and Practical Application in Real Clinical Conditions. In A. Rosiek-Kryszewska & K. Leksowski (Eds.), *Healthcare Administration for Patient Safety and Engagement* (pp. 276–298). Hershey, PA: IGI Global. doi:10.4018/978-1-5225-3946-9.ch015

Konecny, L. T. (2018). Medical School Wellness Initiatives. In C. Smith (Ed.), *Exploring the Pressures of Medical Education From a Mental Health and Wellness Perspective* (pp. 209–228). Hershey, PA: IGI Global. doi:10.4018/978-1-5225-2811-1.ch009

Konieczna, A., & Słomkowski, P. (2016). Confrontation of Human Rights in Daily Clinical Situations. In A. Rosiek & K. Leksowski (Eds.), *Organizational Culture and Ethics in Modern Medicine* (pp. 255–281). Hershey, PA: IGI Global. doi:10.4018/978-1-4666-9658-7.ch011

Kromrei, H., Solomonson, W. L., & Juzych, M. S. (2017). Teaching Residents How to Teach. In J. Stefaniak (Ed.), *Advancing Medical Education Through Strategic Instructional Design* (pp. 164–185). Hershey, PA: IGI Global. doi:10.4018/978-1-5225-2098-6.ch008

Kulkarni, S., Savyanavar, A., Kulkarni, P., Stranieri, A., & Ghorpade, V. (2018). Framework for Integration of Medical Image and Text-Based Report Retrieval to Support Radiological Diagnosis. In M. Kolekar & V. Kumar (Eds.), *Biomedical Signal and Image Processing in Patient Care* (pp. 86–122). Hershey, PA: IGI Global. doi:10.4018/978-1-5225-2829-6.ch006

Kumar, A., & Sarkar, B. K. (2018). Performance Analysis of Nature-Inspired Algorithms-Based Bayesian Prediction Models for Medical Data Sets. In U. Singh, A. Tiwari, & R. Singh (Eds.), *Soft-Computing-Based Nonlinear Control Systems Design* (pp. 134–155). Hershey, PA: IGI Global. doi:10.4018/978-1-5225-3531-7.ch007

Kurtz, R. S. (2016). Fortitude: A Study of African Americans in Surgery in New York City. In U. Thomas & J. Drake (Eds.), *Critical Research on Sexism and Racism in STEM Fields* (pp. 153–169). Hershey, PA: IGI Global. doi:10.4018/978-1-5225-0174-9.ch009

Labbadi, W., & Akaichi, J. (2017). Efficient Algorithm for Answering Fuzzy Medical Requests in Pervasive Healthcare Information Systems. *International Journal of Healthcare Information Systems and Informatics, 12*(2), 46–64. doi:10.4018/IJHISI.2017040103

Lagumdzija, A., & Swing, V. K. (2017). Health, Digitalization, and Individual Empowerment. In F. Topor (Ed.), *Handbook of Research on Individualism and Identity in the Globalized Digital Age* (pp. 380–402). Hershey, PA: IGI Global. doi:10.4018/978-1-5225-0522-8.ch017

Lamey, T. W., & Davidson-Shivers, G. V. (2017). Instructional Strategies and Sequencing. In J. Stefaniak (Ed.), *Advancing Medical Education Through Strategic Instructional Design* (pp. 30–52). Hershey, PA: IGI Global. doi:10.4018/978-1-5225-2098-6.ch002

Leon, G. (2017). The Role of Forensic Medicine in Medical Errors. In M. Riga (Ed.), *Impact of Medical Errors and Malpractice on Health Economics, Quality, and Patient Safety* (pp. 144–170). Hershey, PA: IGI Global. doi:10.4018/978-1-5225-2337-6.ch006

Lidia, B., Federica, G., & Maria, V. E. (2016). The Patient-Centered Medicine as the Theoretical Framework for Patient Engagement. In G. Graffigna (Ed.), *Promoting Patient Engagement and Participation for Effective Healthcare Reform* (pp. 25–39). Hershey, PA: IGI Global. doi:10.4018/978-1-4666-9992-2.ch002

Love, L., & McDowelle, D. (2018). Developing a Comprehensive Wellness Program for Medical Students. In C. Smith (Ed.), *Exploring the Pressures of Medical Education From a Mental Health and Wellness Perspective* (pp. 190–208). Hershey, PA: IGI Global. doi:10.4018/978-1-5225-2811-1.ch008

Lovell, K. L. (2017). Development and Evaluation of Neuroscience Computer-Based Modules for Medical Students: Instructional Design Principles and Effectiveness. In J. Stefaniak (Ed.), *Advancing Medical Education Through Strategic Instructional Design* (pp. 262–276). Hershey, PA: IGI Global. doi:10.4018/978-1-5225-2098-6.ch013

Lubin, R., & Hamlin, M. D. (2018). Medical Student Burnout: A Social Cognitive Learning Perspective on Medical Student Mental Health and Wellness. In C. Smith (Ed.), *Exploring the Pressures of Medical Education From a Mental Health and Wellness Perspective* (pp. 92–121). Hershey, PA: IGI Global. doi:10.4018/978-1-5225-2811-1.ch004

Luk, C. Y. (2018). Moving Towards Universal Health Coverage: Challenges for the Present and Future in China. In B. Fong, A. Ng, & P. Yuen (Eds.), *Sustainable Health and Long-Term Care Solutions for an Aging Population* (pp. 19–45). Hershey, PA: IGI Global. doi:10.4018/978-1-5225-2633-9.ch002

Mahat, M., & Pettigrew, A. (2017). The Regulatory Environment of Non-Profit Higher Education and Research Institutions and Its Implications for Managerial Strategy: The Case of Medical Education and Research. In L. West & A. Worthington (Eds.), *Handbook of Research on Emerging Business Models and Managerial Strategies in the Nonprofit Sector* (pp. 336–351). Hershey, PA: IGI Global. doi:10.4018/978-1-5225-2537-0.ch017

Maitra, I. K., & Bandhyopadhyaay, S. K. (2017). Adaptive Edge Detection Method towards Features Extraction from Diverse Medical Imaging Technologies. In S. Bhattacharyya, S. De, I. Pan, & P. Dutta (Eds.), *Intelligent Multidimensional Data Clustering and Analysis* (pp. 159–192). Hershey, PA: IGI Global. doi:10.4018/978-1-5225-1776-4.ch007

Mangu, V. P. (2017). Mobile Health Care: A Technology View. In C. Bhatt & S. Peddoju (Eds.), *Cloud Computing Systems and Applications in Healthcare* (pp. 1–18). Hershey, PA: IGI Global. doi:10.4018/978-1-5225-1002-4.ch001

Manirabona, A., Fourati, L. C., & Boudjit, S. (2017). Investigation on Healthcare Monitoring Systems: Innovative Services and Applications. *International Journal of E-Health and Medical Communications, 8*(1), 1–18. doi:10.4018/IJEHMC.2017010101

Mannai, M. M., & Karâa, W. B. (2016). Biomedical Image Processing Overview. In W. Karâa & N. Dey (Eds.), *Biomedical Image Analysis and Mining Techniques for Improved Health Outcomes* (pp. 1–12). Hershey, PA: IGI Global. doi:10.4018/978-1-4666-8811-7.ch001

Manzoor, A. (2016). RFID in Health Care-Building Smart Hospitals for Quality Healthcare. *International Journal of User-Driven Healthcare, 6*(2), 21–45. doi:10.4018/IJUDH.2016070102

Marwan, M., Kartit, A., & Ouahmane, H. (2018). A Framework to Secure Medical Image Storage in Cloud Computing Environment. *Journal of Electronic Commerce in Organizations, 16*(1), 1–16. doi:10.4018/JECO.2018010101

Masoud, M. P., Nejad, M. K., Darebaghi, H., Chavoshi, M., & Farahani, M. (2018). The Decision Support System and Conventional Method of Telephone Triage by Nurses in Emergency Medical Services: A Comparative Investigation. *International Journal of E-Business Research, 14*(1), 77–88. doi:10.4018/IJEBR.2018010105

McDonald, W. G., Martin, M., & Salzberg, L. D. (2018). From Medical Student to Medical Resident: Graduate Medical Education and Mental Health in the United States. In C. Smith (Ed.), *Exploring the Pressures of Medical Education From a Mental Health and Wellness Perspective* (pp. 145–169). Hershey, PA: IGI Global. doi:10.4018/978-1-5225-2811-1.ch006

Medhekar, A. (2017). The Role of Social Media for Knowledge Dissemination in Medical Tourism: A Case of India. In R. Chugh (Ed.), *Harnessing Social Media as a Knowledge Management Tool* (pp. 25–54). Hershey, PA: IGI Global. doi:10.4018/978-1-5225-0495-5.ch002

Medhekar, A., & Haq, F. (2018). Urbanization and New Jobs Creation in Healthcare Services in India: Challenges and Opportunities. In U. Benna & I. Benna (Eds.), *Urbanization and Its Impact on Socio-Economic Growth in Developing Regions* (pp. 198–218). Hershey, PA: IGI Global. doi:10.4018/978-1-5225-2659-9.ch010

Mehta, G., Dutta, M. K., & Kim, P. S. (2016). An Efficient and Lossless Cryptosystem for Security in Tele-Ophthalmology Applications Using Chaotic Theory. *International Journal of E-Health and Medical Communications, 7*(4), 28–47. doi:10.4018/IJEHMC.2016100102

Mehta, P. (2017). Framework of Indian Healthcare System and its Challenges: An Insight. In V. Bryan & J. Bird (Eds.), *Healthcare Community Synergism between Patients, Practitioners, and Researchers* (pp. 247–271). Hershey, PA: IGI Global. doi:10.4018/978-1-5225-0640-9.ch011

Metelmann, B., & Metelmann, C. (2016). M-Health in Prehospital Emergency Medicine: Experiences from the EU funded Project LiveCity. In A. Moumtzoglou (Ed.), *M-Health Innovations for Patient-Centered Care* (pp. 197–212). Hershey, PA: IGI Global. doi:10.4018/978-1-4666-9861-1.ch010

Mi, M. (2017). Informal Learning in Medical Education. In J. Stefaniak (Ed.), *Advancing Medical Education Through Strategic Instructional Design* (pp. 225–244). Hershey, PA: IGI Global. doi:10.4018/978-1-5225-2098-6.ch011

Miglioretti, M., Mariani, F., & Vecchio, L. (2016). Could Patient Engagement Promote a Health System Free From Malpractice Litigation Risk? In G. Graffigna (Ed.), *Promoting Patient Engagement and Participation for Effective Healthcare Reform* (pp. 240–264). Hershey, PA: IGI Global. doi:10.4018/978-1-4666-9992-2.ch012

Misra, S. C., & Bisui, S. (2016). Feasibility of Large Scale Implementation of Personalized Medicine in the Current Scenario. *International Journal of E-Health and Medical Communications, 7*(2), 30–49. doi:10.4018/IJEHMC.2016040103

Misra, S. C., Bisui, S., & Fantazy, K. (2016). Identifying Critical Changes in Adoption of Personalized Medicine (PM) in Healthcare Management. *International Journal of E-Health and Medical Communications, 7*(3), 1–15. doi:10.4018/IJEHMC.2016070101

Mohammadian, M., Hatzinakos, D., Spachos, P., & Jentzsh, R. (2016). An Intelligent and Secure Framework for Wireless Information Technology in Healthcare for User and Data Classification in Hospitals. In V. Wang (Ed.), *Handbook of Research on Advancing Health Education through Technology* (pp. 452–479). Hershey, PA: IGI Global. doi:10.4018/978-1-4666-9494-1.ch019

Mokeddem, S., & Atmani, B. (2016). Assessment of Clinical Decision Support Systems for Predicting Coronary Heart Disease. *International Journal of Operations Research and Information Systems, 7*(3), 57–73. doi:10.4018/IJORIS.2016070104

Mokrzycka, A., & Kowalska-Bobko, I. (2016). Public Health Legislation and Patient's Rights: Health2020 Strategy, European Perspective. In A. Rosiek & K. Leksowski (Eds.), *Organizational Culture and Ethics in Modern Medicine* (pp. 298–322). Hershey, PA: IGI Global. doi:10.4018/978-1-4666-9658-7.ch013

Moon, J. D., & Galea, M. P. (2016). Overview of Clinical Decision Support Systems in Healthcare. In J. Moon & M. Galea (Eds.), *Improving Health Management through Clinical Decision Support Systems* (pp. 1–27). Hershey, PA: IGI Global. doi:10.4018/978-1-4666-9432-3.ch001

Mosadeghrad, A. M., & Woldemichael, A. (2017). Application of Quality Management in Promoting Patient Safety and Preventing Medical Errors. In M. Riga (Ed.), *Impact of Medical Errors and Malpractice on Health Economics, Quality, and Patient Safety* (pp. 91–112). Hershey, PA: IGI Global. doi:10.4018/978-1-5225-2337-6.ch004

Moumtzoglou, A. (2017). Digital Medicine: The Quality Standpoint. In A. Moumtzoglou (Ed.), *Design, Development, and Integration of Reliable Electronic Healthcare Platforms* (pp. 179–195). Hershey, PA: IGI Global. doi:10.4018/978-1-5225-1724-5.ch011

Moumtzoglou, A., & Pouliakis, A. (2018). Population Health Management and the Science of Individuality. *International Journal of Reliable and Quality E-Healthcare, 7*(2), 1–26. doi:10.4018/IJRQEH.2018040101

Mpofu, C. (2016). International Medical Experiences Outbound New Zealand: An Economic and Medical Workforce Strategy. In D. Velliaris & D. Coleman-George (Eds.), *Handbook of Research on Study Abroad Programs and Outbound Mobility* (pp. 446–469). Hershey, PA: IGI Global. doi:10.4018/978-1-5225-0169-5.ch018

Mukhtar, W. F., & Abuelyaman, E. S. (2017). Opportunities and Challenges of Big Data in Healthcare. In N. Wickramasinghe (Ed.), *Handbook of Research on Healthcare Administration and Management* (pp. 47–58). Hershey, PA: IGI Global. doi:10.4018/978-1-5225-0920-2.ch004

Munugala, S., Brar, G. K., Syed, A., Mohammad, A., & Halgamuge, M. N. (2018). The Much Needed Security and Data Reforms of Cloud Computing in Medical Data Storage. In M. Lytras & P. Papadopoulou (Eds.), *Applying Big Data Analytics in Bioinformatics and Medicine* (pp. 99–113). Hershey, PA: IGI Global. doi:10.4018/978-1-5225-2607-0.ch005

Narasimhamurthy, A. (2016). An Overview of Machine Learning in Medical Image Analysis: Trends in Health Informatics. In N. Dey & A. Ashour (Eds.), *Classification and Clustering in Biomedical Signal Processing* (pp. 23–45). Hershey, PA: IGI Global. doi:10.4018/978-1-5225-0140-4.ch002

Ngara, R. (2017). Multiple Voices, Multiple Paths: Towards Dialogue between Western and Indigenous Medical Knowledge Systems. In P. Ngulube (Ed.), *Handbook of Research on Theoretical Perspectives on Indigenous Knowledge Systems in Developing Countries* (pp. 332–358). Hershey, PA: IGI Global. doi:10.4018/978-1-5225-0833-5.ch015

O'Connor, Y., & Heavin, C. (2018). Defining and Characterising the Landscape of eHealth. In M. Khosrow-Pour, D.B.A. (Ed.), Encyclopedia of Information Science and Technology, Fourth Edition (pp. 5864-5875). Hershey, PA: IGI Global. doi:10.4018/978-1-5225-2255-3.ch510

Ochonogor, W. C., & Okite-Amughoro, F. A. (2018). Building an Effective Digital Library in a University Teaching Hospital (UTH) in Nigeria. In A. Tella & T. Kwanya (Eds.), *Handbook of Research on Managing Intellectual Property in Digital Libraries* (pp. 184–204). Hershey, PA: IGI Global. doi:10.4018/978-1-5225-3093-0.ch010

Olaniran, B. A. (2016). ICTs, E-health, and Multidisciplinary Healthcare Teams: Promises and Challenges. *International Journal of Privacy and Health Information Management*, *4*(2), 62–75. doi:10.4018/IJPHIM.2016070105

Olaniran, B. A. (2016). ICT Use and Multidisciplinary Healthcare Teams in the Age of e-Health. *International Journal of Reliable and Quality E-Healthcare*, *5*(1), 18–31. doi:10.4018/IJRQEH.2016010102

Olivares, S. L., Cruz, A. G., Cabrera, M. V., Regalado, A. I., & García, J. E. (2017). An Assessment Study of Quality Model for Medical Schools in Mexico. In S. Mukerji & P. Tripathi (Eds.), *Handbook of Research on Administration, Policy, and Leadership in Higher Education* (pp. 404–439). Hershey, PA: IGI Global. doi:10.4018/978-1-5225-0672-0.ch016

Omidoyin, E. O., Opeke, R. O., & Osagbemi, G. K. (2016). Utilization Pattern and Privacy Issues in the use of Health Records for Research Practice by Doctors: Selected Nigerian Teaching Hospitals as Case Study. *International Journal of Privacy and Health Information Management*, *4*(1), 1–11. doi:10.4018/IJPHIM.2016010101

Omoruyi, E. A., & Omidele, F. (2018). Resident Physician and Medical Academic Faculty Burnout: A Review of Current Literature. In C. Smith (Ed.), *Exploring the Pressures of Medical Education From a Mental Health and Wellness Perspective* (pp. 171–189). Hershey, PA: IGI Global. doi:10.4018/978-1-5225-2811-1.ch007

Osop, H., & Sahama, T. (2016). Data-Driven and Practice-Based Evidence: Design and Development of Efficient and Effective Clinical Decision Support System. In J. Moon & M. Galea (Eds.), *Improving Health Management through Clinical Decision Support Systems* (pp. 295–328). Hershey, PA: IGI Global. doi:10.4018/978-1-4666-9432-3.ch014

Ossowski, R., & Izdebski, P. (2016). Ethical Aspects of Talking to a Patient. In A. Rosiek & K. Leksowski (Eds.), *Organizational Culture and Ethics in Modern Medicine* (pp. 203–235). Hershey, PA: IGI Global. doi:10.4018/978-1-4666-9658-7.ch009

Pandian, P. S. (2016). An Overview of Telemedicine Technologies for Healthcare Applications. *International Journal of Biomedical and Clinical Engineering*, *5*(2), 29–52. doi:10.4018/IJBCE.2016070103

Papadopoulou, P., Lytras, M., & Marouli, C. (2018). Bioinformatics as Applied to Medicine: Challenges Faced Moving from Big Data to Smart Data to Wise Data. In M. Lytras & P. Papadopoulou (Eds.), *Applying Big Data Analytics in Bioinformatics and Medicine* (pp. 1–25). Hershey, PA: IGI Global. doi:10.4018/978-1-5225-2607-0.ch001

Paraskou, A., & George, B. P. (2018). An Overview of Reproductive Tourism. In *Legal and Economic Considerations Surrounding Reproductive Tourism: Emerging Research and Opportunities* (pp. 1–17). Hershey, PA: IGI Global. doi:10.4018/978-1-5225-2694-0.ch001

Park, S., & Moon, J. (2016). Strategic Approach towards Clinical Information Security. In J. Moon & M. Galea (Eds.), *Improving Health Management through Clinical Decision Support Systems* (pp. 329–359). Hershey, PA: IGI Global. doi:10.4018/978-1-4666-9432-3.ch015

Pereira, D., Castro, A., Gomes, P., Areias, J. C., Reis, Z. S., Coimbra, M. T., & Cruz-Correia, R. (2016). Digital Auscultation: Challenges and Perspectives. In M. Cruz-Cunha, I. Miranda, R. Martinho, & R. Rijo (Eds.), *Encyclopedia of E-Health and Telemedicine* (pp. 910–927). Hershey, PA: IGI Global. doi:10.4018/978-1-4666-9978-6.ch070

Peters, R. A., & Cruz, M. (2016). The States as Generators of Incremental Change in American Health Care Policy: 1935 to 1965. In R. Gholipour & K. Rouzbehani (Eds.), *Social, Economic, and Political Perspectives on Public Health Policy-Making* (pp. 86–114). Hershey, PA: IGI Global. doi:10.4018/978-1-4666-9944-1.ch005

Phuritsabam, B., & Devi, A. B. (2017). Information Seeking Behavior of Medical Scientists at Jawaharlal Nehru Institute of Medical Science: A Study. In S. Ram (Ed.), *Library and Information Services for Bioinformatics Education and Research* (pp. 177–187). Hershey, PA: IGI Global. doi:10.4018/978-1-5225-1871-6.ch010

Pieczka, B. (2018). Management of Risk and Adverse Events in Medical Entities. In A. Rosiek-Kryszewska & K. Leksowski (Eds.), *Healthcare Administration for Patient Safety and Engagement* (pp. 31–46). Hershey, PA: IGI Global. doi:10.4018/978-1-5225-3946-9.ch003

Poduval, J. (2016). Curriculum Development. In *Optimizing Medicine Residency Training Programs* (pp. 45–102). Hershey, PA: IGI Global. doi:10.4018/978-1-4666-9527-6.ch003

Poduval, J. (2016). Ethics and Professionalism. In *Optimizing Medicine Residency Training Programs* (pp. 103–133). Hershey, PA: IGI Global. doi:10.4018/978-1-4666-9527-6.ch004

Poduval, J. (2016). Management Skills and Leadership. In *Optimizing Medicine Residency Training Programs* (pp. 182–205). Hershey, PA: IGI Global. doi:10.4018/978-1-4666-9527-6.ch008

Poduval, J. (2016). Medicine Residency Training. In *Optimizing Medicine Residency Training Programs* (pp. 1–27). Hershey, PA: IGI Global. doi:10.4018/978-1-4666-9527-6.ch001

Poduval, J. (2016). Personal Issues. In *Optimizing Medicine Residency Training Programs* (pp. 206–221). Hershey, PA: IGI Global. doi:10.4018/978-1-4666-9527-6.ch009

Poduval, J. (2017). Medical Errors: Impact on Health Care Quality. In M. Riga (Ed.), *Impact of Medical Errors and Malpractice on Health Economics, Quality, and Patient Safety* (pp. 33–60). Hershey, PA: IGI Global. doi:10.4018/978-1-5225-2337-6.ch002

Politis, D., Stagiopoulos, P., Aidona, S., Kyriafinis, G., & Constantinidis, I. (2018). Autonomous Learning and Skill Accreditation: A Paradigm for Medical Studies. In A. Kumar (Ed.), *Optimizing Student Engagement in Online Learning Environments* (pp. 266–296). Hershey, PA: IGI Global. doi:10.4018/978-1-5225-3634-5.ch012

Pomares-Quimbaya, A., Gonzalez, R. A., Quintero, S., Muñoz, O. M., Bohórquez, W. R., García, O. M., & Londoño, D. (2016). A Review of Existing Applications and Techniques for Narrative Text Analysis in Electronic Medical Records. In M. Cruz-Cunha, I. Miranda, R. Martinho, & R. Rijo (Eds.), *Encyclopedia of E-Health and Telemedicine* (pp. 796–811). Hershey, PA: IGI Global. doi:10.4018/978-1-4666-9978-6.ch062

Pomares-Quimbaya, A., González, R. A., Sierra, A., Daza, J. C., Muñoz, O., García, A., ... Bohórquez, W. R. (2017). ICT for Enabling the Quality Evaluation of Health Care Services: A Case Study in a General Hospital. In A. Moumtzoglou (Ed.), *Design, Development, and Integration of Reliable Electronic Healthcare Platforms* (pp. 196–210). Hershey, PA: IGI Global. doi:10.4018/978-1-5225-1724-5.ch012

Pouliakis, A., Margari, N., Karakitsou, E., Archondakis, S., & Karakitsos, P. (2018). Emerging Technologies Serving Cytopathology: Big Data, the Cloud, and Mobile Computing. In I. El Naqa (Ed.), *Emerging Developments and Practices in Oncology* (pp. 114–152). Hershey, PA: IGI Global. doi:10.4018/978-1-5225-3085-5.ch005

Queirós, A., Silva, A. G., Ferreira, A., Caravau, H., Cerqueira, M., & Rocha, N. P. (2016). Assessing Mobile Applications Considered Medical Devices. In M. Cruz-Cunha, I. Miranda, R. Martinho, & R. Rijo (Eds.), *Encyclopedia of E-Health and Telemedicine* (pp. 111–127). Hershey, PA: IGI Global. doi:10.4018/978-1-4666-9978-6.ch010

Raffaeli, L., Spinsante, S., & Gambi, E. (2016). Integrated Smart TV-Based Personal e-Health System. *International Journal of E-Health and Medical Communications*, *7*(1), 48–64. doi:10.4018/IJEHMC.2016010103

Rai, A., Kothari, R., & Singh, D. P. (2017). Assessment of Available Technologies for Hospital Waste Management: A Need for Society. In R. Singh, A. Singh, & V. Srivastava (Eds.), *Environmental Issues Surrounding Human Overpopulation* (pp. 172–188). Hershey, PA: IGI Global. doi:10.4018/978-1-5225-1683-5.ch010

Ramamoorthy, S., & Sivasubramaniam, R. (2018). Image Processing Including Medical Liver Imaging: Medical Image Processing from Big Data Perspective, Ultrasound Liver Images, Challenges. In M. Lytras & P. Papadopoulou (Eds.), *Applying Big Data Analytics in Bioinformatics and Medicine* (pp. 380–392). Hershey, PA: IGI Global. doi:10.4018/978-1-5225-2607-0.ch016

Rathor, G. P., & Gupta, S. K. (2017). Improving Multimodality Image Fusion through Integrate AFL and Wavelet Transform. In V. Tiwari, B. Tiwari, R. Thakur, & S. Gupta (Eds.), *Pattern and Data Analysis in Healthcare Settings* (pp. 143–157). Hershey, PA: IGI Global. doi:10.4018/978-1-5225-0536-5.ch008

Rawat, D. B., & Bhattacharya, S. (2016). Wireless Body Area Network for Healthcare Applications. In N. Meghanathan (Ed.), *Advanced Methods for Complex Network Analysis* (pp. 343–358). Hershey, PA: IGI Global. doi:10.4018/978-1-4666-9964-9.ch014

Rea, P. M. (2016). Advances in Anatomical and Medical Visualisation. In M. Pinheiro & D. Simões (Eds.), *Handbook of Research on Engaging Digital Natives in Higher Education Settings* (pp. 244–264). Hershey, PA: IGI Global. doi:10.4018/978-1-5225-0039-1.ch011

Rexhepi, H., & Persson, A. (2017). Challenges to Implementing IT Support for Evidence Based Practice Among Nurses and Assistant Nurses: A Qualitative Study. *Journal of Electronic Commerce in Organizations*, *15*(2), 61–76. doi:10.4018/JECO.2017040105

Reychav, I., & Azuri, J. (2016). Including Elderly Patients in Decision Making via Electronic Health Literacy. In M. Cruz-Cunha, I. Miranda, R. Martinho, & R. Rijo (Eds.), *Encyclopedia of E-Health and Telemedicine* (pp. 241–249). Hershey, PA: IGI Global. doi:10.4018/978-1-4666-9978-6.ch020

Richards, D., & Caldwell, P. H. (2016). Gamification to Improve Adherence to Clinical Treatment Advice: Improving Adherence to Clinical Treatment. In D. Novák, B. Tulu, & H. Brendryen (Eds.), *Handbook of Research on Holistic Perspectives in Gamification for Clinical Practice* (pp. 47–77). Hershey, PA: IGI Global. doi:10.4018/978-1-4666-9522-1.ch004

Rissman, B. (2016). Medical Conditions Associated with NLD. In B. Rissman (Ed.), *Medical and Educational Perspectives on Nonverbal Learning Disability in Children and Young Adults* (pp. 27–66). Hershey, PA: IGI Global. doi:10.4018/978-1-4666-9539-9.ch002

Rosiek, A. (2018). The Assessment of Actions of the Environment and the Impact of Preventive Medicine for Public Health in Poland. In A. Rosiek-Kryszewska & K. Leksowski (Eds.), *Healthcare Administration for Patient Safety and Engagement* (pp. 106–119). Hershey, PA: IGI Global. doi:10.4018/978-1-5225-3946-9.ch006

Rosiek, A., Leksowski, K., Goch, A., Rosiek-Kryszewska, A., & Leksowski, Ł. (2016). Medical Treatment and Difficult Ethical Decisions in Interdisciplinary Hospital Teams. In A. Rosiek & K. Leksowski (Eds.), *Organizational Culture and Ethics in Modern Medicine* (pp. 121–153). Hershey, PA: IGI Global. doi:10.4018/978-1-4666-9658-7.ch006

Rosiek, A., & Rosiek-Kryszewska, A. (2018). Managed Healthcare: Doctor Life Satisfaction and Its Impact on the Process of Communicating With the Patient. In A. Rosiek-Kryszewska & K. Leksowski (Eds.), *Healthcare Administration for Patient Safety and Engagement* (pp. 244–261). Hershey, PA: IGI Global. doi:10.4018/978-1-5225-3946-9.ch013

Rosiek-Kryszewska, A., Leksowski, Ł., Rosiek, A., Leksowski, K., & Goch, A. (2016). Clinical Communication in the Aspect of Development of New Technologies and E-Health in the Doctor-Patient Relationship. In A. Rosiek & K. Leksowski (Eds.), *Organizational Culture and Ethics in Modern Medicine* (pp. 18–51). Hershey, PA: IGI Global. doi:10.4018/978-1-4666-9658-7.ch002

Rosiek-Kryszewska, A., & Rosiek, A. (2018). The Involvement of the Patient and his Perspective Evaluation of the Quality of Healthcare. In A. Rosiek-Kryszewska & K. Leksowski (Eds.), *Healthcare Administration for Patient Safety and Engagement* (pp. 121–144). Hershey, PA: IGI Global. doi:10.4018/978-1-5225-3946-9.ch007

Rosiek-Kryszewska, A., & Rosiek, A. (2018). The Impact of Management and Leadership Roles in Building Competitive Healthcare Units. In A. Rosiek-Kryszewska & K. Leksowski (Eds.), *Healthcare Administration for Patient Safety and Engagement* (pp. 13–30). Hershey, PA: IGI Global. doi:10.4018/978-1-5225-3946-9.ch002

Roşu, S. M., Păvăloiu, I. B., Dragoi, G., Apostol, C. G., & Munteanu, D. (2016). Telemedicine Based on LMDS in the Urban/Metropolitan Area: A Romanian Case Study. In M. Cruz-Cunha, I. Miranda, R. Martinho, & R. Rijo (Eds.), *Encyclopedia of E-Health and Telemedicine* (pp. 96–109). Hershey, PA: IGI Global. doi:10.4018/978-1-4666-9978-6.ch009

Rouzbehani, K. (2017). Health Policy Implementation: Moving Beyond Its Barriers in United States. In N. Wickramasinghe (Ed.), *Handbook of Research on Healthcare Administration and Management* (pp. 541–552). Hershey, PA: IGI Global. doi:10.4018/978-1-5225-0920-2.ch032

Ruiz, I. M., Cohen, D. S., & Marco, Á. M. (2016). ISO/IEEE11073 Family of Standards: Trends and Applications on E-Health Monitoring. In M. Cruz-Cunha, I. Miranda, R. Martinho, & R. Rijo (Eds.), *Encyclopedia of E-Health and Telemedicine* (pp. 646–660). Hershey, PA: IGI Global. doi:10.4018/978-1-4666-9978-6.ch050

Sam, S. (2017). Mobile Phones and Expanding Human Capabilities in Plural Health Systems. In K. Moahi, K. Bwalya, & P. Sebina (Eds.), *Health Information Systems and the Advancement of Medical Practice in Developing Countries* (pp. 93–114). Hershey, PA: IGI Global. doi:10.4018/978-1-5225-2262-1.ch006

Sankaranarayanan, S., & Ganesan, S. (2016). Applications of Intelligent Agents in Health Sector-A Review. *International Journal of E-Health and Medical Communications*, *7*(1), 1–30. doi:10.4018/IJEHMC.2016010101

Santos-Trigo, M., Suaste, E., & Figuerola, P. (2018). Technology Design and Routes for Tool Appropriation in Medical Practices. In M. Khosrow-Pour, D.B.A. (Ed.), Encyclopedia of Information Science and Technology, Fourth Edition (pp. 3794-3804). Hershey, PA: IGI Global. doi:10.4018/978-1-5225-2255-3.ch329

Sarivougioukas, J., Vagelatos, A., Parsopoulos, K. E., & Lagaris, I. E. (2018). Home UbiHealth. In M. Khosrow-Pour, D.B.A. (Ed.), Encyclopedia of Information Science and Technology, Fourth Edition (pp. 7765-7774). Hershey, PA: IGI Global. doi:10.4018/978-1-5225-2255-3.ch675

Sarkar, B. K. (2017). Big Data and Healthcare Data: A Survey. *International Journal of Knowledge-Based Organizations*, *7*(4), 50–77. doi:10.4018/IJKBO.2017100104

Saxena, K., & Banodha, U. (2017). An Essence of the SOA on Healthcare. In R. Bhadoria, N. Chaudhari, G. Tomar, & S. Singh (Eds.), *Exploring Enterprise Service Bus in the Service-Oriented Architecture Paradigm* (pp. 283–304). Hershey, PA: IGI Global. doi:10.4018/978-1-5225-2157-0.ch018

Schmeida, M., & McNeal, R. (2016). Consulting Online Healthcare Information: E-Caregivers as Knowledgeable Decision Makers. *International Journal of Computers in Clinical Practice*, *1*(1), 42–52. doi:10.4018/IJCCP.2016010104

Sen, K., & Ghosh, K. (2018). Incorporating Global Medical Knowledge to Solve Healthcare Problems: A Framework for a Crowdsourcing System. *International Journal of Healthcare Information Systems and Informatics*, *13*(1), 1–14. doi:10.4018/IJHISI.2018010101

Shakdher, A., & Pandey, K. (2017). REDAlert+: Medical/Fire Emergency and Warning System using Android Devices. *International Journal of E-Health and Medical Communications*, *8*(1), 37–51. doi:10.4018/IJEHMC.2017010103

Shekarian, E., Abdul-Rashid, S. H., & Olugu, E. U. (2017). An Integrated Fuzzy VIKOR Method for Performance Management in Healthcare. In M. Tavana, K. Szabat, & K. Puranam (Eds.), *Organizational Productivity and Performance Measurements Using Predictive Modeling and Analytics* (pp. 40–61). Hershey, PA: IGI Global. doi:10.4018/978-1-5225-0654-6.ch003

Shi, J., Erdem, E., & Liu, H. (2016). Expanding Role of Telephone Systems in Healthcare: Developments and Opportunities. In A. Dwivedi (Ed.), *Reshaping Medical Practice and Care with Health Information Systems* (pp. 87–131). Hershey, PA: IGI Global. doi:10.4018/978-1-4666-9870-3.ch004

Shijina, V., & John, S. J. (2017). Multiple Relations and its Application in Medical Diagnosis. *International Journal of Fuzzy System Applications*, *6*(4), 47–62. doi:10.4018/IJFSA.2017100104

Shipley, N., & Chakraborty, J. (2018). Big Data and mHealth: Increasing the Usability of Healthcare Through the Customization of Pinterest – Literary Perspective. In J. Machado, A. Abelha, M. Santos, & F. Portela (Eds.), *Next-Generation Mobile and Pervasive Healthcare Solutions* (pp. 46–66). Hershey, PA: IGI Global. doi:10.4018/978-1-5225-2851-7.ch004

Singh, A., & Dutta, M. K. (2017). A Reversible Data Hiding Scheme for Efficient Management of Tele-Ophthalmological Data. *International Journal of E-Health and Medical Communications*, *8*(3), 38–54. doi:10.4018/IJEHMC.2017070103

Sivaji, A., Radjo, H. K., Amin, M., & Abu Hashim, M. A. (2016). Design of a Hospital Interactive Wayfinding System: Designing for Malaysian Users. In N. Mohamed, T. Mantoro, M. Ayu, & M. Mahmud (Eds.), *Critical Socio-Technical Issues Surrounding Mobile Computing* (pp. 88–123). Hershey, PA: IGI Global. doi:10.4018/978-1-4666-9438-5.ch005

Skourti, P. K., & Pavlakis, A. (2017). The Second Victim Phenomenon: The Way Out. In M. Riga (Ed.), *Impact of Medical Errors and Malpractice on Health Economics, Quality, and Patient Safety* (pp. 197–222). Hershey, PA: IGI Global. doi:10.4018/978-1-5225-2337-6.ch008

Smith, C. R. (2018). Medical Students' Quest Towards the Long White Coat: Impact on Mental Health and Well-Being. In C. Smith (Ed.), *Exploring the Pressures of Medical Education From a Mental Health and Wellness Perspective* (pp. 1–42). Hershey, PA: IGI Global. doi:10.4018/978-1-5225-2811-1.ch001

Smith, S. I., & Dandignac, M. (2018). Perfectionism: Addressing Lofty Expectations in Medical School. In C. Smith (Ed.), *Exploring the Pressures of Medical Education From a Mental Health and Wellness Perspective* (pp. 68–91). Hershey, PA: IGI Global. doi:10.4018/978-1-5225-2811-1.ch003

Sobrinho, Á. A., Dias da Silva, L., Perkusich, A., Cunha, P., Pinheiro, M. E., & Melo de Medeiros, L. (2016). Towards Medical Systems to Aid the Detection and Treatment of Chronic Diseases. In D. Fotiadis (Ed.), *Handbook of Research on Trends in the Diagnosis and Treatment of Chronic Conditions* (pp. 50–69). Hershey, PA: IGI Global. doi:10.4018/978-1-4666-8828-5.ch003

Soczywko, J., & Rutkowska, D. (2018). The Patient/Provider Relationship in Emergency Medicine: Organization, Communication, and Understanding. In A. Rosiek-Kryszewska & K. Leksowski (Eds.), *Healthcare Administration for Patient Safety and Engagement* (pp. 74–105). Hershey, PA: IGI Global. doi:10.4018/978-1-5225-3946-9.ch005

Sołtysik-Piorunkiewicz, A., Furmankiewicz, M., & Ziuziański, P. (2016). Web Healthcare Applications in Poland: Trends, Standards, Barriers and Possibilities of Implementation and Usage of E-Health Systems. In I. Deliyannis, P. Kostagiolas, & C. Banou (Eds.), *Experimental Multimedia Systems for Interactivity and Strategic Innovation* (pp. 258–283). Hershey, PA: IGI Global. doi:10.4018/978-1-4666-8659-5.ch013

Soni, P. (2018). Implications of HIPAA and Subsequent Regulations on Information Technology. In M. Gupta, R. Sharman, J. Walp, & P. Mulgund (Eds.), *Information Technology Risk Management and Compliance in Modern Organizations* (pp. 71–98). Hershey, PA: IGI Global. doi:10.4018/978-1-5225-2604-9.ch004

Spinelli, R., & Benevolo, C. (2016). From Healthcare Services to E-Health Applications: A Delivery System-Based Taxonomy. In A. Dwivedi (Ed.), *Reshaping Medical Practice and Care with Health Information Systems* (pp. 205–245). Hershey, PA: IGI Global. doi:10.4018/978-1-4666-9870-3.ch007

Srivastava, A., & Aggarwal, A. K. (2018). Medical Image Fusion in Spatial and Transform Domain: A Comparative Analysis. In M. Anwar, A. Khosla, & R. Kapoor (Eds.), *Handbook of Research on Advanced Concepts in Real-Time Image and Video Processing* (pp. 281–300). Hershey, PA: IGI Global. doi:10.4018/978-1-5225-2848-7.ch011

Srivastava, S. K., & Roy, S. N. (2018). Recommendation System: A Potential Tool for Achieving Pervasive Health Care. In J. Machado, A. Abelha, M. Santos, & F. Portela (Eds.), *Next-Generation Mobile and Pervasive Healthcare Solutions* (pp. 111–127). Hershey, PA: IGI Global. doi:10.4018/978-1-5225-2851-7.ch008

Stancu, A. (2016). Correlations and Patterns of Food and Health Consumer Expenditure. In A. Jean-Vasile (Ed.), *Food Science, Production, and Engineering in Contemporary Economies* (pp. 44–101). Hershey, PA: IGI Global. doi:10.4018/978-1-5225-0341-5.ch003

Stanimirovic, D. (2017). Digitalization of Death Certification Model: Transformation Issues and Implementation Concerns. In S. Saeed, Y. Bamarouf, T. Ramayah, & S. Iqbal (Eds.), *Design Solutions for User-Centric Information Systems* (pp. 22–43). Hershey, PA: IGI Global. doi:10.4018/978-1-5225-1944-7.ch002

Stavros, A., Vavoulidis, E., & Nasioutziki, M. (2016). The Use of Mobile Health Applications for Quality Control and Accreditational Purposes in a Cytopathology Laboratory. In A. Moumtzoglou (Ed.), *M-Health Innovations for Patient-Centered Care* (pp. 262–283). Hershey, PA: IGI Global. doi:10.4018/978-1-4666-9861-1.ch013

Sukkird, V., & Shirahada, K. (2018). E-Health Service Model for Asian Developing Countries: A Case of Emergency Medical Service for Elderly People in Thailand. In M. Khosrow-Pour (Ed.), *Optimizing Current Practices in E-Services and Mobile Applications* (pp. 214–232). Hershey, PA: IGI Global. doi:10.4018/978-1-5225-5026-6.ch011

Sygit, B., & Wąsik, D. (2016). Patients' Rights and Medical Personnel Duties in the Field of Hospital Care. In A. Rosiek & K. Leksowski (Eds.), *Organizational Culture and Ethics in Modern Medicine* (pp. 282–297). Hershey, PA: IGI Global. doi:10.4018/978-1-4666-9658-7.ch012

Sygit, B., & Wąsik, D. (2016). The Idea of Human Rights in Conditions of Hospital Treatment. In A. Rosiek & K. Leksowski (Eds.), *Organizational Culture and Ethics in Modern Medicine* (pp. 236–254). Hershey, PA: IGI Global. doi:10.4018/978-1-4666-9658-7.ch010

Talbot, T. B. (2017). Making Lifelike Medical Games in the Age of Virtual Reality: An Update on "Playing Games with Biology" from 2013. In B. Dubbels (Ed.), *Transforming Gaming and Computer Simulation Technologies across Industries* (pp. 103–119). Hershey, PA: IGI Global. doi:10.4018/978-1-5225-1817-4.ch006

Tamposis, I., Pouliakis, A., Fezoulidis, I., & Karakitsos, P. (2016). Mobile Platforms Supporting Health Professionals: Need, Technical Requirements, and Applications. In A. Moumtzoglou (Ed.), *M-Health Innovations for Patient-Centered Care* (pp. 91–114). Hershey, PA: IGI Global. doi:10.4018/978-1-4666-9861-1.ch005

Tiago, M. T., Tiago, F., Amaral, F. E., & Silva, S. (2016). Healthy 3.0: Healthcare Digital Dimensions. In A. Dwivedi (Ed.), *Reshaping Medical Practice and Care with Health Information Systems* (pp. 287–322). Hershey, PA: IGI Global. doi:10.4018/978-1-4666-9870-3.ch010

Tripathy, B. (2016). Application of Rough Set Based Models in Medical Diagnosis. In S. Dash & B. Subudhi (Eds.), *Handbook of Research on Computational Intelligence Applications in Bioinformatics* (pp. 144–168). Hershey, PA: IGI Global. doi:10.4018/978-1-5225-0427-6.ch008

Turcu, C. E., & Turcu, C. O. (2017). Social Internet of Things in Healthcare: From Things to Social Things in Internet of Things. In C. Reis & M. Maximiano (Eds.), *Internet of Things and Advanced Application in Healthcare* (pp. 266–295). Hershey, PA: IGI Global. doi:10.4018/978-1-5225-1820-4.ch010

Unwin, D. W., Sanzogni, L., & Sandhu, K. (2017). Developing and Measuring the Business Case for Health Information Technology. In K. Moahi, K. Bwalya, & P. Sebina (Eds.), *Health Information Systems and the Advancement of Medical Practice in Developing Countries* (pp. 262–290). Hershey, PA: IGI Global. doi:10.4018/978-1-5225-2262-1.ch015

Urooj, S., & Singh, S. P. (2016). Wavelet Transform-Based Soft Computational Techniques and Applications in Medical Imaging. In P. Saxena, D. Singh, & M. Pant (Eds.), *Problem Solving and Uncertainty Modeling through Optimization and Soft Computing Applications* (pp. 339–363). Hershey, PA: IGI Global. doi:10.4018/978-1-4666-9885-7.ch016

Vasant, P. (2018). A General Medical Diagnosis System Formed by Artificial Neural Networks and Swarm Intelligence Techniques. In U. Kose, G. Guraksin, & O. Deperlioglu (Eds.), *Nature-Inspired Intelligent Techniques for Solving Biomedical Engineering Problems* (pp. 130–145). Hershey, PA: IGI Global. doi:10.4018/978-1-5225-4769-3.ch006

Waegemann, C. P. (2016). mHealth: History, Analysis, and Implementation. In A. Moumtzoglou (Ed.), M-Health Innovations for Patient-Centered Care (pp. 1-19). Hershey, PA: IGI Global. doi:10.4018/978-1-4666-9861-1.ch001

Watfa, M. K., Majeed, H., & Salahuddin, T. (2016). Computer Based E-Healthcare Clinical Systems: A Comprehensive Survey. *International Journal of Privacy and Health Information Management*, *4*(1), 50–69. doi:10.4018/IJPHIM.2016010104

Wietholter, J. P., Coetzee, R., Nardella, B., Kincaid, S. E., & Slain, D. (2016). International Healthcare Experiences: Caring While Learning and Learning While Caring. In D. Velliaris & D. Coleman-George (Eds.), *Handbook of Research on Study Abroad Programs and Outbound Mobility* (pp. 470–496). Hershey, PA: IGI Global. doi:10.4018/978-1-5225-0169-5.ch019

Witzke, K., & Specht, O. (2017). M-Health Telemedicine and Telepresence in Oral and Maxillofacial Surgery: An Innovative Prehospital Healthcare Concept in Structurally Weak Areas. *International Journal of Reliable and Quality E-Healthcare*, *6*(4), 37–48. doi:10.4018/IJRQEH.2017100105

Wójcik, A., & Chojnacki, M. (2016). Behavior and Ethical Problems in the Functioning of the Operating Theater (Case Study). In A. Rosiek & K. Leksowski (Eds.), *Organizational Culture and Ethics in Modern Medicine* (pp. 154–179). Hershey, PA: IGI Global. doi:10.4018/978-1-4666-9658-7.ch007

Wolk, K., & Marasek, K. P. (2016). Translation of Medical Texts using Neural Networks. *International Journal of Reliable and Quality E-Healthcare*, *5*(4), 51–66. doi:10.4018/IJRQEH.2016100104

Wong, A. K., & Lo, M. F. (2018). Using Pervasive Computing for Sustainable Healthcare in an Aging Population. In B. Fong, A. Ng, & P. Yuen (Eds.), *Sustainable Health and Long-Term Care Solutions for an Aging Population* (pp. 187–202). Hershey, PA: IGI Global. doi:10.4018/978-1-5225-2633-9.ch010

Wu, W., Martin, B. C., & Ni, C. (2017). A Systematic Review of Competency-Based Education Effort in the Health Professions: Seeking Order Out of Chaos. In K. Rasmussen, P. Northrup, & R. Colson (Eds.), *Handbook of Research on Competency-Based Education in University Settings* (pp. 352–378). Hershey, PA: IGI Global. doi:10.4018/978-1-5225-0932-5.ch018

Yadav, N., Aliasgari, M., & Poellabauer, C. (2016). Mobile Healthcare in an Increasingly Connected Developing World. *International Journal of Privacy and Health Information Management*, *4*(2), 76–97. doi:10.4018/IJPHIM.2016070106

Yadav, S., Ekbal, A., Saha, S., Pathak, P. S., & Bhattacharyya, P. (2017). Patient Data De-Identification: A Conditional Random-Field-Based Supervised Approach. In S. Saha, A. Mandal, A. Narasimhamurthy, S. V, & S. Sangam (Eds.), Handbook of Research on Applied Cybernetics and Systems Science (pp. 234-253). Hershey, PA: IGI Global. doi:10.4018/978-1-5225-2498-4.ch011

Yu, B., Wijesekera, D., & Costa, P. C. (2017). Informed Consent in Healthcare: A Study Case of Genetic Services. In A. Moumtzoglou (Ed.), *Design, Development, and Integration of Reliable Electronic Healthcare Platforms* (pp. 211–242). Hershey, PA: IGI Global. doi:10.4018/978-1-5225-1724-5.ch013

Yu, B., Wijesekera, D., & Costa, P. C. (2017). Informed Consent in Electronic Medical Record Systems. In I. Management Association (Ed.), Healthcare Ethics and Training: Concepts, Methodologies, Tools, and Applications (pp. 1029-1049). Hershey, PA: IGI Global. doi:10.4018/978-1-5225-2237-9.ch049

Yu, M., Li, J., & Wang, W. (2017). Creative Life Experience among Students in Medical Education. In C. Zhou (Ed.), *Handbook of Research on Creative Problem-Solving Skill Development in Higher Education* (pp. 158–184). Hershey, PA: IGI Global. doi:10.4018/978-1-5225-0643-0.ch008

Zarour, K. (2017). Towards a Telehomecare in Algeria: Case of Diabetes Measurement and Remote Monitoring. *International Journal of E-Health and Medical Communications*, *8*(4), 61–80. doi:10.4018/IJEHMC.2017100104

Zavyalova, Y. V., Korzun, D. G., Meigal, A. Y., & Borodin, A. V. (2017). Towards the Development of Smart Spaces-Based Socio-Cyber-Medicine Systems. *International Journal of Embedded and Real-Time Communication Systems*, *8*(1), 45–63. doi:10.4018/IJERTCS.2017010104

Zeinali, A. A. (2018). Word Formation Study in Developing Naming Guidelines in the Translation of English Medical Terms Into Persian. In M. Khosrow-Pour, D.B.A. (Ed.), Encyclopedia of Information Science and Technology, Fourth Edition (pp. 5136-5147). Hershey, PA: IGI Global. doi:10.4018/978-1-5225-2255-3.ch446

Ziminski, T. B., Demurjian, S. A., Sanzi, E., & Agresta, T. (2016). Toward Integrating Healthcare Data and Systems: A Study of Architectural Alternatives. In T. Iyamu & A. Tatnall (Eds.), *Maximizing Healthcare Delivery and Management through Technology Integration* (pp. 270–304). Hershey, PA: IGI Global. doi:10.4018/978-1-4666-9446-0.ch016

Ziminski, T. B., Demurjian, S. A., Sanzi, E., Baihan, M., & Agresta, T. (2016). An Architectural Solution for Health Information Exchange. *International Journal of User-Driven Healthcare*, *6*(1), 65–103. doi:10.4018/IJUDH.2016010104

Zineldin, M., & Vasicheva, V. (2018). Reducing Medical Errors and Increasing Patient Safety: TRM and 5 Q's Approaches for Better Quality of Life. In *Technological Tools for Value-Based Sustainable Relationships in Health: Emerging Research and Opportunities* (pp. 87–115). Hershey, PA: IGI Global. doi:10.4018/978-1-5225-4091-5.ch005

About the Contributors

Dakshina Ranjan Kisku is currently an Assistant Professor in the Department of Computer Science and Engineering at National Institute of Technology Durgapur, India. He obtained his Bachelor of Engineering, Master of Engineering and Ph.D. (Engg) degrees all in Computer Science and Engineering from Jadavpur University, India. He is an active researcher in the areas of biometrics and computer vision. From March 2006 to March 2007, he worked as a Researcher in the Computer Vision Laboratory at University of Sassari, Italy. From August 2005 to March 2006, Dr. Kisku was also involved as a Project Associate with the Biometrics and Computer Vision Group at Indian Institute of Technology Kanpur, India. Besides, he has worked as a faculty member for several years at various engineering colleges in India. Dr. Kisku has published more than 55 scientific articles in international conferences, journals and edited books. He has co-authored and edited three books on biometrics and homeland security. He is a recipient of the IEI Young Engineers Award. Dr. Kisku is a Senior Member of IEEE (USA). He serves as a reviewer for many refereed journals, conferences and edited books.

Phalguni Gupta did his Ph.D. from IIT Kharagpur and started his career in 1983 by joining in Space Applications Centre (ISRO) Ahmedabad, India, as a Scientist. In 1987, he joined the Department of Computer Science and Engineering, Indian Institute of Technology Kanpur, India. Currently he is director at National Institute of Technical Teachers' Training & Research Kolkata and also a Professor in the department of CSE IIT Kanpur. He works in the field of Data Structures, Sequential Algorithms, Parallel algorithms, Online Algorithms, Image Analysis, and Biometrics. He has published more than 250 papers in International Journals and Conferences. He has dealt with several sponsored and consultancy projects which are funded by the Government of India. Some of these projects are in the area of Biometrics, System Solver, Grid Computing, Image Processing, Mobile Computing, and Network Flow. During this period he has proved himself a well known researchers in theoretical computer science; especially in the field of Biometrics.

Jamuna Kanta Sing received his B.E. (Computer Science & Engineering) degree from Jadavpur University in 1992, M.Tech. (Computer & Information Technology) degree from Indian Institute of Technology (IIT) Kharagpur in 1994 and Ph.D. (Engineering) degree from Jadavpur University in 2006. Prof. Sing has been a faculty member of the Department of Computer Science & Engineering, Jadavpur University since March 1997. He is a recipient of the BOYSCAST Fellowship of the Department of Science & Technology, Govt. of India in 2006 and the UGC Research Award in 2014. He has served as principal investigator for 4 R&D projects sponsored by AICTE and UGC, New Delhi. He is a senior member of the IEEE, USA. He has published more than 30 papers in refereed international journals and more than 60 papers in international conferences. His research interest includes face recognition/ detection, medical image processing, and pattern recognition.

* * *

Ferdous Ahmed is a graduate research assistant in the Biometrics Laboratory at the University of Calgary. He has completed his bachelor's degree in Computer Science from the Islamic University of Technology, Bangladesh. He is also a former lecturer at his graduating University. His current research work includes emotion recognition using body movement analysis.

Geetika Arora is currently a Ph.D. candidate at Birla Institute of Technology and Science Pilani, India. She received M. Tech. degree in Computer Science and Engineering from the Department of Computer Science and Engineering, at Banasthali Vidyapith Rajasthan, India, in 2016. Her research interest includes Machine Learning, Biometrics, System Securities, and Image Processing.

Marina L. Gavrilova is a Full Professor in the Department of Computer Science, University of Calgary, a head of the Biometric Technologies Laboratory and a Board Member of ISPIA. Her publications include over 200 journal and conference papers, edited special issues, books and book chapters in the areas of image processing, pattern recognition, machine learning, biometric and online security. She has founded ICCSA – an international conference series with LNCS/IEEE, co-chaired a number of top international conferences, and is Founding Editor-in-Chief of LNCS Transactions on Computational Science Journal. Dr. Gavrilova is on the Editorial Boards of the Visual Computer, International Journal of Biometrics, and six other journals. She has given over 50 keynotes, invited lectures and tutorials at major scientific gatherings and industry research centers, including at Stanford University, SERIAS Center at Purdue, Microsoft Research USA, Oxford University UK, Samsung Research South Korea and others.

Kapil Kumar Gupta is an assistant professor in the department of Computer Science and Engineering at Shri Ramswaroop Memorial University, Barabanki, India. He obtained his B. Tech (IT) degree in 2009 from JSS Academy of Technical Education, Noida, Uttar Pradesh, India and M.Tech. (CSE) from Integral University Lucknow in 2013. He is currently pursuing Ph.D. from Amity University, UP. His main research interest is in the field of Medical Image Processing, Design and analysis of algorithms, etc.. He published many research papers in international journals and conferences. He is a Member of CSTA and IEEE.

Bin Hao received the Ph.D. degree from Beijing University of Posts and Telecommunications, China, in 2012. He is now a Postdoctoral Fellow at School of Computing and Informatics, University of Louisiana at Lafayette. His research interests focus on acoustical channel based access control, wireless device security, key agreement protocol, and trusted computing.

Xiali (Sharon) Hei is an assistant professor in the School of Computing and Informatics at the University of Louisiana at Lafayette. Prior to joining the University of Louisiana at Lafayette, she was an assistant professor at Delaware State University from 2015-2017 and Frostburg State University 2014-2015. Sharon received his Ph.D. in computer science from Temple University in 2014, focusing on computer security. She was awarded NSF CRII grant and Delaware DEDO grant. Also, she earned an M.S. in Software Engineering from Tsinghua University and a B.S. in Electrical Engineering from Xi'an Jiaotong University.

Aman Kamboj received his M.Tech. Degree in Computer Science and Engineering from NIT Jalandhar in 2014-15. He is currently a research scholar in the Department of Computer Science and Engineering, NIT Jalandhar doing Ph.D in Biometrics using deep learning under the guidance of (Dr. Rajneesh Rani, Assistant Professor at NITJ) and (Dr. Aditya Nigam, Assistant Professor at IIT Mandi). His research interest includes Biometrics, Pattern Recognition, Deep Learning and Image Processing.

Ramgopal Kashyap's areas of interest are image processing, pattern recognition and machine learning. He has published more than 20 research papers, and book chapters in international journals and conferences like Springer, Inderscience, Elsevier, ACM and IGI-Global indexed by Science Citation Index (SCI) and Scopus (Elsevier).He has Reviewed Research Papers in the Science Citation Index Expanded, Springer Journals and Editorial Board Member and conferences programme committee member of the IEEE, Springer international conferences and journals held in countries: Czech Republic, Switzerland, UAE, Australia, Hungary, Poland, Taiwan, Denmark, India, USA, UK, Austria, and Turkey. He has written many book chapters published by IGI Global, USA.

Gurjot Kaur graduated in Electronics and Communication Engineering from Guru Nanak Dev University, Amritsar and did her Masters from UIET, Panjab University, Chandigarh. She is currently pursuing her PhD from Dr. B.R. Ambedkar, NIT Jalandhar in Electronics and Communication department. Her current area of research focuses on Biomedical Signal Processing.

Upendra Kumar has done schoolings (10th and 10+2) from Govt. Inter College, Allahabad, India with First Honors. He graduated in B. Tech. (Information Technology) from UPTU Lucknow, India with First Honors (75.60% marks) and received M. Tech. (Information Technology) from Indian Institute of Information Technology, Allahabad, India with 9.19/10 CGPI. Currently, he has completed his Ph. D. in the faculty of Computer Science & Engineering under supervision of Dr. Tapobrata Lahiri, IIIT Allahabad, India. During M.Tech program, he got the placement opportunity in Tata Consultancy Services Limited and joined as an Assistant Engineer in 2008. He worked in various corporate projects for 4 years in TCS as a Java/ J2EE developer: i) Automation System for Ministry of Corporate Affairs from Dec, 2008 to Sep, 2009, ii) Passport Seva Project for India from Sep, 2009 to June, 2011 iii) SMB Education Project from July, 2011 to Aug, 2012. Owing to his research interest and also coding skill, during his post-graduation, he worked on Biomedical Signal Processing, especially ECG to develop automatic diagnosis of heart diseases. In continuation with this he got the opportunity to be registered in Ph. D. program from UPTU Lucknow and research working place at IIIT Allahabad. He also got opportunity to work as a Research Engineer, in Knowledge Flow Limited, Indian Institute of Technology, Delhi on Real Time Image Processing projects offered from Dr. Anirban Das, Head & Founder Member, from Aug, 2012 to till now. He developed various industrial projects like Home Automation and Security Surveillance System, Perimeter Protection and Security System and Sensor based Interactive Robot. He also developed some android programming based applications for robotics. As for his teaching experience, he worked as a Lecturer in Department of Computer Science and Information Technology, SIET, Allahabad from August 2004 to June 2006. He attended and completed various training programs and workshops during his Ph. D work: i) Training Course on "Robotics and Embedded in C" by Robosapians India Pvt. Ltd., Noida (01-15, Dec, 2012), ii) Training on "Initial Learning Program" by TCS, Tiruvanatpuram, Kerala (Aug- Sept, 2008), iii) "5th IEEE Conference on Wireless Communication and Sensor Networks" (Held at IIIT-Allahabad 2008), iv) Workshop on "Network and Information Security: Emerging Challenges and Practices", held at Ansal Technical Campus, Lucknow (05-08 Dec, 2013). Currently he is working as Assistant Professor also deputy dean of student welfare in the Department of Computer Science and Engineering of Institute of Engineering & Technology, Lucknow, India from 07 September, 2017 to till now.

Aditya Nigam received Ph.D. and M.Tech. degrees in Computer Science and Engineering from Indian Institute of Technology Kanpur (IITK), India, in 2015 and 2009 respectively. He is currently working as an Assistant Professor in the School of Computing and Electrical Engineering at Indian Institute of Technology Mandi, India. His research interest includes Biometrics, Pattern Recognition, Computer Vision, Deep Learning and Image Processing.

Rajneesh Rani has received his B.Tech and M.Tech degrees in Computer Science and Engineering, from Punjab Technical University, Jalandhar, India in 2001 and Punjabi University Patiala, India in 2003 respectively. She has done her PhD in computer Science and Engineering from NIT Jalandhar in 2015. From 2003 to 2005, she was a lecturer in Guru Nanak Dev Engineering College, Ludhiana. Currently, she has been working as an assistant professor in NIT Jalandhar since 2007. Her teaching and research include areas like Image Processing, Biometrics, Pattern Recognition, Machine Learning, Computer Programming and Document Analysis and Recognition.

Lakhan Dev Sharma received his B.E. in electronics and communication engineering from Rajiv Gandhi Proudyogiki Vishwavidyalaya, Bhopal in year 2007. He has completed his M. Tech. from ABV-Indian Institute of Information Technology and Management, Gwalior in year 2012 and Ph. D. from Dr. B. R. Ambedkar National Institute of Technology, Jalandhar, India in year 2018. He has teaching and industrial experience. Currently he is working at MLV Textile and Engineering College, Bhilwara, Rajasthan, India as Assistant Professor. His research interests include Biomedical Signals and Image Processing.

Brandon Sieu is a BSc student pursuing degree in Computer Science at the University of Calgary, and member of the University of Calgary BTLabs.

Tanuja Sudhakar is a Computer Science student at the University of Calgary and also has two years of work experience in the IT Sector. Her interests are Bio-metric technology, Deep Learning and AI. She holds a Bachelors degree in IT from Anna University, India.

Ramesh Kumar Sunkaria was born in Kot Khalsa, Amritsar, on February 20, 1967. He received B.Tech. degree in Electronics Engg. from Department of Electronics Technology, Guru Nanak Dev University Amritsar, and M.Tech. (Hons.) Degree in Electronics Engg. From Guru Nanak Dev Engineering College, Ludhiana (Punjab Technical University, Jalandhar) in 1990, and 2004 respectively. He did his Ph.D. degree from Electrical Engineering Department, Indian Institute of Technology Roorkee, Roorkee in June, 2010. He worked as Engineer in Bharat Heavy Electricals Limited, Haridwar up to June 2004. Thereafter he joined Indian Broadcasting (Engineers) Service, Govt. of India and worked as Assistant Director(Engineering) until August 1996. He directed the installation of numerous radio and TV projects all-over India, namely, VLPT, Joginder Nagar, VLPT Ahjufort, VLPT Kalakot etc. Afterwards he joined academics and at present he is Assistant Professor in Department of Electronics & Communication Engineering, Dr. B. R. Ambedkar National Institute of Technology, where he has been teaching UG/PG courses related to signal processing and communication engineering. His current research interests are in bio-electric signal and image processing, filter banks and non-linear heart variability studies. He has numerous publications in referred international journals and conference proceedings. At present seven PhD thesis are in progress under his guidance, 18 M.Tech thesis has already been completed and more than thirty B.Tech projects have been completed. He is member of IEEE Engineering in Medicine and Biology Society and IEEE Computer society.

Kamlesh Tiwari is working as an assistant professor in the department of Computer Science and Information Systems at Birla Institute of Technology and Science Pilani. He has earned his PhD degree from the department of Computer Science and Engineering of Indian Institute of Technology Kanpur in 2016. His research interests include Machine Learning, Computer Vision, Multimodal Biometric (Fingerprint, Face, Palmprint, knuckleprint) and Security. His PhD work was in the area of biometrics, entitled as "Some Efficient Techniques To Improve Fingerprint Based Recognition System". He is a member of IEEE and Signal Processing Society (SPS). He is heads AI-ML lab and is co-in charge of Advanced Data Analytics & Parallel Technologies Lab at BITS Pilani. He is also an active member of Multimedia & HCI Laboratory.

Esha Tripathi is currently a Ph.D. Research Scholar selected through Homi Bhabha Teaching Assistant Fellowship program 2017-2018 in Computer Science and Engineering department at Institute of Engineering & Technology, Lucknow under the supervision of Prof. S.P. Tripathi and Dr. Upendra Kumar. My research activity mainly focuses on Digital Image Processing. I obtained the M.Tech (Computer Science & Engineering) Degree from Banasthali Vidyapith, Rajasthan in 2012 and completed thesis work under the supervision of Dr. T. V. Prabhakar at IIT Kanpur.

Surya Tripathi is a Professor in Department of Computer Science & Engineering of Institute of Engineering & Technology, Lucknow, India.

Tanu Wadhera received the B.tech degree in Electronics and Communication from Guru Nanak Engineering college, Ludhiana, India and M.tech degree in Electronics & Communication from Punjabi University, Patiala, India. Currently, she is a Research Scholar in National institute of Technology, Jalandhar, India. Her research interests are cognitive neuroscience, behavioral understanding, technology in Autism Spectrum Disorder, biomedical signal processing, signal classification and prediction of neurodevelopmental disorders.

Index

H

I

M

O

P

R

S

T

U

V

W

www.ingramcontent.com/pod-product-compliance
Lightning Source LLC
LaVergne TN
LVHW062145300926
849161LV00010B/2633